Biosensors for Monitoring and Diagnostics

Biosensors for Monitoring and Diagnostics

Guest Editors

Radivoje Prodanović
Dalibor M. Stanković

MDPI

Basel • Beijing • Wuhan • Barcelona • Belgrade • Novi Sad • Cluj • Manchester

Guest Editors

Radivoje Prodanović
Faculty of Chemistry
University of Belgrade
Belgrade
Serbia

Dalibor M. Stanković
Faculty of Chemistry
University of Belgrade
Belgrade
Serbia

Editorial Office
MDPI AG
Grosspeteranlage 5
4052 Basel, Switzerland

This is a reprint of the Special Issue, published open access by the journal *Biosensors* (ISSN 2079-6374), freely accessible at: https://www.mdpi.com/journal/biosensors/special_issues/5ZQX616KO3.

For citation purposes, cite each article independently as indicated on the article page online and as indicated below:

Lastname, A.A.; Lastname, B.B. Article Title. *Journal Name* **Year**, *Volume Number*, Page Range.

ISBN 978-3-7258-7961-8 (Hbk)
ISBN 978-3-7258-7962-5 (PDF)
https://doi.org/10.3390/books978-3-7258-7962-5

Contents

About the Editors

Radivoje Prodanović

Radivoje Prodanović is a Full Professor at the Department of Biochemistry, Faculty of Chemistry, University of Belgrade, Serbia. He obtained his Ph.D. and M.Sc. in Chemistry from the University of Belgrade. His research interests include protein engineering, directed evolution, high-throughput screening, enzyme immobilization, nonaqueous enzymology, and the application of enzymes in organic synthesis, as well as hydrogels and biocomposite materials. His research focuses on enzyme engineering, microfluidic screening systems, biosensors, and enzyme–material hybrid materials for biotechnological and analytical applications. He was awarded the Alexander von Humboldt Fellowship, the Fulbright Fellowship, and the Institute of Advanced Studies Visiting Fellowship. His current work is focused on biosensors, enzyme immobilization, protein engineering, bioinformatics, and functional biomaterials.

Dalibor M. Stanković

Dalibor M. Stanković is an Associate Professor at the Faculty of Chemistry, University of Belgrade, Serbia, and a Research Professor at the Vinča Institute of Nuclear Sciences, University of Belgrade, Serbia. His research interests include electrochemical sensors and biosensors, nanomaterials, electrode modification, and electroanalytical methods. His work is focused on the development of electrochemical sensing platforms based on modified carbon electrodes and nanomaterials for environmental monitoring and biomedical applications. He is the coordinator of an international CEEPUS network and principal investigator of several bilateral and international research projects, including a Eureka project. He has authored numerous scientific publications in the fields of electrochemistry, analytical chemistry, and sensor development.

Preface

The Reprint entitled "Biosensors for Monitoring and Diagnostics" presents a collection of research and review articles published in the Special Issue of the journal *Biosensors*. The aim of this Reprint is to provide an overview of recent developments in biosensor technologies and their applications in monitoring and diagnostic systems. The contributions included in this Reprint cover a wide range of topics, including electrochemical and optical biosensors, nanomaterials for biosensing, protein and nucleic acid engineering, enzyme immobilization, microfluidic devices, wearable sensors, and point-of-care diagnostic systems. The papers demonstrate how interdisciplinary research combining biotechnology, analytical chemistry, materials science, nanotechnology, and bioinformatics enables the development of advanced biosensing platforms for medical diagnostics, environmental monitoring, and real-time analysis. This Reprint is intended for researchers, engineers, and students working in biosensors, biotechnology, analytical chemistry, materials science, and biomedical diagnostics, and aims to provide insight into current research trends and future directions in biosensor technology.

Radivoje Prodanović and Dalibor M. Stanković
Guest Editors

biosensors

MDPI

Editorial

Recent Advances and Future Perspectives in Biosensors for Monitoring and Diagnostics

Radivoje Prodanović * and Dalibor Stanković *

University of Belgrade-Faculty of Chemistry, Studentski Trg 12, 11000 Belgrade, Serbia
* Correspondence: radivojep@chem.bg.ac.rs (R.P.); dalibors@chem.bg.ac.rs (D.S.)

1. Introduction

Biosensors have become some of the most vital analytical tools in modern biotechnology, medicine, environmental monitoring, food safety, and industrial process control. They combine biological recognition elements, such as enzymes, antibodies, nucleic acids, or cells, with physical and chemical transducers, such as electrochemical, optical, piezoelectric, and electronic detection systems. These systems allow for selective detection of biologically relevant compounds and are used in medical diagnostics, environmental monitoring, food analysis, and biotechnology.

Over the past few decades, biosensor technology has evolved from simple enzyme electrodes to advanced integrated analytical systems that combine biological recognition, nanomaterials, microelectronics, microfluidics, and data processing. Modern biosensors are increasingly integrated into portable electronics and wearable devices, enabling continuous monitoring and point-of-care diagnostics.

2. Recent Developments in Biosensors

In recent years, biosensor research has shifted toward smaller, portable, and intelligent sensing systems capable of real-time monitoring and data analysis. Advances in biosensors have been driven by progress in nanomaterials, microfluidics, wearable electronics, molecular diagnostics, and bioinformatics. Nanomaterials such as metal nanoparticles, carbon nanomaterials, quantum dots, metal–organic frameworks, and polymer nanocomposites have enhanced the sensitivity, stability, and miniaturization of biosensors. These nanostructured materials provide a high surface area, improved electron transfer, and better immobilization of biological molecules, significantly enhancing biosensor performance.

Advances in molecular diagnostics, including clustered regularly interspaced short palindromic repeats (CRISPR)-based detection systems, nucleic acid sensors, and immunoassays, have enabled fast and multiplexed detection of biomarkers and pathogens. Microfluidic systems have supported the development of smaller analytical devices that use less sample and provide fast results. Wearable biosensors are another rapidly growing area, particularly for continuous monitoring of physiological parameters such as glucose, lactate, and oxygen saturation, as well as other biomarkers.

Artificial intelligence and data analysis tools are also becoming increasingly essential for biosensor signal processing, pattern recognition, and the creation of smart diagnostic systems.

3. Challenges and Knowledge Gaps

Despite notable progress, several challenges persist in the field of biosensors. These include maintaining the long-term stability of biological recognition elements, ensuring

 https://doi.org/10.3390/bios16050237

reproducibility during biosensor fabrication, integrating biosensors with electronic systems, enabling real-time data processing, and transforming laboratory biosensor systems into practical and commercial devices.

Another key challenge is developing biosensors for early disease detection, monitoring environmental pollutants at low levels, and creating continuous monitoring systems for personalized medicine. Developing new materials, improved immobilization techniques, and sophisticated signal-processing methods will be essential for addressing these challenges.

Besides technological challenges, there is also a gap between biosensor research in labs and real-world use. Many biosensor systems show excellent performance in controlled lab conditions but face issues with stability, reproducibility, cost, and large-scale manufacturing. Additionally, integrating biosensors with data processing, wireless communication, and artificial intelligence is an important area for future research. Therefore, interdisciplinary approaches that combine materials science, biotechnology, electronics, and data science are crucial for developing next-generation biosensor systems.

An important research area and knowledge gap concerns the design and engineering of biological recognition elements for biosensors. While enzymes, antibodies, and nucleic acids are commonly used as sensing elements, their stability, selectivity, and electron-transfer properties are often not ideal for biosensor design. Protein engineering, directed evolution, and synthetic biology provide powerful tools for creating biological components with better stability, catalytic activity, and electron-transfer efficiency. However, combining protein engineering with biosensor design, materials science, and electronic systems is still a developing field and an important direction for future biosensor advancements.

Addressing these challenges will require strong interdisciplinary collaboration between chemists, biochemists, materials scientists, engineers, and data scientists.

4. Overview of the Special Issue

This Special Issue, titled "Biosensors for Monitoring and Diagnostics", contains research and review articles exploring key areas of biosensor research, including molecular diagnostics, electrochemical and optical biosensors, nanomaterials and composites for biosensing, wearable and physiological monitoring systems, bioelectronic sensors, and protein and nanomaterial engineering for biosensors.

Several papers in this Special Issue focus on molecular diagnostics and biomarker detection, including molecular detection systems, CRISPR-based assays for nucleic acid detection, and point-of-care viral biosensors and diagnostic platforms. These studies highlight the increasing importance of rapid, portable, and multiplexed diagnostic systems.

Another set of papers discusses wearable biosensors and physiological monitoring systems, including sensors for tracking lactate in sweat, tissue oxygenation, and arterial pulse strain, as well as ultrasound-based monitoring devices. These technologies highlight the growing importance of wearable, non-invasive biosensors for continuous monitoring of physiological parameters.

Nanomaterials and advanced materials for biosensing are prominently featured in this Special Issue, including nanomaterial-based sensing platforms such as surface-enhanced Raman spectroscopy (SERS) sensors, quantum dot-based detection methods, and carbon nanotube electrochemistry. These materials improve the sensitivity, selectivity, and overall performance of biosensors.

Bioelectronic sensing systems and the monitoring of biological systems are also discussed, including platforms for tracking neuronal network activity with bioelectronic measurement tools. These studies highlight the vital role of biosensors in researching biological systems and cellular networks.

https://doi.org/10.3390/bios16050237

Additionally, this Special Issue features review articles on protein, nucleic acid, and nanomaterial engineering for biosensors, emphasizing the importance of protein engineering, enzyme immobilization, and nanomaterials in the development of modern biosensing platforms.

The papers published in this Special Issue address many of these challenges by introducing new biosensor materials, improved sensing platforms, wearable monitoring systems, and molecular diagnostic tools, as well as review articles summarizing recent progress in biosensor engineering and analytical methods. Together, these contributions help reduce the gap between fundamental biosensor research and practical monitoring and diagnostic applications.

5. Future Perspectives

The future development of biosensors will likely be heavily impacted by advances in nanotechnology, synthetic biology, protein engineering, microfluidics, wearable electronics, and artificial intelligence. Combining biosensors with artificial intelligence and data analysis tools will enable the creation of smart diagnostic systems and personalized health monitoring.

Advances in protein engineering and synthetic biology will allow the development of new biological recognition elements with better stability, specificity, and catalytic efficiency. Innovative materials such as nanocomposites, metal–organic frameworks, and functional polymers will enable the development of more sensitive and durable biosensors.

Wearable and implantable biosensors will facilitate continuous monitoring of physiological parameters, which is vital for personalized medicine and preventive healthcare. Environmental monitoring and food safety will also remain key applications for biosensors. In the future, biosensors are expected to become fully integrated smart analytical systems capable of continuous monitoring, real-time data processing, and autonomous decision-making, which will greatly influence healthcare, environmental monitoring, and industrial biotechnology.

6. Conclusions

This Special Issue highlights recent advances in biosensor technologies and emphasizes the importance of interdisciplinary research that combines biotechnology, analytical chemistry, materials science, nanotechnology, electronics, and data science to develop modern biosensors for monitoring and diagnostic applications. We hope this Special Issue will contribute to further progress in biosensor technologies and inspire new research in this rapidly growing field.

Author Contributions: Writing—original draft preparation, R.P.; writing—review and editing, R.P. and D.S. All authors have read and agreed to the published version of the manuscript.

Funding: This research was funded by the Ministry of Science, Technological Development, and Innovation of the Republic of Serbia, grant numbers 451-03-136/2025-03/200168.

Conflicts of Interest: The authors declare no conflicts of interest.

https://doi.org/10.3390/bios16050237

biosensors

MDPI

Review

Protein, Nucleic Acid, and Nanomaterial Engineering for Biosensors and Monitoring

Milica Crnoglavac Popović [1], Vesna Stanković [2], Dalibor Stanković [1] and Radivoje Prodanović [1,*]

[1] Faculty of Chemistry, University of Belgrade, Studentski trg 12, 11000 Belgrade, Serbia; milicac@chem.bg.ac.rs (M.C.P.); dalibors@chem.bg.ac.rs (D.S.)
[2] IHTM, Njegoševa 12, 11000 Belgrade, Serbia; vesna.stankovic@ihtm.bg.ac.rs
* Correspondence: rprodano@chem.bg.ac.rs; Tel.: +381-63-870-8018

Abstract

The engineering of proteins, nucleic acids, and nanomaterials has significantly advanced the development of biosensors for the monitoring of rare diseases. These innovative biosensing technologies facilitate the early detection and management of conditions that often lack adequate diagnostic solutions. By utilizing engineered proteins and functional nucleic acids, such as aptamers and nucleic acid sensors, these biosensors can achieve high specificity in identifying the biomarkers associated with rare diseases. The incorporation of nanomaterials, like nanoparticles and nanosensors, enhances sensitivity and allows for the real-time monitoring of biochemical changes, which is critical for timely intervention. Moreover, integrating these technologies into wearable devices provides patients and healthcare providers with continuous monitoring capabilities, transforming the landscape of healthcare for rare diseases. The ability to detect low-abundance biomarkers in varied sample types, such as blood or saliva, can lead to breakthroughs in understanding disease pathways and personalizing treatment strategies. As the field continues to evolve, the combination of protein, nucleic acid, and nanomaterial engineering will play a crucial role in developing next-generation biosensors that are not only cost-effective but also easy to use, ultimately improving outcomes and the quality of life for individuals affected by rare diseases.

Keywords: directed evolution; DNA; RNA; nanocomposites; MOF; biosensor; rare diseases

1. Introduction

In this review, we will cover the most recent studies concerning protein, nucleic acid, and nanomaterial engineering for biosensors that can detect and be used for the monitoring of rare diseases, as detailed in Figure 1.

The rapid advancement of biosensor technology has emerged as a transformative force in diagnostics, enabling the detection of rare diseases through the engineering of proteins, nucleic acids, and nanomaterials. Rare diseases often present significant challenges in terms of diagnosis, given the complexity and low prevalence of the associated biomarkers. To address these challenges, this review focuses on the engineering strategies applied to enhance the performance of biosensors, thus facilitating the timely detection and monitoring of rare illnesses.

The optical and electrochemical biosensors discussed in this article operate on distinct, yet complementary, principles for detecting biological interactions. Optical biosensors detect target analytes by measuring changes in optical properties, such as absorbance, fluorescence, luminescence, or the refractive index, resulting from biomolecular recognition events on the sensor's surface. Techniques like surface plasmon resonance (SPR), fluorescence spectroscopy, and interferometry are commonly employed to monitor these changes, often enabling label-free and real-time analyses with high sensitivity. In contrast, electrochemical biosensors convert biochemical interactions into electrical signals—such as current, voltage, or impedance—using recognition elements, like enzymes, antibodies, or aptamers. Amperometric, potentiometric, and impedimetric approaches are used to quantify these signals, offering advantages such as a low cost, portability, and suitability for point-of-care diagnostics.

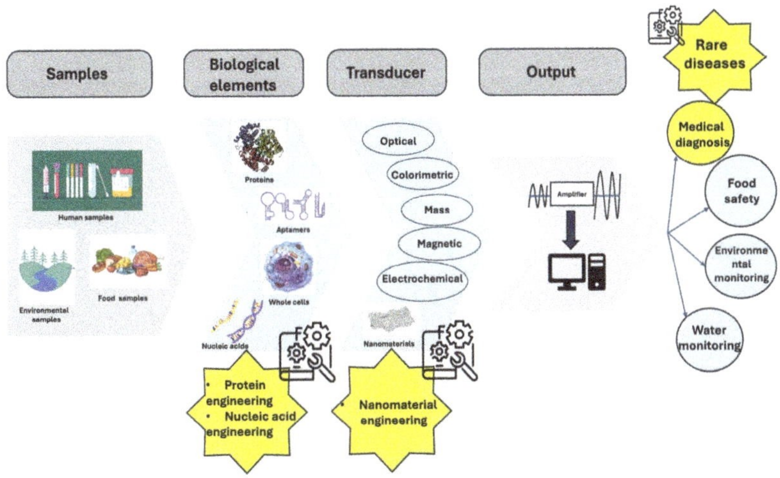

Figure 1. Biomolecular and nanomaterial engineering in the development of biosensors for monitoring.

Section 2 of this review addresses protein engineering for biosensors, including directed evolution, semi-rational design, and rational design approaches, each contributing to the development of highly targeted biosensors with improved binding capabilities. In Section 3, we further explore the engineering of nucleic acids, particularly RNA and DNA, which have revolutionized biosensor specificity and sensitivity by developing novel aptamers and genetic circuits.

Section 4 discusses the engineering of nanomaterials and highlights their critical role in amplifying sensor signals and improving the overall biosensor performance. Finally, Section 5 examines the application of these engineered biosensors for monitoring rare diseases, emphasizing their potential impact on personalized medicine and patient outcomes. Through this synthesis, the review emphasizes the critical intersections of these fields, paving the way for future advancements in biosensing technologies to enhance the management of rare diseases.

2. Protein Engineering for Biosensors

2.1. Directed Evolution of Proteins for Biosensors

Directed evolution has rapidly emerged as a transformative strategy for developing biosensor proteins, enabling researchers to create highly tailored biomolecules that respond to specific analytes with improved sensitivity and selectivity. By mimicking natural selection principles through iterative mutation, selection, and amplification rounds, scientists can evolve proteins that outperform their natural counterparts. For instance, in the development of biosensors for detecting the R248Q mutant of the tumor suppressor protein p53, Pellerano et al. effectively utilized directed evolution to generate a fluorescent peptide biosensor that demonstrated a significant interaction with the aggregation-prone variant of p53, showing the potential of this strategy in clinical applications and biomarker detection [1]. This approach highlights how engineered proteins can provide critical insights into specific mutations linked to disease states.

Additionally, Liang et al. developed a lead (Pb) whole-cell biosensor based on the Pb resistance transcriptional regulatory factor and green fluorescent protein and improved its performance through directed evolution in conjunction with fluorescence-activated cell sorting (FACS) [2], Figure 2.

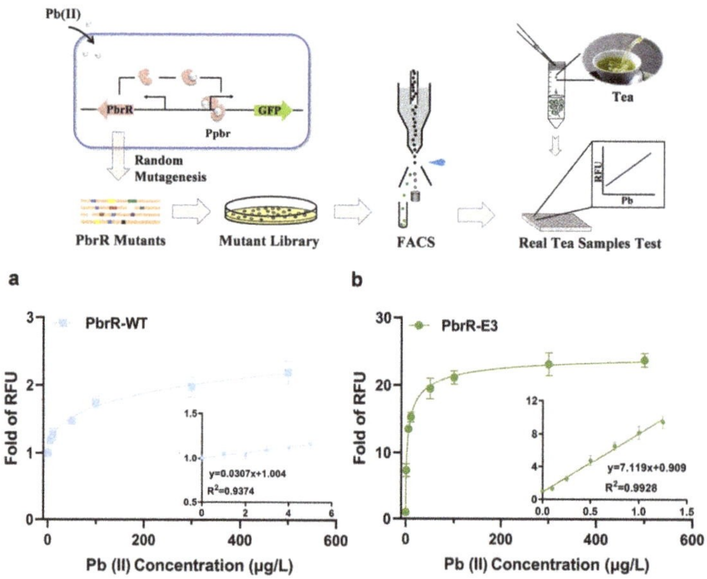

Figure 2. A strategy for the development of a highly sensitive, PbrR-based biosensor via directed evolution and its application for lead detection. (**a**) The dose-dependent response of PbrR-WT WCB detected by flow cytometer. (**b**) The dose-dependent response of PbrR-E3 WCB detected by flow cytometer. Reprinted from ref. [2] with permission. Copyright © 2025 Elsevier B.V.

The adaptability of directed evolution is further illustrated by its utilization in the development of transcription-factor-based biosensors, which are highly effective in monitoring intricate biological systems. Feng et al. [3] proposed a comprehensive framework for the construction of small-molecule biosensors within eukaryotic contexts, utilizing a ligand-binding domain (LBD) fused to a transcriptional activator, Figure 3.

In this system, the binding of the ligand serves to stabilize the LBD, thereby preventing its degradation. By fusing this destabilized LBD to an appropriate reporter protein—such as an enzyme, a fluorescent protein, or a transcription factor—the resulting construct becomes conditionally stable and produces a measurable sensor response. When conditionally stabilized LBDs are linked to transcription factors (TFs), the resulting biosensors exhibit a higher degree of signal induction in response to target ligands compared to those fused with fluorescent proteins. Utilizing these TF-biosensors, the authors successfully enhanced the biosynthetic yield of progesterone in yeast, demonstrating the practical applications of this biosensing technology [3]. Researchers can create biosensors with the precise response characteristics needed for various small molecules or environmental conditions by systematically mutating transcription factors and utilizing selection systems.

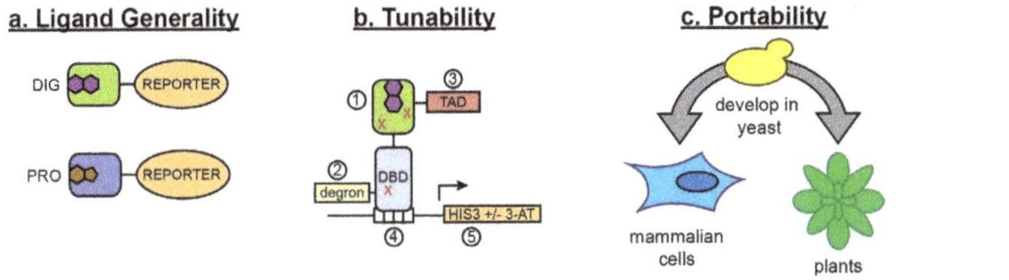

Figure 3. A schematic of the biosensor platform. (**a**) Biosensors for small molecules are modularly constructed by replacing the LBD with proteins possessing altered substrate preferences. (**b**) The activity of the biosensor can be tuned by (1) introducing destabilizing mutations (red Xs), (2) adding a degron, (3) altering the strength of the TAD or DNA binding affinity of the TF, (4) changing the number of TF-binding sites or sequences, and (5) titrating 3-aminotriazole, an inhibitor of His3. (**c**) Yeast provides a genetically tractable chassis for biosensor development before implementation in more complex eukaryotes, such as mammalian cells and plants. Reprinted from ref. [3]; licensed under a Creative Commons Attribution license (CC-BY).

In a study by Paulmurugan et al. [4], a novel protein folding molecular imaging biosensor is introduced to monitor the drug effects that restore the structure and function of the mutant p53 protein in glioblastoma cells. This innovative assay enables rapid high-throughput screening for compounds that promote proper p53 folding, with potential applications for non-invasive studies in living subjects using optical bioluminescence imaging in small animal models. The research highlights the biosensor's effectiveness in screening drugs that reactivate p53 activity, representing a promising advancement in cancer therapy targeting p53 mutations. This technique not only streamlines the biosensor design process but also accelerates the timeline from concept to functioning sensor in dynamic biological systems.

Additionally, integrating novel strategies, such as Förster resonance energy transfer (FRET), into directed evolution allows for a more refined biosensor architecture that can yield more nuanced insights into biological processes. In the publication by Taylor et al. [5] a significant advancement in protein engineering is demonstrated through the development of allosteric transcription factors (aTFs) that can be tailored to respond to novel inducers. The authors employed the *Escherichia coli* lac repressor, LacI, and modified its structure to achieve responsiveness to new small molecules, such as fucose and gentiobiose. By integrating computational design with directed mutagenesis techniques, they successfully generated LacI variants that maintain specificity and induction efficiencies comparable to the native protein when activated by the traditional inducer IPTG. This engineering strategy not only enables the precise control of gene expression in synthetic biology applications but also opens up avenues for designing biosensors capable of detecting specific biomolecules in complex environments. The ability to create customized biosensors based on engineered proteins represents a significant leap forward in both biosensing technology and metabolic engineering, enabling the monitoring and manipulation of cellular processes with unprecedented specificity and sensitivity.

Furthermore, in a study by Gohil et al. [6] the authors present a comprehensive evaluation of biosensor optimization utilizing Förster resonance energy transfer (FRET) pairs based on the mScarlet red fluorescent protein and a derived green fluorescent protein. The research explores two primary classes of biosensors: one that employs intensiometric measurements, where a conformational change in the recognition domain is relayed to a single fused fluorescent protein, resulting in altered fluorescence intensity; and another that utilizes the recognition-domain-dependent modulation of FRET between two fluorescent proteins of differing colors, allowing for ratiometric changes in fluorescence emission. The optimization strategies detailed in the study aim to enhance the sensitivity and specificity of FRET-based biosensors, which have been successfully engineered to monitor various intracellular processes, including enzyme activity, ion dynamics, post-translational modifications, and protein–protein interactions. This work highlights the potential of utilizing engineered fluorescent protein pairs to develop advanced biosensors that provide real-time insights into complex biological systems, thereby facilitating a deeper understanding of cellular dynamics and interactions.

Cai et al. [7] developed an advanced biosensor for the sensitive and specific detection of cadmium in environmental samples by integrating an engineered cadmium resistance transcriptional regulator (CadR) with a green fluorescent protein (GFP) reporter, Figure 4.

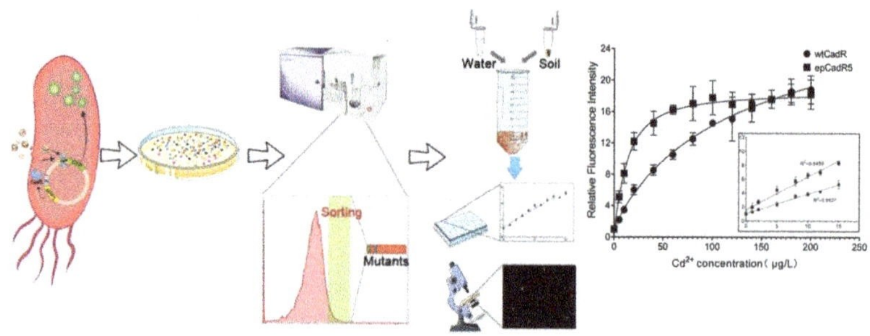

Figure 4. Strategy for evolving bacterial biosensor for measuring cadmium concentration. Dose-dependent response of bacterial biosensors to cadmium in *E. coli* strain JudeI. Reprinted from ref. [7] with permission. Copyright © 2022 American Chemical Society.

Using directed evolution, the researchers generated mutant libraries via error-prone PCR and optimized the biosensor through multiple rounds of fluorescence-activated cell sorting (FACS). This process led to the identification of the variant epCadR5, which exhibited significantly improved performance. The evolved biosensor achieved a remarkable detection limit of 0.45 μg/L for cadmium, demonstrating 6.8 times greater specificity compared to the wild-type sensor. This study highlights the potential of directed evolution in enhancing biosensor functionality for environmental monitoring, thereby facilitating the more reliable detection of hazardous heavy metals, such as cadmium.

Guo et al. [8] explored the potential of using the anthocyanin biosynthetic pathway to develop a bacterial biosensor for detecting lead toxicity. By engineering the metalloregulator PbrR to regulate anthocyanin biosynthesis in response to lead (Pb) levels, the researchers created a visually detectable biosensor that changes color upon exposure to this toxic heavy metal. Their findings highlight the effectiveness of metabolic engineering in utilizing natural pigments for the development of sensitive, cost-effective biosensors for monitoring environmental toxins. This approach highlights the ecological significance of tracking lead pollution and demonstrates the potential of biosensors in assessing the bioavailability and ecotoxicity of heavy metals. Overall, the study represents a significant advancement in biosensing technology, contributing to the development of practical tools for environmental monitoring. This application extends beyond basic detection; it aims to identify potential toxicity issues in aquatic ecosystems early, thereby emphasizing the societal relevance of such biosensor technologies.

In a study by Lee et al. [9], the authors explore the development and application of genetic enzyme screening systems (GESSs) specifically designed for the detection of C1 compounds—formate, formaldehyde, and methanol—using genetically engineered biosensors. The research presents the FA-GESS, Frm-GESS, and MeOH-GESS biosensors, which incorporate enzymatic pathways that enable the specific detection of these compounds through their respective enzymatic activities. Each biosensor demonstrates a linear response to its target compound—formate (1.0–250 mM), formaldehyde (1.0–50 μM), and methanol (5–400 mM)—showing the high specificity that is crucial for practical applications. By harnessing the capabilities of bacterial oxidoreductases within these biosensing techniques, the study significantly broadens the spectrum of detectable ligands available to genetic biosensor systems. Overall, their work underlines the potential of engineered biosensors in environmental monitoring and metabolic engineering, offering valuable tools for the real-time analysis of key C1 pollutants.

Fana et al. [10] applied directed evolution to enhance the thermostability of cholesterol oxidase (ChOx), a key enzyme in cholesterol biosensors. Using error-prone PCR, they generated a diverse library of ChOx variants and identified mutants with improved stability, crucial for maintaining biosensor reliability in dynamic environments. Their study underscores the power of directed evolution in enzyme engineering, demonstrating its potential to create more sensitive and durable biosensors for cholesterol monitoring. By optimizing ChOx, the research advances medical diagnostics and metabolic monitoring, highlighting the critical role of engineered enzymes in biosensor development.

Despite the successes observed in the directed evolution of proteins for biosensing applications, challenges remain in optimizing the protein-engineering process to meet specific assay requirements. The increasing complexity of biological systems necessitates continuously refining biosensors to ensure robust performance under variable conditions. Current research indicates that a combination of computational modeling alongside empirical evolution techniques will be crucial for future research in protein engineering for biosensors. This approach will enable biosensor development and advance our understanding of the underlying molecular dynamics that govern protein–analyte interactions.

2.2. *Semi-Rational Design of Proteins for Biosensors*

The semi-rational design of proteins for biosensors combines computational modeling and empirical experimentation to optimize the performance of these biological sensors. This strategy enables targeted modifications of proteins or receptor elements, leveraging structural information to enhance binding affinity and specificity toward analytes. For example, Ding et al. used a semi-rational approach to fine-tune cross-ribosome-binding sites in a glucarate biosensor, significantly improving its dynamic range [11]. It represents a rational design strategy for fine-tuning the dynamic range of bacterial biosensors by engineering cross-ribosome-binding sites. The study focuses on optimizing biosensor performance for detecting specific target molecules by precisely adjusting expression levels, enabling a more controlled response over a broader concentration range. By strategically modifying ribosome-binding sites, the researchers enhance the balance between sensitivity and specificity, thereby improving the applicability of biosensors in fields such as environmental monitoring and metabolic engineering. Their findings demonstrate the potential of this approach to enhance biosensor efficacy and practicality, offering a framework for refining existing biosensors and guiding the development of next-generation biosensing technologies with tailored response characteristics. Such methodologies illustrate that semi-rational design can efficiently utilize existing biological and computational knowledge to create more-effective biosensors, making it a valuable tool in synthetic biology and biotechnological applications.

In addition to improving binding affinities, semi-rational design strategies have been employed to enhance the selectivity of biosensors for complex mixtures. Moussa et al. [12] reported engineering d-amino acid oxidase enzymes to increase their selectivity for d-serine and d-alanine, demonstrating the potential for the more sensitive detection of these amino acids in biological samples, Figure 5.

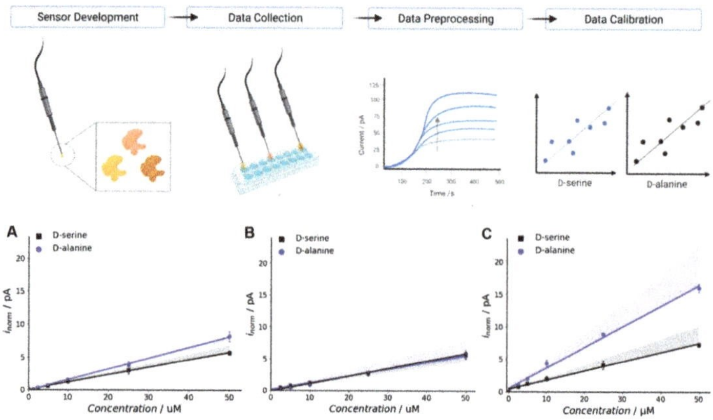

Figure 5. Strategy for enhancing electrochemical biosensor selectivity with engineered d-amino acid oxidase enzymes for d-serine and d-alanine quantification. RgDAAO WT (**A**), RgDAAO M213G (**B**), and hDAAO W209R (**C**) biosensor calibration in standard solutions of 0–50 μM d-serine and d-alanine. Error bars represent ± S.E.M, (n = 3). The shaded areas represent 95% confidence interval regions. Reprinted from ref. [12] with permission. Copyright © 2022 American Chemical Society.

Given the crucial roles of these d-amino acids in neurotransmission and metabolic regulation, accurate detection is essential for biomedical research. Using targeted mutagenesis, the authors developed enzyme variants with enhanced catalytic efficiency and substrate specificity, enabling the selective detection of targets in complex biological matrices. The resulting electrochemical biosensor exhibited a low limit of detection and quantification, making it a valuable tool for applications in neuroscience and metabolic studies. The study employed rational design techniques to introduce point mutations in the DAAO enzyme, which enhance its substrate specificity. The two engineered variants, human DAAO W209R and yeast DAAO M213G, exhibit significant improvements in selectivity and activity compared to the wild-type enzyme. The authors immobilized these variants to facilitate practical biosensor application, demonstrating that the resulting electrochemical biosensors achieve selectivity levels comparable to those observed with free DAAO enzymatic activity. Their findings emphasize the importance of integrating engineered biocatalysts into biosensor design to enhance performance in terms of selectivity and sensitivity. It also highlights the potential for creating sensitive tools for measuring D-amino acids, which are crucial in various biological processes and hold promise for potential therapeutic applications.

Furthermore, as semi-rational design continues to evolve, integrating high-throughput screening techniques facilitates the rapid evaluation of protein variants, identifying optimal candidates for biosensing applications. Approaches like domain-insertion profiling, as employed by Nadler et al., rapidly construct metabolite biosensors by assessing diverse protein libraries with high-throughput assays [13]. The authors introduce a methodology for the rapid development of single-fluorescent protein biosensors (SFPBs) designed to detect specific metabolites in living cells. Traditional biosensor design has been hindered by the challenge of achieving effective allosteric coupling between a green fluorescent protein (GFP) and a ligand-binding domain, limiting biosensor efficiency and functionality. To overcome this limitation, the researchers employed domain-insertion profiling, a systematic strategy that accelerates the identification of optimal insertion sites within the ligand-binding domain, streamlining the biosensor engineering process. The resulting SFPBs demonstrated exceptional performance, enabling the real-time quantification of metabolites, such as trehalose, at the single-cell level. Their study represents a significant advancement in biosensing technology, offering a scalable and efficient approach for developing high-performance biosensors applicable to diverse biological and medical contexts, ultimately improving our ability to monitor metabolite dynamics in complex environments. This integration of computational predictions and experimental datasets supports the dynamic profiling of protein-based biosensors.

Zhou et al. [14] described a novel approach to enhancing metabolite biosensing in *Saccharomyces cerevisiae* by integrating DNA barcoding with machine learning techniques. Their study focused on constructing a biosensor system using six distinct promoters (pGPD, pENO2, pHSP12, pEXG1, pCYC1, pULI1) with varying activity levels to monitor biochemical reactions based on transcription factor (TF) dosages. By employing a minimal synthetic promoter with a shortened sequence length, high robustness, and strong orthogonality, the authors drove the expression of the fluorescent protein YPet. A key challenge in biosensor development is gene expression noise at the single-cell level, arising from both extrinsic factors, such as cell morphology, plasmid copy number, and cell cycle stage, and intrinsic factors, including fluctuations in DNA, RNA, enzymes, and TF levels. The researchers addressed this issue by integrating machine learning into response curve profiling, which enables the improved deconvolution of noise factors and a more precise characterization of biosensor performance. Their study improves biosensing technology by demonstrating how computational approaches can enhance the accuracy and reliability of biosensors. Using machine learning for data-driven optimization, the work provides a way to develop more sophisticated and high-precision biosensors.

Jiang et al. [15] enhanced the thermostability of *Culex pipiens* acetylcholinesterase (AChE) using a semi-rational design approach, combining computational modeling and site-directed mutagenesis to identify stabilizing mutations. The engineered AChE exhibits improved thermal stability, making it more suitable for biosensing applications. Additionally, a sensitivity analysis confirmed its effectiveness in detecting acephate, an organophosphate pesticide, highlighting its potential for environmental monitoring. This study advances enzyme engineering for robust biosensors capable of detecting environmental contaminants under challenging conditions.

Balaž et al. [16] enhanced the stability of cellobiose dehydrogenase (CDH) using a semi-rational design approach, improving its resilience to peroxide, a common inhibitor in biological systems. By integrating structural insights with targeted mutagenesis, they identified key mutations that significantly increased the enzyme's thermal stability and oxidative resistance. Three mutants with increased activity and stability in the presence of peroxide were found: M65F, M738S, and M685Y. These advancements are particularly valuable for biosensors that detect carbohydrates in complex samples, ensuring reliable performance under challenging conditions. This study highlights the potential of targeted enzyme engineering to optimize biocatalysts for biosensing applications, driving progress in biotechnology and environmental monitoring.

Collectively, the semi-rational design of proteins serves as the methodology of choice in improving biosensors that should meet the increasing demands in medical diagnostics, environmental monitoring, and synthetic biology [17].

2.3. Rational Design of Proteins for Biosensors

The rational design of proteins for biosensors is a sophisticated approach that uses an understanding of protein structure and function to optimize performance in biosensing applications. This method involves the precise engineering of protein scaffolds to enhance their binding affinity and specificity for target analytes, thereby improving the sensitivity of biosensors.

For instance, Pedotti et al. demonstrated the rational modification of estrogen receptor proteins to detect endocrine disruptors (EDCs) with increased sensitivity and specificity, showcasing the potential for tailored receptor designs in environmental monitoring [18]. It represents an innovative strategy for developing biosensing techniques to detect endocrine-disrupting chemicals (EDCs) in environmental samples by leveraging rationally engineered estrogen receptor (ER) proteins as biorecognition elements. Through protein engineering, the researchers designed recombinant ER variants with enhanced binding affinities for specific EDCs, significantly improving assay sensitivity and specificity compared to traditional detection methods. These engineered receptors enable the detection of estradiol, a well-characterized EDC, and offer a more sensitive and convenient alternative to conventional mass spectrometry techniques. This study highlights the potential of modifying natural protein receptors to enhance biosensor performance, opening new avenues for environmental monitoring and pollutant detection using biosensing technology.

Integrating computational modeling with empirical data enhances this design process, enabling researchers to predict how changes in amino acid sequences affect protein interactions and activity, thereby guiding the rational design effort.

This level of control is crucial for applications that require the precise quantification of metabolites or experimental conditions. Additionally, synthetic biosensors have been developed to monitor metabolic states in real-time by engineering transcription factors that respond to specific ligands, as illustrated by Rogers et al. [19], who explored transcription-factor-based synthetic biosensors for precise gene regulation and real-time metabolite monitoring in microbial and mammalian systems. Luminescent, protein-based sensors enable dynamic transcriptional control in response to environmental or intracellular signals. Key regulators, such as MphR (responsive to macrolide antibiotics) and TtgR (modulating multidrug efflux pumps via flavonoids), demonstrate broad applicability in gene expression modulation. This study highlights the potential of synthetic biology to enhance biosensing technology, advancing real-time metabolic monitoring and genetic control in diverse research and medical fields.

Hiraka et al. [20] optimized *Aerococcus viridans* l-lactate oxidase (LOX) using rational design to enhance lactate biosensor performance by facilitating quasi-direct electron transfer. Through targeted mutations, Ala96Leu and Asn212Lys, they enhanced the lactate binding affinity and electron transfer efficiency, resulting in increased sensitivity and a lower detection limit, critical for medical applications, such as lactic acid monitoring in patients. The study highlights the potential of enzyme engineering in biosensor development and underscores the role of immobilization techniques in enhancing stability and reproducibility, advancing clinical and biotechnological biosensing technologies.

Golynskiy et al. emphasized the utility of engineered protein switches that translate biochemical signals into measurable outputs, a key attribute for developing efficient biosensing systems [21]. This trend is further supported by efforts to create multicolored biosensors through synthetic biology techniques, which enable high-throughput and multiparametric analyses of biological systems.

Moreover, ongoing research highlights the importance of incorporating advanced techniques, such as Förster resonance energy transfer (FRET) and the engineering of allosteric proteins, into the rational design of biosensors. Hellweg et al. [22] for example, address the limitations of traditional fluorescent biosensors by developing a novel family of FRET pairs with near-quantitative efficiencies. Utilizing reversible interactions between fluorescent proteins and a labeled HaloTag, this approach enables biosensors with significantly expanded dynamic ranges. The engineered FRET pairs enable the precise monitoring of key cellular metabolites, such as calcium, ATP, and NAD+, while facilitating multicolor, simultaneous imaging in live cells. This method enhances biosensor versatility, advancing research on metabolic processes and cellular signaling with greater accuracy and sensitivity.

For instance, Kimura et al. [23] demonstrated this technique by evolving the LuxR protein to modify its responsiveness to signaling molecules, resulting in a more stringent biosensor capable of detecting lower concentrations of target analytes.

Similarly, directed evolution strategies have been employed to fine-tune the functionalities of transcription factors, resulting in significant improvements in the recognition of environmental pollutants and biological markers. The ability to systematically adjust biosensor properties makes directed evolution a powerful tool in biosensor engineering.

In a study by Du et al., the researchers explored the engineering of biosensors responsive to 4-hydroxyphenylpyruvate (HPP) using a novel dual-selection system. They employed the transcription factor PobR from *Acinetobacter* ADP1, which activates the PpobA promoter upon binding its natural ligand, 4-hydroxybenzoic acid (4HB). The aim was to create a biosensor capable of detecting HPP by generating mutant variants of PobR that are specifically responsive to this substrate. By implementing a selection protocol in *Escherichia coli*, the study successfully identified mutant PobR variants that exhibit enhanced sensitivity and specificity toward HPP, thereby demonstrating the potential of directed evolution and rational design strategies to optimize biosensor performance and expand their applicability in synthetic biology [24].

In their 2016 study, Chong and Ching focused on the development of a colorimetric-based, whole-cell biosensor specifically designed to detect organophosphorus compounds, such as those found in pesticides, Figure 6 [25].

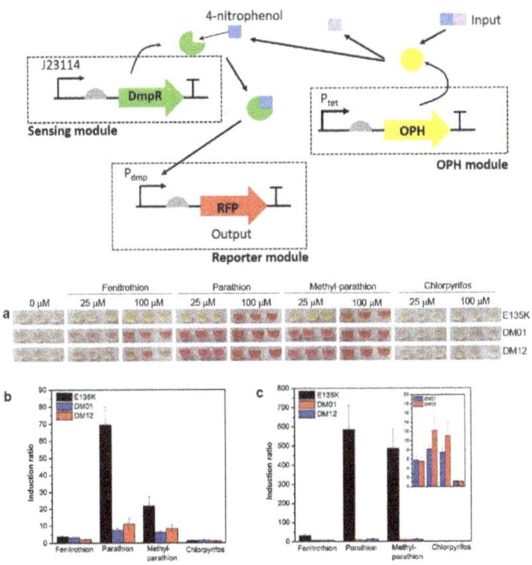

Figure 6. Strategy for development of colorimetric-based, whole-cell biosensor for organophosphorus compounds by engineering transcription regulator DmpR. (**a**) Color of liquid cultures of DmpR mutants after overnight exposure to various OP pesticides (fenitrothion, parathion, methyl-parathion, and chlorpyrifos) at concentrations of 25 and 100 μM at 30 °C at 225 rpm. (**b**) Induction ratios of DmpR mutants after overnight induction to 25 μM OP pesticides at 30 °C at 225 rpm. (**c**) Induction ratios of DmpR mutants after overnight induction to 100 μM OP pesticides at 30 °C at 225 rpm. Reprinted from ref. [25] with permission. Copyright © 2022 American Chemical Society.

They employed directed evolution techniques on the transcription regulator DmpR to optimize the system's responsiveness to these compounds. By modifying the DmpR regulator, they aimed to enhance the expression of a promoter fused to the red fluorescent protein (mRFP1), enabling a visual red coloration as an output signal upon detecting organophosphate pesticides with a phenolic group. The researchers addressed the inherent limitation of colorimetric biosensors, which typically exhibit higher detection limits compared to fluorescence or bioluminescence methods. Their results demonstrated improved sensitivity and effectiveness in utilizing colorimetric signals for detecting hazardous substances, thereby advancing the field of biosensing technology in environmental monitoring [25].

As biotechnological applications grow, the rational design of proteins for biosensors will drive significant innovations in detecting and quantifying various biological targets, enhancing our capabilities across biomedicine and environmental science.

3. Nucleic Acid Engineering for Biosensors

Nucleic acids can be utilized as biosensors to analyze biomarkers within living cells, which is crucial for understanding cellular dynamics and achieving accurate disease diagnoses. These biosensors utilize specific DNA sequences as probes to detect a wide range of biomarkers, providing significant insights for disease diagnostics and drug development. As these methodologies continue to evolve, they will expand the capabilities of DNA biosensors, making them indispensable for applications in disease diagnostics, environmental monitoring, and biotechnology [26,27].

The combination of directed evolution with state-of-the-art screening methods promises to drive innovation in biosensor technology, making it a vibrant area of research.

Directed evolution has emerged as a powerful strategy in the field of RNA engineering, particularly for the development of biosensors with enhanced performance characteristics. By mimicking natural selection processes, researchers can create molecular libraries containing vast variations of RNA sequences, allowing for the identification of variants that possess desired traits, such as increased affinity for target analytes or enhanced specificity. This approach has been particularly effective in optimizing transcription factors and riboswitches, which serve as critical components of RNA biosensors. For instance, directed evolution has been utilized to enhance the responsiveness of the PcaV allosteric transcription factor to aromatic aldehydes, leading to the development of biosensors that can effectively detect small molecules involved in microbial catabolic pathways, thereby facilitating environmental monitoring and bioremediation applications [28].

Also, in a recent study by Huo et al. [29], a riboswitch biosensor specifically responsive to a sesquiterpene amorpha-4,11-diene—was developed through SELEX-SMB that represents a biosensor and a powerful tool for the detection and regulation of small molecules, with broad applications in strain engineering, pathway regulation, and metabolic optimization in synthetic biology.

Recent advancements have highlighted the potential of utilizing large RNA aptamer libraries to generate biosensors capable of detecting diverse small molecules, extending the capabilities of RNA-based sensing technologies (Townshend et al., 2022) [30]. Townshend et al. investigated RNA-based biosensors for detecting small molecules, which is crucial for applications in synthetic biology, healthcare, and environmental monitoring. Using the DRIVER (Diversity-Rich In Vitro Evolution with RNA) method, the study enabled the multiplexed selection of RNA aptamers against unmodified small molecules, mimicking high-throughput drug screening conditions. The approach yields diverse RNA aptamers with strong binding affinities, advancing the development of genetically encoded biosensors. This work accelerates biosensor development for detecting and quantifying small metabolites in complex biological environments by demonstrating the feasibility of multiplexed selections. As the field evolves, combining computational techniques for predicting RNA structure and function alongside experimentally directed evolution will likely yield new solutions in biosensing, leading to novel diagnostic tools that can address critical needs in environmental monitoring, healthcare, and biological research.

Aptamers, which are single-stranded DNA or RNA molecules, can be isolated from combinatorial libraries using the Systematic Evolution of Ligands by Exponential Enrichment (SELEX) process. This strategy mimics natural evolutionary principles to yield high-affinity and high-specificity nucleic acid receptors, effectively serving as recognition elements in biosensors. Recent innovations in SELEX methodologies, including Capture-SELEX and integrated expression platforms, have further refined the directed evolution of nucleic acids for biosensor development. For example, Capture-SELEX techniques have been successfully employed to select DNA aptamers for aminoglycoside antibiotics by directly isolating sequences that not only bind their targets but also exhibit the signaling properties required for biosensor function (Stoltenburg et al., 2012) [31]. Similarly, strategies that integrate an expression platform into the SELEX cycle have enabled the selection of aptamers optimized for binding disease biomarkers, thereby enhancing both the functional performance and the robustness of the resulting biosensor (Ao et al., 2022) [32].

Despite these advances, challenges remain in bridging the gap between laboratory selection and real-world biosensing applications. The integration of evolved aptamers into diverse detection platforms, such as fluorescence, electrochemical, and paper-based biosensors, has demonstrated improved analytical performance; for instance, RNA aptamers have been tailored for detecting specific antibiotics, like levofloxacin, with relatively few selection cycles (Kramat et al., 2024) [33]. This study presents the development and characterization of an RNA aptamer-based biosensor for the detection of levofloxacin, a fluoroquinolone antibiotic widely used in human and veterinary medicine. The study describes the selection of a high-affinity RNA aptamer through a SELEX-based approach and its integration into a sensing platform using fluorescence quenching as a transduction mechanism. The aptamer exhibited strong specificity for levofloxacin over other structurally similar antibiotics, with a dissociation constant (Kd) in the low micromolar range, indicating good binding affinity. The sensor was capable of detecting levofloxacin concentrations in the nanomolar to micromolar range, making it suitable for applications in pharmaceutical quality control and potentially in environmental monitoring, where antibiotic contamination is a concern.

Moreover, ongoing improvements in selection techniques, including the refinement of screening conditions and amplification fidelity, are critical to overcoming practical limitations, such as high detection limits and reproducibility issues (Jaeger et al., 2019) [34]. The work presents the development of a paper-based biosensor utilizing an RNA aptamer specific to ciprofloxacin, a widely used antibiotic. The researchers characterized the binding affinity and specificity of the RNA aptamer to ciprofloxacin, confirming its potential as a biorecognition element. They employed inkjet printing technology to immobilize the aptamer onto paper substrates, creating a cost-effective and user-friendly platform for antibiotic detection. This approach offers a promising avenue for the development of portable and accessible diagnostic tools, particularly beneficial in resource-limited settings where the rapid and reliable detection of antibiotics is crucial for monitoring environmental contamination and ensuring food safety.

A summary of the biosensors improved with protein and nucleic acid engineering, as well as their characteristics, is given in Table 1.

Table 1. A summary of the most representative examples of biosensors improved using protein/nucleic acid engineering and their characteristics.

Biosensor	Detection	LOD	Linear Range	Sample	References
Protein engineering for biosensors					
TP2-Rho P53 peptide biosensor	R248Q mutant of protein p53	Nanomolar concentrations of p53 mutant in lung cancer cell extract	NR	Lung cancer cells	[1]
Pb whole-cell biosensor	Detection of Pb	0.045 μg/L	NR	Food products, water samples	[2]
Bacterial Cd biosensor (CadR)	Detection of Cd	0.45 μg/L	NR	Soil and water	[7]
Bacterial biosensor for detecting lead toxicity	Detection of Pb	0.012 μM Pb	0.012–3.125 μM Pb	Artificially polluted water	[8]
Bacterial biosensor for C1 molecules	Detection of formate, formaldehyde, and methanol	NR	1.0–250 mM formate 1.0–50 μM formaldehyde 5–400 mM methanol	Cell culture	[9]
Bacterial biosensor	Detection of organophosphorus compounds	10 μM parathion	NR	Artificial samples	[25]
Nucleic acid engineering for biosensors					
Riboswitch biosensor	Detection of sesquiterpene	NR	10–100 mg/mL amorpha-4,11—diene	Samples for high throughput screening and for metabolic engineering	[29]
Nucleotide aptamer	Detection of levofloxacin	100 μM	NR	Artificial samples	[33]
Nucleotide aptamer	Detection of ciprofloxacin	The maximum residue limit (MRL)	NR	Food samples	[34]

4. Nanomaterials Engineering for Biosensors

Nanomaterials have transformed the development of biosensors, providing distinct properties that improve biosensor performance, including enhanced stability, repeatability, and sensitivity [35]. Nanomaterials possess a high surface-area-to-volume ratio, great electronic conductivity, and excellent magnetic and physicochemical properties [36,37], and they provide enhanced accessibility for the analyte to the sensing element. Different kinds of nanomaterials, including carbon nanomaterials, metal and metal oxide nanoparticles, magnetic nanoparticles, and quantum dots, etc., have been successfully applied to develop biosensors. Numerous studies have investigated the use of different carbon nanomaterials in biosensors, demonstrating their ability to produce highly sensitive biosensors. These materials have proven to be effective transducers in electrochemical processes due to their exceptional properties, including chemical stability, high electrical conductivity, a large surface-to-volume ratio, and robust mechanical strength [38]. Carbon nanotubes, carbon nanofibers, quantum dots, graphene, and graphene oxide are commonly used carbon materials for biosensor development. The easy modification and functionalization of carbon-based nanomaterials have allowed their diverse integration with sensing elements in biotechnology [39].

For instance, Ansari et al. (see Table 2) utilized a nanocomposite consisting of two carbon nanomaterials: graphitic carbon nanosheets and carboxyl-functionalized multi-walled carbon nanotubes, which were modified with zirconium oxide nanoparticles, for the development of a biosensor [40]. They reported a label-free electrochemical aptasensor for the ultrasensitive detection of matrix metalloproteinase-9 (MMP-9), a biomarker for various pathological conditions. The proposed modification approach for carbon nanomaterials resulted in a significant increase in the electroactive surface area, by 205%, enabling the better immobilization of the matrix metalloproteinase aptamer. As a result, the sensor demonstrated an impressively broad detection range from 50 to 1250 pg/mL, with an outstandingly low limit of detection of 10.51 pg/mL. Additionally, the aptasensor exhibited real-world applicability when tested with human serum and saliva samples.

Similarly, Shi et al. conducted a study where they utilized multiwalled carbon nanotubes that were modified with gold and manganese dioxide nanoparticles to develop an electrochemical biosensor for the detection of serum homocysteine (Hcy): a reliable biomarker for cardiovascular, cerebrovascular, and neurological diseases [41]. The researchers applied the proposed composite to screen-printed carbon electrodes, resulting in the creation of a portable, disposable sensor with exceptional electrochemical activity and catalytic ability. The biosensor demonstrated a wide linear range, high sensitivity, and a low detection limit for homocysteine, making it a promising tool for the accurate determination of serum homocysteine levels. This development holds potential for enabling convenient home health monitoring for individuals at risk of these diseases.

du Plooy et al. developed a highly sensitive, label-free electrochemical immunosensor for SARS-CoV-2 IgG antibody detection through the modification of screen-printed carbon paste electrodes with graphene quantum dots (GQDs) [42]. Graphene quantum dots are a zero-dimensional (0D) member of the carbon family, consisting of a single to a few layers of graphene sheets with lateral dimensions of <10 nm [43]. GQDs possess extraordinary physicochemical characteristics, including edge effects, non-zero band gaps, and quantum confinement effects, as a result of which they hold great potential in the energy, electronic, and optical industries, as well as biosensing [44]. du Plooy et al. confirmed in their work that GQDs were instrumental in raising the detection sensitivity and expanding the active surface area of the working electrode, resulting in the improved binding efficiency of the immunocomplex. The sensing mechanism depends on the disruption of the redox reactions and electron transfer kinetics of the $Fe(CN)_6^{3-/4-}$ redox probe, due to the formation of an immunocomplex between the specific antigen and the IgG antibody. The high sensitivity of the sensor was confirmed by achieving a low limit of detection (LOD) of 2028 ng/mL, while remarkable selectivity was proven by the minimal interference of other antibodies. Additionally, the authors examined the possibility of using other highly conductive and low-oxygen-content carbon nanomaterials (electrochemically reduced graphene oxide and carbon quantum dots) for sensor development. Their results indicate that the performance of the electrochemical immunosensor is primarily determined by its binding affinity, rather than its electrode conductivity, which may be enhanced by utilizing carbon–carbon conjugated carbon nanostructures.

Nanostructured conducting polymers (nCPs) are receiving increasing interest, with regard to the development of biosensors, due to their biocompatibility in cellular environments [45]. Mirzaei et al. reported an aptasensor for *H. pylori* (Hsp60) detection using reduced graphene oxide (RGO) decorated with gold nanoparticles (Au NPs) [46]. The main role of the RGO was as a medium, assembling different parts of the biosensor, while the Au NPs improved the system's conductivity and provided functional groups for aptamer attachment. In addition, the authors used a conductive polymer, PTP, in order to fill the surface of the RGO at sites where there were no Au NPs, and this method prevented the attachment of the analyte to the surface and any false responses from the biosensor. The determination of Hsp60 was studied in a hexacyanoferrate medium, and the sensitivity of the prepared sensor was investigated by several electrochemical techniques (CV, SWV, and EIS analysis) to check the effect of the used material on the performance of the sensor after each synthesis step. Their results indicate that the prepared aptasensor has suitable selectivity for Hsp60, among other biological materials.

Jaradat et al. have developed an innovative electrochemical impedimetric immunosensor that can detect *H. pylori* bacteria directly by using the HopQ protein as a biomarker [47]. The biosensor fabrication process involves electropolymerizing a layer of polypyrrole nanotubes (PPy-NTs) onto SPCE, followed by the drop-casting of multiwalled carbon nanotubes (MWCNT-COOH). The researchers chose PPy-NTs because they have excellent conductivity, biocompatibility, and redox capabilities. This simplifies the sample preparation process by eliminating the need to add redox probes during measurement. MWCNT-COOH provides covalent binding sites for HopQ antibodies (HopQ-Ab) on the biosensor surface. The results demonstrate excellent selectivity and a dynamic linear range of 5 pg/mL to 1.063 ng/mL. The calculated limit of detection was 2.06 pg/mL.

High-entropy nanomaterials have gained significant attention in research due to their unique synergy effect from multiple metals and exceptional material properties. These materials provide increased active sites, optimized electronic structures, and improved electronic conductivity in electrochemical immunoassays [48]. Lv et al. have developed an innovative, label-free electrochemical immunosensor using a self-supported, PtPdMnCoFe, high-entropy alloy with nanochain-like internetworks (HEAINN) for the highly sensitive detection of the biomarker neuron-specific enolase (NSE) [49]. The proposed material was created using a simple, one-pot, wet-chemical, co-reduction strategy and has been shown to be effective for the reduction of H_2O_2, acting as a signal amplifier for a novel, label-free, electrochemical, amperometric immunosensor for the highly sensitive determination of NSE. Under optimized analytical conditions, the sensor has demonstrated a broad linear range (0.1 pg mL^{-1} to 200 ng mL^{-1}) and a low limit of detection (LOD = 0.0036 pg mL^{-1}). Furthermore, it has successfully quantified NSE in human serum samples, yielding highly satisfactory results. Similarly, Tang et al. prepared well-defined, PtRhMoCoFe, high-entropy, alloyed nanodendrites (HEANDs), by a wet-chemical, co-reduction method and adopted it to build an electrochemical, label-free biosensor for an ultrasensitive bioassay of the biomarker cardiac troponin I (cTnI) [50]. By using the developed method, the authors achieved a low detection limit of 0.0095 pg mL^{-1} and further explored the practicability of the method in serum samples with satisfactory recovery (102.0%).

Recent breakthroughs in nanotechnology have paved the way for the development of functional nanomaterials with inherent enzyme-like activity. These innovative nanomaterials, referred to as "nanozymes", effectively replicate the catalytic action of natural enzymes. Nanozymes demonstrate comparable kinetic behaviors to natural enzymes and effectively catalyze the conversion of substrates into oxidized products, offering promising applications in biosensing [51].

Metal–organic frameworks have great potential in developing nanozymes, due to their ability to precisely identify uniformly distributed active sites at the atomic scale in MOF-based nanozymes, offering valuable insights into catalytic mechanisms and enabling the design of more-efficient catalytic systems [52,53]. Huang et al. developed a high-performance electrochemical biosensor based on a highly active, dual-nanozyme-amplified system [54]. For this purpose, they used conductive MOF nanosheets decorated with high-density, ultrafine gold nanoparticles (Au-NPs). The MOFs provide a large surface area and abundant active, open metal sites ($Cu–O_4$), which could improve the catalytic activity towards H_2O_2. The large number of exposed oxygen atoms also provided additional sites for the deposition of gold nanoparticles without agglomeration. The combination of the high catalytic activity of MOF and Au-NP, along with their unique structural and electrical properties, resulted in a nanohybrid-modified electrode with an excellent sensing performance for H_2O_2. It achieved an extremely low detection limit of 5.6 nM and a high sensitivity. Furthermore, this nanoscale electrochemical biosensor was used for the real-time monitoring of H_2O_2 released from different human colon cells to distinguish colon cancer cells from normal colon epithelial cells. This demonstrates great potential for the early diagnosis and management of various cancer diseases.

An innovative 3D-printed device enhanced with an iron-based metal–organic framework (Fe(II)-MOF) has been created for measuring glucose levels in artificial sweat [55]. In this study, Koukouviti et al. developed an Fe(II)-MOF that is water-insoluble and non-toxic, serving as a nanozyme for detecting glucose through differential pulse voltammetry (DPV). The electrochemical device, consisting of three electrodes and a holder, was fabricated using 3D printing technology in a single step. The Fe(II)-MOF was applied to the working electrode, followed by trapping it with a Nafion film. Unlike most glucose sensors, the DPV detection of glucose was carried out in a slightly acidic environment. The Fe(II)-MOF exhibited the capability to capture glucose on its surface and convert it to gluconolactone through the redox activity of Fe centers during the electrode's pre-polarization step and subsequent anodic potential scan. With 3D printing allowing the creation of wearable electrochemical sensors using flexible filaments, the proposed Fe(II)-MOF/3D-printed device opens up possibilities for on-skin glucose monitoring.

A summary of the most representative examples of biosensors improved using nanomaterial engineering and their characteristics is given in Table 2.

Table 2. A summary of the most representative examples of biosensors improved using nanomaterial engineering and their characteristics.

Electrode Nanomaterial	Analyte	Linear Range	LOD	Real Matrices	Reference
Graphitic carbon nitride/zirconium dioxide/multiwalled carbon nanotubes	Metalloproteinase-9	50 to 1250 pg/mL	10.51 pg/mL	Human serum and saliva	[40]
Gold/manganese (IV) oxide/multiwalled carbon nanotubes	Homocysteine	5–125 μM/L	0.6173 μM/L	Serum homocysteine	[41]
Graphene-oxide-doped gold nanoparticles conjugated with polythiophene	*H. pylori*	10–900 nM/L	0.0080 μM/L with impedance; 0.0067 μM/L for SWV method	/	[46]
Polypyrrole nanotubes and carbon nanotubes nanocomposite	*H. pylori* outer membrane protein (HopQ)	5 pg/mL 1.063 ng/mL	2.06 pg/mL	Spiked drinking water	[47]
Self-supported PtPdMnCoFe	Neuron-specific enolase (NSE)	0.1 pg/mL–200 ng/mL	0.0036 pg/mL	Human serum samples	[49]
Dendritic quinary PtRhMoCoFe	cTnI	0.0001–200 ng/mL	0.0095 pg/mL	Serum samples	[50]
MOF@Pt@MOF nanozyme	Exosomal miRNA	1 fM/L–1 nM/L	0.29 fM/L	MCF-7 cells, MCF-10 A cells, and serums of a breast cancer patient and a healthy person were tested	[53]
2D conductive MOF nanosheets/gold nanoparticles	H_2O_2	50 nM/L–16.4 mM/L	5.6 nM/L	Human colon cells	[54]
Fe(II)-MOF nanozyme	Glucose	100–600 μmol/L	17.6 μmol/L	Artificial sweat samples	[55]

5. Biosensors for Rare Diseases Monitoring

5.1. Biosensors for the Detection of Aβ42, a Biomarker for Alzheimer's Disease Diagnosis

Alzheimer's disease (AD) is a slowly progressing degenerative condition with an extended initial period, during which effective therapies are lacking. Timely identification and intervention during the early phases of AD are widely acknowledged to enhance treatment outcomes. Utilizing body fluid biomarkers is a valuable approach for the prompt detection of early-stage AD. Amyloid β (1–42) (Aβ42) is a crucial biomarker for monitoring AD development. Electrochemical biosensors provide a rapid, portable, and field-capable method for analyzing Aβ42 levels, offering significant benefits in this context. Because of all of the above, a rapid increase in research on this topic has been observed.

Yu and co-authors introduced a novel approach for the highly sensitive detection of soluble β-amyloid peptides (Aβ1–40/1–42) using an electrochemical nanoprobe composed of HRP–Au–gelsolin. Instead of relying on antibody-based detection methods focusing on peptide termini recognition, the study highlights the potential of the specific interaction between gelsolin and Aβ for precise and sensitive peptide evaluation. By developing the HRP–Au–gelsolin nanohybrid through Au nanoparticle functionalization, this group of authors established a sandwich-type sensor array capturing soluble Aβ1–40/1–42 with surface-bound gelsolin and enhancing electrochemical signals via HRP catalysis. This innovative methodology demonstrated high sensitivity and a broad linear range towards Aβ1–40/1–42, achieving a remarkable detection limit of 28 pM suitable for monitoring Aβ levels in both normal and Alzheimer's disease rat brains. The experimental results indicated significant decreases in soluble β-amyloid levels in the cerebrospinal fluid (CSF) and specific brain tissues of AD model rats compared to the normal group, showing the potential of this new method for early AD diagnosis [56].

A study conducted by Diba Sharmin and co-authors [57] showed a demonstration of an electrochemical immunosensor by using a surface sandwich complex on a gold nanoparticle (NP)-modified, screen-printed carbon electrode for the highly sensitive detection of the amyloid-beta 1–42 peptide (A) at femtomolar levels in serum and plasma. This involved the use of selective antibodies (antiA (12F4) and (1E11)) with distinct binding sites for the amyloid-beta 1–42 peptide and the formation of surface sandwich complexes of antiA (12F4)/A/antiA (1E11)-ALP employing a sequential adsorption method. The immobilized ALP enzyme, upon reaction with the substrate, 4-amino phenyl phosphate, generated voltammetric signals that increased linearly with the A concentration. The applied differential pulse voltammetry established the lowest detectable concentration of 100 fM of amyloid-beta 1–42 peptide within a linear response range from 100 fM to 25 pM. After optimization, the immunoassay platform was successfully used to determine the native concentration of the amyloid-beta 1–42 peptide in diluted human serum and plasma samples, with the results validated using a commercially available ELISA test.

A study by Ding and co-authors [58] focused on the use of the amyloid-β peptide (1–42) for the early diagnosis of Alzheimer's disease, emphasizing its potential for high diagnostic accuracy. The researchers designed an immunosensor based on the peroxidase-like activity of heme-Aβ42 for the purpose of detecting Aβ42. Through a series of modifications to the glassy carbon electrode, including the electrodeposition of Au particles and the successive addition of polythionine–methylene blue (PTH−MB), AuNPs, monoclonal antibodies (anti-Aβ42), and bovine serum albumin (BSA), a competitive recognition capability for Aβ42 and heme-Aβ42 was achieved. The anchored heme-Aβ42 exhibited robust electrocatalytic activity in the presence of H_2O_2 and ferrocenemethanol. The study determined detection limits of 17.3 pM in a PBS of pH 7, 25.2 pM in serum, and 23.8 pM in saliva, with a satisfactory recovery range of 71.0% to 117.8% under optimized conditions.

A study done by Hsu and co-workers [59] reported the development of an Alzheimer's disease biosensor for the early detection of amyloid-beta 1–42 protein by employing a simple nanorestructuring technique involving the oxidation–reduction cycle (ORC) via an electrochemical system to modify a Au sheet plate. The ORC technique was utilized to enhance the topology of Au substrates through the roughening and growth of Au grains in a KCl electrolyte solution, resulting in approximately 15% larger electroactive sites compared to polished Au without ORC. The roughened substrate was further functionalized with the specific antibody β-amyloid Aβ (1–28) through HS-PEG-NHS modification, enabling the detection of Aβ (1–42) peptide. The ORC structure provided the detection of the Aβ (1–42) peptide, with a low detection limit of 10.4 fg/mL and a wide linear range of 10^{-2} to 10^{6} pg/mL. This innovative structure demonstrated promising potential as an early-stage Alzheimer's disease (AD) detection platform, offering low-cost fabrication and ease of operation.

A study reported by Le and co-authors [60] applied a capacitive biosensor for the non-Faradaic detection of the aβ 1–42 peptide. This approach involves the immobilization of a specific anti-aβ antibody onto a self-assembled, monolayer-functionalized, interdigitated, chain-shaped electrode. What sets this research apart from previous studies is its direct detection of the aβ peptide without the use of a redox probe ($(Fe(CN)_6)^{3-}/^{4-}$), eliminating the potential protein denaturation caused by metallization. The non-Faradaic capacitive measurement technique employed for the direct detection of aβ without a redox probe represents a significant departure from the Faradaic measurement approach, avoiding the charge transfer resistance of the redox probe. By measuring the relative change in electrode interfacial capacitance due to specific antibody-aβ binding, the biosensor demonstrated a linear detection range of between 10 pg/mL and 104 pg/mL in human serum. The high binding affinity of the anti-aβ/SAM/ICE biosensor is evidenced by the small dissociation constant, Kd, of the antibody–antigen interaction, indicating its potential for the label-free and direct measurement of the aβ 1–42 peptide for point-of-care Alzheimer's disease diagnosis without the need for a redox probe.

A study done by Ribeiro and co-workers [61] introduces a potentiometric biosensor designed for the point-of-care analysis of amyloid β-42 (Aβ-42). This method utilizes the molecular imprint polymer technique, where covalently immobilized Aβ-42 is utilized to create specific detection sites on the surface of single-walled carbon nanotubes. The biosensor fabrication involved binding Aβ-42 to the single-walled carbon nanotubes' surface and imprinting it with acrylamide, N,N′-methylene-bis-acrylamide, and ammonium persulphate. The presence of imprinting sites was verified by comparing a molecular-imprint-polymer-modified surface with a negative control. The evaluation of the sensing material's ability to rebind Aβ-42 was carried out using Raman spectroscopy and Fourier transform infrared spectroscopy, followed by an analytical performance assessment via potentiometric transduction: cationic slopes of 75 mV-decade^{-1} in a buffer of pH 8.0 and a detection limit of 0.72 µg/mL. The molecular imprint polymer material demonstrated favorable cationic slopes and a low detection limit, affirming its capability to discriminate Aβ-42 in the presence of other biomolecules in a given solution. This innovative biosensor presents a promising approach for the detection of Aβ-42, with potential applications in point-of-care analysis.

5.2. *Biosensors for the Monitoring of Hepatitis B*

Monitoring and treating hepatitis B are essential due to the potential development of acute or chronic hepatitis, which can result in serious health complications. Regular monitoring enables the early detection of infection and the timely implementation of appropriate treatment, significantly improving the disease outcome. Continuous hepatitis

B treatment, including vaccination and therapy, plays a key role in suppressing the HBV virus and preventing potential hepatitis complications.

A study conducted by Chen and co-authors [62] aimed to develop an efficient, sensitive, precise, and cost-effective method for genotyping hepatitis B viruses (HBVs). This involved the use of mercapto-modified B1-, B2-, C1-, and C2-specific genotyping probes, each comprising two probes for every HBV genotype, creating a dual verification system. These probes were immobilized on four gold electrodes via Au-S bonds, allowing for the distinction of different genotypes based on the differential charge generated by RuHex binding to DNA phosphate groups before and after hybridization. The results demonstrated distinct changes in the detected charges during hybridization with genotypes B and C, as well as a significant increase in charges during hybridization with mixed genotypes (B and C) at all four electrodes. The method exhibited a wide linear detection range (10^{-10}–10^{-7} M) and a high sensitivity for detecting mixed B or C HBV genotypes, proving to be a simple, specific, and cost-effective approach for rapid HBV genotyping via electrochemical sensing. This innovative method offers the sensitive detection of mixed B and C HBV genotypes, with significant potential for practical application.

An innovative, wireless-based detection system utilizing an electrochemical card (eCard) controlled by a smartphone was created to target the hepatitis B surface antigen (HBsAg) by Teengam and co-workers [63]. This label-free electrochemical platform provides user-friendly operation for convenient point-of-care diagnosis. The modification of a disposable, screen-printed carbon electrode with chitosan and glutaraldehyde, followed by antibody immobilization, established a simple, yet efficient and stable, method verified by electrochemical techniques. The smartphone-controlled eCard sensor quantified HBsAg by detecting the change in the current response of the $[Fe(CN)_6]^{3-/4-}$ redox couple upon HBsAg presence, achieving a linear calibration curve from 10 to 100,000 IU/mL with a detection limit of 9.55 IU/mL. Successfully tested on 500 chronically HBV-infected serum samples, the HBsAg eCard sensor exhibited high sensitivity (97.75%) and specificity (93%), highlighting its practicality and accuracy.

In a study from Fan and co-authors [64], the researchers focused on preparing amino-functionalized Ti_3C_2Tx MXene and fabricating a highly sensitive electrochemical DNA biosensor for the swift detection of HBV-DNA. By regulating the surface's−NH2 concentration and anchoring gold nanoparticles, they successfully controlled the density of p-DNA, leading to the creation of a robust, stable, and specific $Ti_3C_2NH_2$ MXene@Au nanocomposite DNA sensor. This sensor exhibited a wide detection range of 10^{-17}–10^{-7} M and a low detection limit of 1.05×10^{-14} M, enabling the detection of HBV-DNA fragments, even in an artificial serum environment. This innovative biosensor holds promise for advancing the rapid and accurate detection of HBV-DNA, supporting the overall management and prevention of hepatitis B.

In a study by Li and co-authors [65], the research introduces a straightforward paper-based electrochemical sensor, created through paper folding, capable of detecting a 30-base nucleotide sequence specific to DNA from the hepatitis B virus with a remarkable sensitivity of 85 pM. This innovative device was developed based on previous design principles used for protein detection via a metalloimmunoassay. It boasts several key features: 1. the design integrates a simple origami assembly with a hollow-channel paper structure, ideal for accommodating micrometer-scale particles, along with a practical slip layer for precise incubation timing; 2. the sensor achieves two amplification stages: using silver nanoparticle labels for a potent amplification factor of up to 250,000 and strategically placing magnetic microbeads for an additional ~25-fold amplification at the detection electrode; 3. the assay does not require enzymes or antibodies, enhancing its speed, stability, and reliability significantly; 4. a single sample incubation step initiates the detection process, streamlining the workflow effectively.

5.3. Biosensors for the Monitoring of Human Papillomavirus

Monitoring for HPV is crucial because certain types of the virus, such as HPV-16, are linked to the development of cervical cancer, which can be prevented if caught at an early stage. Regular monitoring can help identify changes in cervical cells caused by HPV, allowing for timely treatment to prevent the progression to cancer. Additionally, monitoring for HPV enables healthcare providers to assess the infection status and provide appropriate care, contributing to better overall health outcomes for individuals.

Parrek and co-authors [66] constructed a new device that uses DNA to detect human papillomavirus-16 (HPV-16), which is linked to cervical cancer. To build this tool, the scientists modified a glass electrode with indium tin oxide (ITO) and added graphene oxide and silver-coated gold nanoparticles to sense HPV-16. They then attached specific DNA probes for HPV-16 to the modified electrode. Using special techniques, the scientists examined the characteristics of these materials at a microscopic level. By using various methods, like cyclic voltammetry, they analyzed how the DNA probes interacted with the target DNA by measuring electrical changes. This new biosensor displayed significant sensitivity, detecting HPV-16 effectively with a detailed range of concentrations. The successful development of this biosensor indicates a promising future for quick and accurate point-of-care diagnostic tools, potentially enabling the early detection of HPV-16 infections. The developed biosensor showed a sensitivity as high as 0.54 mA/aM with a linear range of 100 aM to 1 μM and a 100 aM LOD; the proposed biosensor exhibited an excellent performance with a rapid diagnosis.

A study conducted by Yang and co-workers [67] introduced a unique biosensor by combining a modified gold electrode and a super sandwich structure using 3-aminopropyltriethoxysilane (APTES), resulting in a biosensor with a lower detection limit (5.475×10^{-16} mol/L) and a wide linear range (1.0×10^{-13} mol/L to 1.0×10^{-6} mol/L). The experimental results demonstrated the biosensor's good stability, and further testing revealed its potential as a reliable clinical tool for disease diagnosis with strong interference resistance in complex human serum samples. This innovative biosensor has promising implications for the early detection and management of cervical cancer.

The development of an electrochemical sensor for detecting HPV 16 by using dual-signal amplification is a significant step in improving the early diagnosis and prevention of cervical cancer. This kind of sensor utilized an APTES-modified glassy carbon electrode to enhance its stability and incorporated gold nanoparticles and a chain amplification reaction for signal amplification [68]. With a limit of detection (LOD) of 1.731×10^{-16} mol/L and a wide linear response range, the sensor demonstrated great potential, exhibiting the good recovery of serum samples and strong anti-interference properties. The emphasis on low cost, high sensitivity, and stability aligns with the future prospects of developing effective cervical cancer prevention methods and electrochemical biosensors.

A study done by Rasouli and co-authors [69] presents a label-free biosensor designed to detect HPV-16 using a straightforward and environmentally friendly method. The screen-printed carbon electrodes were coated with Fe_3O_4-Au core-shell nanoparticles to enhance their sensing capabilities. Then, the electrode surfaces were functionalized with thiolated single-strand DNA (ssDNA) probes specific to HPV DNA sequences. The hybridization process was monitored using cyclic voltammetry and differential pulse voltammetry with $[Fe(CN)_6]^{3-/4-}$ as the redox indicator, successfully distinguishing the absence and presence of immobilized probe DNA. The biosensor demonstrated its optimal performance at a probe DNA concentration of 5 μM, achieving a limit of detection of 0.1 nM and a sensitivity of

2.4 μA/nM, providing a promising avenue for the development of effective HPV detection tools.

A very interesting study was done by Chaibun and his co-workers [70]. In contrast to previous studies, where the focus was on the detection of a specific analyte—HPV 16 peptide—in this study with a sandwich assay, the concentration of two types of analytes—HPV 16 and HPV 18 peptides—was determined at the femtomolar detection limit. The method utilizes a sandwich hybridization process involving the HPV target, a silica-redox dye reporter probe, and a capture probe, with subsequent electrochemical detection. This biosensor exhibits high specificity and sensitivity, detecting HPV-16 at a limit of 22 fM and HPV-18 at 20 fM, within a wide range of 1 fM to 1 μM. Assessment using oral and cervical samples demonstrated the biosensor's consistency with nested PCR/gel electrophoresis detection. Its quick completion time of 90 min, simplicity, and high sensitivity make it suitable for HPV detection in clinical laboratories and epidemiological studies, offering an alternative to existing detection methods. This innovative biosensor holds great promise for efficient HPV detection.

A summary of the techniques and biomarkers used in some recent research articles to improve biosensors is presented in Table 3.

Table 3. Summary of techniques and biomarkers used in recent biosensor research.

Field	Technique	Biosensor Type	Biomarkers	Advancements	Refs.
Protein Engineering	Directed evolution	Optical, Electrochemical	Mutant p53, Pb, Cd, C1 compounds, cholesterol, transcription factors	High-specificity biosensors via protein mutation and screening	[1–10]
	Semi-rational design	Optical, Electrochemical	Glucarate, D-serine, D-alanine, metabolites, acephate	Enhanced enzyme selectivity/stability using structural data	[11–17]
	Rational design	Optical, Electrochemical	Estrogen, lactate, calcium, ATP, NAD+, HPP, pesticides	Targeted mutations for ligand binding, allosteric sensor design	[18–25]
Nucleic Acid Engineering	SELEX and aptamer engineering	Fluorescent, Electrochemical	Aminoglycosides, levofloxacin	Improved aptamer affinity and specificity for antibiotics and proteins	[26,27,31–33]
	Riboswitch development	Fluorescent	Amorpha-4,11-diene, aromatic aldehydes	Riboswitches for metabolite detection in synthetic biology	[28,29]
	RNA aptamer screening (DRIVER)	Fluorescent, Paper-based	Small molecules	High-throughput aptamer generation with multiplex selection	[30]
Nanomaterials Engineering	Carbon nanomaterials (CNTs, graphene)	Electrochemical	MMP-9, *H. pylori*, SARS-CoV-2 IgG	Enhanced conductivity and immobilization for pathogen detection	[38,40–42,46,47]
	MOF-based nanozymes	Electrochemical	Exosomal miRNA, H_2O_2, glucose	Ultrasensitive catalytic detection via MOF-Au composites	[52–55]
	High-entropy alloy nanomaterials	Electrochemical	NSE, cTnI	Multi-metallic platforms with ultralow LOD for disease markers	[48–50]

6. Conclusions

The engineering of proteins, nucleic acids, and nanomaterials plays a key role in the development of advanced biosensors that enable the early detection and monitoring of rare diseases. Using innovative engineering techniques, including protein evolution, aptamer design, and nanomaterial engineering, these biosensors offer high specificity and sensitivity in detecting the specific biomarkers of rare diseases, even at very low concentrations, which is vital for timely intervention, thereby significantly improving treatment, while at the same time reducing the cost and simplifying the treatment of the disease itself. The individual or group integration of these technologies into wearable devices opens new opportunities for continuous patient monitoring, advancing personalized medicine and improving treatment outcomes. Currently, these systems are still in the development and research phase, but their application could have a significant impact on the future of rare disease diagnostics, providing healthcare professionals with better tools for assessing and managing patient conditions. The importance of this research is also reflected in the increased number of papers in this area, some of which are summarized in this review.

Proteins and nucleic acids offer several distinct advantages over nanomaterials in the development of biosensors, particularly in terms of biorecognition, specificity, and compatibility with biological systems. Engineered proteins, such as enzymes, transcription factors, and antibodies, exhibit high catalytic efficiency and target selectivity, enabling the sensitive detection of specific analytes under physiological conditions. Nucleic-acid-based elements, including aptamers and DNAzymes, provide programmable and highly tunable recognition capabilities through predictable base-pair interactions, allowing the design of versatile and robust sensing platforms. Unlike many synthetic nanomaterials, which may require complex functionalization and raise concerns about biocompatibility or environmental toxicity, biomolecules are inherently biocompatible. They can be sustainably produced via recombinant or synthetic methods. Furthermore, their molecular adaptability supports integration into various transduction systems, enhancing the modularity and precision of biosensor designs.

From a commercial perspective, integrating engineered proteins, nucleic acids, and nanomaterials into biosensors presents promising avenues for developing scalable, cost-effective, and highly sensitive diagnostic platforms. Advances in synthetic biology and nanotechnology have accelerated the development of the next-generation biosensors suitable for real-time, point-of-care, and wearable applications. With a growing market demand across healthcare, environmental monitoring, and food safety, these bioengineered systems will drive innovation and commercialization, bridging the gap between laboratory research and practical deployment.

In the coming years, further intensive work in this field is expected, expanding research capacities, and further progress in reducing costs, improving accuracy, and expanding the availability of these technologies is expected, contributing to the global improvement of healthcare and the quality of life of patients with rare diseases.

Author Contributions: Conceptualization R.P. & D.S., writing—original draft preparation M.C.P. & V.S., writing—review and editing R.P. & D.S. All authors have read and agreed to the published version of the manuscript.

Funding: This research was funded by the Ministry of Science, Technological Development, and Innovation of the Republic of Serbia, grant numbers 451-03-136/2025-03/200168 and 451-03-136/2025-03/200026.

Conflicts of Interest: The authors declare no conflicts of interest.

References

1. Pellerano, M.; Naud-Martin, D.; Mahuteau-Betzer, F.; Morille, M.; Morris, M.C. Fluorescent Biosensor for Detection of the R248Q Aggregation-Prone Mutant of p53. *ChemBioChem* **2019**, *20*, 605–613. [CrossRef]
2. Shen, L.; Chen, Y.W.; Pan, J.J.; Yu, X.; Zhang, Y.B.; Guo, B.X.; Wang, J.Q.; Liu, Y.; Xiao, X.; Chen, S.P.; et al. Development of a highly sensitive PbrR-based biosensor via directed evolution and its application for lead detection. *J. Hazard. Mater.* **2025**, *488*, 137489. [CrossRef]
3. Feng, J.; Jester, B.W.; Tinberg, C.E.; Mandell, D.J.; Antunes, M.S.; Chari, R.; Morey, K.J.; Rios, X.; Medford, J.I.; Church, G.M.; et al. A general strategy to construct small molecule biosensors in eukaryotes. *eLife* **2015**, *4*, e10606. [CrossRef]
4. Paulmurugan, R.; Afjei, R.; Sekar, T.V.; Babikir, H.A.; Massoud, T.F. A protein folding molecular imaging biosensor monitors the effects of drugs that restore mutant p53 structure and its downstream function in glioblastoma cells. *Oncotarget* **2018**, *9*, 21495–21511. [CrossRef]
5. Taylor, N.D.; Garruss, A.S.; Moretti, R.; Chan, S.; Arbing, M.A.; Cascio, D.; Rogers, J.K.; Isaacs, F.J.; Kosuri, S.; Baker, D.; et al. Engineering an allosteric transcription factor to respond to new ligands. *Nat. Methods* **2016**, *13*, 177–183. [CrossRef] [PubMed]
6. Gohil, K.; Wu, S.Y.; Takahashi-Yamashiro, K.; Shen, Y.; Campbell, R.E. Biosensor Optimization Using a Forster Resonance Energy Transfer Pair Based on mScarlet Red Fluorescent Protein and an mScarlet-Derived Green Fluorescent Protein. *ACS Sens.* **2023**, *8*, 587–597. [CrossRef] [PubMed]
7. Cai, Y.; Zhu, K.; Shen, L.; Ma, J.; Bao, L.; Chen, D.; Wei, L.; Wei, N.; Liu, B.; Wu, Y.; et al. Evolved Biosensor with High Sensitivity and Specificity for Measuring Cadmium in Actual Environmental Samples. *Environ. Sci. Technol.* **2022**, *56*, 10062–10071. [CrossRef] [PubMed]
8. Guo, Y.; Huang, Z.L.; Zhu, D.L.; Hu, S.Y.; Li, H.; Hui, C.Y. Anthocyanin biosynthetic pathway switched by metalloregulator PbrR to enable a biosensor for the detection of lead toxicity. *Front. Microbiol.* **2022**, *13*, 975421. [CrossRef]
9. Lee, J.Y.; Sung, B.H.; Oh, S.H.; Kwon, K.K.; Lee, H.; Kim, H.; Lee, D.H.; Yeom, S.J.; Lee, S.G. C1 Compound Biosensors: Design, Functional Study, and Applications. *Int. J. Mol. Sci.* **2019**, *20*, 2253. [CrossRef]
10. Ebrahimi Fana, S.; Fazaeli, A.; Aminian, M. Directed evolution of cholesterol oxidase with improved thermostability using error-prone PCR. *Biotechnol. Lett.* **2023**, *45*, 1159–1167. [CrossRef]
11. Ding, N.; Zhou, S.; Yuan, Z.; Zhang, X.; Chen, J.; Deng, Y. Fine-tuning biosensor dynamic range based on rational design of cross-ribosome-binding sites in bacteria. *bioRxiv* **2020**, 2020.01.27.922302.
12. Moussa, S.; Murtas, G.; Pollegioni, L.; Mauzeroll, J. Enhancing Electrochemical Biosensor Selectivity with Engineered D-Amino Acid Oxidase Enzymes for D-Serine and D-Alanine Quantification. *ACS Appl. Bio Mater.* **2021**, *4*, 5598–5604. [CrossRef]
13. Nadler, D.C.; Morgan, S.A.; Flamholz, A.; Kortright, K.E.; Savage, D.F. Rapid construction of metabolite biosensors using domain-insertion profiling. *Nat. Commun.* **2016**, *7*, 12266. [CrossRef]
14. Zhou, Y.K.; Yuan, Y.M.; Wu, Y.A.; Li, L.; Jameel, A.; Xing, X.H.; Zhang, C. Encoding Genetic Circuits with DNA Barcodes Paves the Way for Machine Learning-Assisted Metabolite Biosensor Response Curve Profiling in Yeast. *ACS Synth. Biol.* **2022**, *11*, 977–989. [CrossRef] [PubMed]
15. Jiang, S.Q.; Zhang, Z.W.; Gu, Q.Y.; Yu, X.B. Semi-rational design for enhancing thermostability of acetylcholinesterase and sensitivity analysis of acephate. *Sci. Total Environ.* **2024**, *934*, 173282. [CrossRef]
16. Balaz, A.M.; Stevanovic, J.; Ostafe, R.; Blazic, M.; Durdic, K.I.; Fischer, R.; Prodanovic, R. Semi-rational design of cellobiose dehydrogenase for increased stability in the presence of peroxide. *Mol. Divers.* **2020**, *24*, 593–601. [CrossRef] [PubMed]
17. Qiu, C.X.; Zhai, H.T.; Hou, J. Biosensors design in yeast and applications in metabolic engineering. *Fems Yeast Res.* **2019**, *19*, foz082. [CrossRef]
18. Pedotti, M.; Ferrero, V.E.V.; Lettieri, T.; Colpo, P.; Follonier, S.; Calzolai, L.; Varani, L. Rationally Modified Estrogen Receptor Protein as a Bio-Recognition Element for the Detection of EDC Pollutants: Strategies and Opportunities. *Int. J. Environ. Res. Public Health* **2015**, *12*, 2612–2621. [CrossRef]
19. Rogers, J.K.; Guzman, C.D.; Taylor, N.D.; Raman, S.; Anderson, K.; Church, G.M. Synthetic biosensors for precise gene control and real-time monitoring of metabolites. *Nucleic Acids Res.* **2015**, *43*, 7648–7660. [CrossRef]
20. Hiraka, K.; Kojima, K.; Tsugawa, W.; Asano, R.; Ikebukuro, K.; Sode, K. Rational engineering of L-lactate oxidase for the mediator modification to achieve quasi-direct electron transfer type lactate sensor. *Biosens. Bioelectron.* **2020**, *151*, 111974. [CrossRef]
21. Golynskiy, M.V.; Koay, M.S.; Vinkenborg, J.L.; Merkx, M. Engineering Protein Switches: Sensors, Regulators, and Spare Parts for Biology and Biotechnology. *ChemBiochem* **2011**, *12*, 353–361. [CrossRef]
22. Hellweg, L.; Edenhofer, A.; Barck, L.; Huppertz, M.C.; Frei, M.S.; Tarnawski, M.; Bergner, A.; Koch, B.; Johnsson, K.; Hiblot, J. A general method for the development of multicolor biosensors with large dynamic ranges. *Nat. Chem. Biol.* **2023**, *19*, 1147–1157. [CrossRef]
23. Kimura, Y.; Kawai-Noma, S.; Saito, K.; Umeno, D. Directed Evolution of the Stringency of the LuxR Quorum Sensor without OFF-State Selection. *Acs Synth. Biol.* **2020**, *9*, 567–575. [CrossRef]

24. Du, H.X.; Liang, Y.Y.; Li, J.N.; Yuan, X.Y.; Tao, F.L.; Dong, C.J.; Shen, Z.K.; Sui, G.C.; Wang, P.C. Directed Evolution of 4-Hydroxyphenylpyruvate Biosensors Based on a Dual Selection System. *Int. J. Mol. Sci.* **2024**, *25*, 1533. [CrossRef] [PubMed]
25. Chong, H.Q.; Ching, C.B. Development of Colorimetric-Based Whole-Cell Biosensor for Organophosphorus Compounds by Engineering Transcription Regulator DmpR. *Acs Synth. Biol.* **2016**, *5*, 1290–1298. [CrossRef] [PubMed]
26. Dong, Z.Z.; Su, R.T.; Fu, Y.; Wang, Y.P.; Chang, L.Q. Recent Progress in DNA Biosensors for Detecting Biomarkers in Living Cells. *ACS Biomater. Sci. Eng.* **2024**, *10*, 5595–5608. [CrossRef] [PubMed]
27. Stephen, B.J.; Suchanti, S.; Jain, D.; Dhaliwal, H.; Sharma, V.; Kaur, R.; Mishra, R.; Singh, A. DNA biosensor based detection for neglected tropical disease: Moving towards smart diagnosis. *Sens. Rev.* **2022**, *42*, 517–525. [CrossRef]
28. Machado, L.F.M.; Currin, A.; Dixon, N. Directed evolution of the PcaV allosteric transcription factor to generate a biosensor for aromatic aldehydes. *J. Biol. Eng.* **2019**, *13*, 91. [CrossRef]
29. Huo, Y.Y.; Zhang, S.D.; Bi, H.R.; Wang, K.; Fang, Y.M.; Wang, M.; Tan, T.W. Development of a specific biosensor for sesquiterpene based on SELEX and directed evolution platforms. *Talanta* **2025**, *283*, 127186. [CrossRef]
30. Townshend, B.; Kaplan, M.; Smolke, C.D. Highly multiplexed selection of RNA aptamers against a small molecule library. *PLoS ONE* **2022**, *17*, e0273381. [CrossRef]
31. Stoltenburg, R.; Nikolaus, N.; Strehlitz, B. Capture-SELEX: Selection of DNA Aptamers for Aminoglycoside Antibiotics. *J. Anal. Methods Chem.* **2012**, *2012*, 415697. [CrossRef]
32. Ao, Y.Q.; Duan, A.Q.; Chen, B.F.; Yu, X.M.; Wu, Y.Y.; Zhang, X.J.; Li, S.S. Integration of an Expression Platform in the SELEX Cycle to Select DNA Aptamer Binding to a Disease Biomarker. *ACS Omega* **2022**, *7*, 10804–10811. [CrossRef] [PubMed]
33. Kramat, J.; Kraus, L.; Gunawan, V.J.; Smyej, E.; Froehlich, P.; Weber, T.E.; Spiehl, D.; Koeppl, H.; Blaeser, A.; Suess, B. Sensing Levofloxacin with an RNA Aptamer as a Bioreceptor. *Biosensors* **2024**, *14*, 56. [CrossRef]
34. Jaeger, J.; Groher, F.; Stamm, J.; Spiehl, D.; Braun, J.; Dörsam, E.; Suess, B. Characterization and Inkjet Printing of an RNA Aptamer for Paper-Based Biosensing of Ciprofloxacin. *Biosensors* **2019**, *9*, 7. [CrossRef]
35. Holzinger, M.; Le Goff, A.; Cosnier, S. Synergetic Effects of Combined Nanomaterials for Biosensing Applications. *Sensors* **2017**, *17*, 1010. [CrossRef] [PubMed]
36. El-Kady, M.M.; Ansari, I.; Arora, C.; Rai, N.; Soni, S.; Verma, D.K.; Singh, P.; Mahmoud, A.D. Nanomaterials: A comprehensive review of applications, toxicity, impact, and fate to environment. *J. Mol. Liq.* **2023**, *370*, 121046. [CrossRef]
37. Mekuye, B.; Abera, B. Nanomaterials: An overview of synthesis, classification, characterization, and applications. *Nano Sel.* **2023**, *4*, 486–501. [CrossRef]
38. Hwang, H.S.; Jeong, J.W.; Kim, Y.A.; Chang, M. Carbon Nanomaterials as Versatile Platforms for Biosensing Applications. *Micromachines* **2020**, *11*, 814. [CrossRef]
39. Eivazzadeh-Keihan, R.; Noruzi, E.B.; Chidar, E.; Jafari, M.; Davoodi, F.; Kashtiaray, A.; Gorab, M.G.; Hashemi, S.M.; Javanshir, S.; Cohan, R.A.; et al. Applications of carbon-based conductive nanomaterials in biosensors. *Chem. Eng. J.* **2022**, *442*, 136183. [CrossRef]
40. Ansari, M.A.; Mohd-Naim, N.F.; Ahmed, M.U. Electrochemical Nanoaptasensor Based on Graphitic Carbon Nitride/Zirconium Dioxide/Multiwalled Carbon Nanotubes for Matrix Metalloproteinase-9 in Human Serum and Saliva. *ACS Appl. Bio Mater.* **2024**, *7*, 1579–1587. [CrossRef]
41. Shi, Z.H.; Zhao, H.L.; Zhou, F.F.; Mao, Y.; Gong, Z.; Lan, M.B. Sensitive Electrochemical Biosensor Using Gold and Manganese (IV) Oxide Nanomaterials on Multiwalled Carbon Nanotubes for the Determination of Homocysteine With a Portable Potentiostat. *Anal. Lett.* **2025**, *58*, 492–505. [CrossRef]
42. du Plooy, J.; Kock, B.; Jahed, N.; Iwuoha, E.; Pokpas, K.; Hou, P.X.; He, M.S. Carbon Nanostructured Immunosensing of Anti-SARS-CoV-2 S-Protein Antibodies. *Molecules* **2023**, *28*, 8022. [CrossRef]
43. Younis, M.R.; He, G.; Lin, J.; Huang, P. Recent Advances on Graphene Quantum Dots for Bioimaging Applications. *Front. Chem.* **2020**, *8*, 424. [CrossRef] [PubMed]
44. Kansara, V.; Shukla, R.; Bahadur, P.; Tiwari, S.; Flora, S.J.S. Graphene quantum dots: Synthesis, optical properties and navigational applications against cancer. *Mater. Today Commun.* **2022**, *31*, 103359. [CrossRef]
45. El-Said, W.A.; Abdelshakour, M.; Choi, J.H.; Choi, J.W. Application of Conducting Polymer Nanostructures to Electrochemical Biosensors. *Molecules* **2020**, *25*, 307. [CrossRef] [PubMed]
46. Mirzaei, S.S.; Mehrdadi, N.; Bidhendi, G.N.; Pourmadadi, M.; Ahmadi, M.; Meknatkhah, S. Novel detection of *H. pylori* using ultrasensitive electrochemical aptasensor based on surface modified graphene oxide doped gold nanoparticles conjugated polythiophene(vol 200, 110279, 2024). *Microchem. J.* **2024**, *204*, 110572. [CrossRef]
47. Jaradat, H.; Hryniewicz, B.M.; Pasti, I.A.; Valerlo, T.L.; Al -Hamry, A.; Marchesi, L.F.; Vidotti, M.; Kanoun, O. Detection of H. pylori outer membrane protein (HopQ) biomarker using electrochemical impedimetric immunosensor with polypyrrole nanotubes and carbon nanotubes nanocomposite on screen-printed carbon electrode. *Biosens. Bioelectron.* **2024**, *249*, 115937. [CrossRef]

48. Yu, T.T.; Xu, H.T.; Jin, Z.Y.; Zhang, Y.Y.; Qiu, H.J. Noble metal-free high-entropy oxide/Co-N-C bifunctional electrocatalyst enables highly reversible and durable Zn-air batteries. *Appl. Surf. Sci.* **2023**, *610*, 155624. [CrossRef]
49. Lv, C.L.; Tang, C.; Zhou, H.Y.; Wang, A.J.; Feng, J.J.; Cheang, T.Y. Self-supported PtPdMnCoFe high-entropy alloy with nanochain-like internetworks for ultrasensitive electrochemical immunoassay of biomarker. *Sens. Actuators B-Chem.* **2024**, *401*, 135041. [CrossRef]
50. Tang, C.; Lv, C.L.; Chen, P.F.; Wang, A.J.; Feng, J.J.; Cheang, T.Y.; Xia, H.M. Dendritic quinary PtRhMoCoFe high-entropy alloy as a robust immunosensing nanoplatform for ultrasensitive detection of biomarker. *Bioelectrochemistry* **2024**, *157*, 108639. [CrossRef]
51. Kurup, C.P.; Ahmed, M.U. Nanozymes towards Personalized Diagnostics: A Recent Progress in Biosensing. *Biosensors* **2023**, *13*, 461. [CrossRef] [PubMed]
52. Keum, C.; Park, S.; Kim, H.; Kim, H.; Lee, K.H.; Jeong, Y. Modular conductive MOF-gated field-effect biosensor for sensitive discrimination on the small molecular scale. *Chem. Eng. J.* **2023**, *456*, 141079. [CrossRef]
53. Li, X.Y.; Li, X.M.; Li, D.D.; Zhao, M.; Wu, H.P.; Shen, B.; Liu, P.; Ding, S.J. Electrochemical biosensor for ultrasensitive exosomal miRNA analysis by cascade primer exchange reaction and MOF@Pt@MOF nanozyme. *Biosens. Bioelectron.* **2020**, *168*, 112554. [CrossRef] [PubMed]
54. Huang, W.; Xu, Y.; Wang, Z.P.; Liao, K.; Zhang, Y.; Sun, Y.M. Dual nanozyme based on ultrathin 2D conductive MOF nanosheets intergraded with gold nanoparticles for electrochemical biosensing of HO in cancer cells. *Talanta* **2022**, *249*, 123612. [CrossRef]
55. Koukouviti, E.; Plessas, A.K.; Pagkali, V.; Economou, A.; Papaefstathiou, G.S.; Kokkinos, C. 3D-printed electrochemical glucose device with integrated Fe(II)-MOF nanozyme. *Microchim. Acta* **2023**, *190*, 274. [CrossRef]
56. Yu, Y.Y.; Sun, X.Y.; Tang, D.Q.; Li, C.L.; Zhang, L.; Nie, D.X.; Yin, X.X.; Shi, G.Y. Gelsolin bound β-amyloid peptides: Electrochemical evaluation of levels of soluble peptide associated with Alzheimer's disease. *Biosens. Bioelectron.* **2015**, *68*, 115–121. [CrossRef]
57. Diba, F.S.; Kim, S.; Lee, H.J. Electrochemical immunoassay for amyloid-beta 1-42 peptide in biological fluids interfacing with a gold nanoparticle modified carbon surface. *Catal. Today* **2017**, *295*, 41–47. [CrossRef]
58. Ding, M.L.; Shu, Q.; Zhang, N.; Yan, C.R.; Niu, H.Z.; Li, X.Q.; Guan, P.; Hu, X.L. Electrochemical Immunosensor for the Sensitive Detection of Alzheimer's Biomarker Amyloid-β (1-42) Using the Heme-amyloid-β (1-42) Complex as the Signal Source. *Electroanalysis* **2022**, *34*, 263–274. [CrossRef]
59. Hsu, C.H.; Gupta, A.K.; Purwidyantri, A.; Prabowo, B.A.; Chen, C.H.; Chuang, C.C.; Tian, Y.C.; Lu, Y.J.; Lai, C.S. Sensing Alzheimer's Disease Utilizing Au Electrode by Controlling Nanorestructuring. *Chemosensors* **2022**, *10*, 94. [CrossRef]
60. Le, H.T.N.; Park, J.; Cho, S. A Probeless Capacitive Biosensor for Direct Detection of Amyloid Beta 1-42 in Human Serum Based on an Interdigitated Chain-Shaped Electrode. *Micromachines* **2020**, *11*, 791. [CrossRef]
61. Ribeiro, S.C.; Fernandes, R.; Moreira, F.T.C.; Sales, M.G.F. Potentiometric Biosensor Based on Artificial Antibodies for an Alzheimer Biomarker Detection. *Appl. Sci.* **2022**, *12*, 3625. [CrossRef]
62. Chen, J.Y.; Weng, S.H.; Chen, Q.Q.; Liu, A.L.; Wang, F.Q.; Chen, J.; Yi, Q.; Liu, Q.C.; Lin, X.H. Development of an Electrochemical Sensing Technique for Rapid Genotyping of Hepatitis B Virus. *Sensors* **2014**, *14*, 5611–5621. [CrossRef]
63. Teengam, P.; Tangkijvanich, P.; Chuaypen, N.; Chailapakul, O. An innovative wireless electrochemical card sensor for field-deployable diagnostics of Hepatitis B surface antigen. *Sci. Rep.* **2023**, *13*, 3523. [CrossRef]
64. Fan, P.Y.; Ma, E.Y.; Liu, C.; Zhao, Y.; Wen, X.W.; Wang, L.; Li, L.; Qu, Q. A new electrochemical DNA biosensor based on the density control strategy of TiCNH MXene@Au nanocomposites for the detection of hepatitis B virus-DNA. *Ionics* **2024**, *30*, 541–552. [CrossRef]
65. Li, X.; Scida, K.; Crooks, R.M. Detection of Hepatitis B Virus DNA with a Paper Electrochemical Sensor. *Anal. Chem.* **2015**, *87*, 9009–9015. [CrossRef] [PubMed]
66. Pareek, S.; Jain, U.; Bharadwaj, M.; Saxena, K.; Roy, S.; Chauhan, N. An ultrasensitive electrochemical DNA biosensor for monitoring (HPV-16) using graphene oxide/Ag/Au nano-biohybrids. *Anal. Biochem.* **2023**, *663*, 115015. [CrossRef]
67. Yang, Y.X.; Qing, Y.; Hao, X.D.; Fang, C.X.; Ouyang, P.; Li, H.Y.; Wang, Z.C.; Liao, Y.Z.; Fang, H.B.; Du, J. APTES-Modified Remote Self-Assembled DNA-Based Electrochemical Biosensor for Human Papillomavirus DNA Detection. *Biosensors* **2022**, *12*, 449. [CrossRef]
68. Yang, Y.X.; Liao, Y.Z.; Qing, Y.; Li, H.Y.; Du, J. Electrochemical DNA Biosensors with Dual-Signal Amplification Strategy for Highly Sensitive HPV 16 Detection. *Sensors* **2023**, *23*, 7380. [CrossRef]

69. Rasouli, E.; Basirun, W.J.; Johan, M.R.; Rezayi, M.; Mahmoudian, M.R.; Poenar, D.P. Electrochemical DNA-nano biosensor for the detection of cervical cancer-causing HPV-16 using ultrasmall Fe_3O_4-Au core-shell nanoparticles. *Sens. Bio-Sens. Res.* **2023**, *40*, 100562. [CrossRef]
70. Chaibun, T.; Thanasapburachot, P.; Chatchawal, P.; Yin, L.S.; Jiaranuchart, S.; Jearanaikoon, P.; Promptmas, C.; Buajeeb, W.; Lertanantawong, B. A Multianalyte Electrochemical Genosensor for the Detection of High-Risk HPV Genotypes in Oral and Cervical Cancers. *Biosensors* **2022**, *12*, 290. [CrossRef]

biosensors

MDPI

Article

CampyTube: Seamless Integration of a Molecular Test and Lateral Flow Detection of *Campylobacter* in a Single Vial

Natalia Sandetskaya [1,*], Andreas Kölsch [1], Kai Mattern [1], Vanessa Vater [2], Dirk Kuhlmeier [1] and Florian Priller [2]

[1] MicroDiagnostics Group, Fraunhofer Institute for Cell Therapy and Immunology IZI, Perlickstrasse 1, 04103 Leipzig, Germany
[2] Hygiena Diagnostics GmbH, Hermannswerder 17, 14473 Potsdam, Germany
* Correspondence: natalia.sandetskaya@izi.fraunhofer.de

Abstract

Background: The efficient control of hygiene and *Campylobacter*'s contamination status at various steps of poultry meat production is essential for the prevention of *Campylobacter* transmission to humans. Microbiological methods are laborious and time-consuming, and molecular methods of detection are often too skill- and infrastructure-demanding. Methods: We have developed CampyTube, a simple and user-friendly format for the integration of isothermal DNA amplification with embedded instrument-free detection on a miniaturized lateral flow test in a single vial. All test components, from the dry amplification reagents to the mini lateral flow tests, are incorporated into a standard single vial, which is closed after the addition of the liquid sample and never has to be opened again. This ensures the absolute prevention of carry-over contamination and makes the system very safe and simple to use in point-of-need settings. Results: As few as 60 *Campylobacter* genome copies per reaction could be successfully detected with CampyTube. We have primarily developed and evaluated CampyTube for the detection of *Campylobacter* in chicken neck skin samples and could reach 100% sensitivity and 100% specificity in the samples exceeding the regulatory limit of 1000 CFU/g confirmed microbiologically, while the sensitivity in all samples that tested positive using qPCR (1.4×10^2–2.5×10^6 genome copies/g) was 71.1%. We discuss the impact of sample preparation on CampyTube performance and suggest further options for test optimization. Conclusions: CampyTube is a highly versatile and efficient, yet simple, affordable, and material-saving system that can be adapted for other targets and sample types.

Keywords: *Campylobacter*; LAMP; lateral flow detection; point-of-care test; integrated diagnostics

1. Introduction

Campylobacteriosis, an acute diarrhoeal disease in humans, remains one of the most frequent foodborne infections worldwide [1–3]. According to the 2019 report by the European Food Safety Authority (EFSA), a total of 220,682 human cases of Campylobacteriosis were confirmed in the European Union, with an average notification rate of 59.7 cases per 100,000 individuals [4,5]. The consumption of insufficiently thermally processed poultry meat or the contamination of cooking utensils and surfaces during meat processing are the most common origins of infection [6].

Campylobacter jejuni and *C. coli* account for almost 90% of the reported cases of human Campylobacteriosis [6]. However, other species, including *C. lari, C. upsaliensis, and C. concisus*, have also been identified as potential causes of infection [6,7].

https://doi.org/10.3390/bios15080497

Campylobacter is commonly found in the intestinal tracts of various wild and domestic animals, as well as birds, and is prevalent throughout the poultry production environment. Chickens serve as a natural reservoir for *Campylobacter* species and are the primary source of transmission to humans [8]. Currently, no single control method effectively eliminates *Campylobacter* contamination in the poultry industry [9]. Therefore, the strict and efficient monitoring and control of hygiene and contamination status at various steps of poultry meat production must be maintained for the prevention of *Campylobacter* transmission to humans.

Current methods for *Campylobacter* detection in meat production facilities rely on laborious and time-consuming microbiological methods [3,10]. More rapid, advanced, and sensitive methods like nucleic acid amplification, DNA hybridization and fingerprinting techniques, mass spectrometry techniques, and biosensors are being developed, yet they demand special equipment, adequate lab infrastructure, and skilled personnel [6,11–14]. A user-friendly molecular detection format suitable for various test throughputs (i.e., compact and cost-effective) could significantly facilitate hygiene control for *Campylobacter*.

We have developed CampyTube—a test format that seamlessly integrates molecular detection with an instrument-free qualitative readout. CampyTube is based on the molecular isothermal amplification of *Campylobacter* DNA with integrated visual detection on a miniaturized lateral flow test (LFT) in a single closed vial. All test components are incorporated in a standard 0.5 mL vial with a screw cap, which remains closed after the addition of the sample and never has to be opened again, thus eliminating any cross-contamination risks. These characteristics make CampyTube especially beneficial for the point-of-need settings.

We have manufactured and evaluated the CampyTube tests for the detection of *Campylobacter* spp. (further referred to as *Campylobacter*) in chicken neck skin samples. In the current manuscript, we describe the principle of the test, its advantages over the State-of-the-Art methods, and propose ideas for the further development of the CampyTube-based detection of pathogens.

While the regulatory framework currently only allows for cultural determination in accordance with EN ISO 10272-1 in Europe [10], rapid molecular tests such as CampyTube could represent a future alternative. The new test platform aims to complement hygiene monitoring with a modern rapid testing method, thereby ensuring higher quality for the end consumer while also simplifying quality assurance in the meat processing industry. We also envisage the CampyTube principle as a versatile and extremely helpful test format for various applications in infectious disease diagnostics.

2. Materials and Methods

The majority of the chemicals for the buffers (PBS and Tris-HCl components, potassium monophosphate, potassium chloride, polyethylene glycol (PEG), bovine serum albumin (BSA), casein, Tween 20, and trehalose), as well as streptavidin, were purchased from Carl Roth, Karlsruhe, Germany. Moreover, 1-ethyl-3-(3-dimethylaminopropyl)carbodiimid-hydrochlorid (EDC) and N-hydroxysulfosuccinimide (sulfo-NHS) were bought from ThermoFisher, Waltham, MA, USA, and hydroxylamine from Sigma Aldrich, Schnelldorf, Germany.

The 150 nm gold nanoshells were obtained from NanoComposix, San Diego, CA, USA (product number GSXR150). The nitrocellulose membrane CN95 was ordered from Sartorius, Göttingen, Germany; the conjugate and absorbent pads from Ahlstrom, Helsinki, Finland; and the backing cards DCN 0.010"from DCN, Carlsbad, CA, USA. The manufactures of the antibodies and their specific product numbers are indicated in the protocols below. The two-component silicone for the manufacturing of the tube inserts was purchased from S. u. K. Hock, Regen, Germany.

The reagents for DNA extraction and qPCR were produced by Hygiena Diagnostics, Potsdam, Germany, as also referred to in the specific procedure steps.

Deoxyribonucleotides (dNTPs), magnesium sulfate, and Bst 2.0 polymerase for the loop-mediated isothermal amplification (LAMP) stemmed from New England Biolabs, Frankfurt am Main, Germany. The primers were ordered from Eurofins, Ebersberg, Germany.

The integration of the CampyTube test was performed in 0.5 mL screw cap tubes from Bio-Rad, Feldkirchen, Germany.

The chicken skin samples for the test evaluation were kindly provided by Plukon Storkow GmbH, Storkow, Germany.

The analysis of the sensitivity and specificity was performed in GraphPad Prism for Windows, Version 6.07 (GraphPad Software, Boston, MA, USA).

2.1. *Manufacturing of Mini LFT*

The conjugate for the mini LFT was produced via the conjugation of 35 µg of mouse anti-FITC antibodies (Abcam, Cambridge, UK, product number ab10257) to 1 mL of gold nanoshells via EDC-NHS cross-linking following the EDC product instructions [15]. Ten µL of hydroxylamine was added and incubated for 10 min to quench the reaction. The final conjugates were stored at 4 °C in conjugate diluent (0.5× PBS, 0.5% BSA, 0.5% casein, 1% Tween 20, 0.05% sodium azide, and pH 8.0).

For the manufacturing of the 25 mm mini LFTs, nitrocellulose membrane CN95 was mounted on 25 mm-wide backing cards DCN 0.010″. Conjugate and absorbent pads were cut to an 8 mm width. Conjugates were dispensed at a 25 µL/cm ratio using the Automated Lateral Flow Reagent Dispenser (ALFRD, Claremont BioSolutions, Upland, CA, USA). Streptavidin and goat anti-mouse IgG (Sigma Aldrich, product number M8642, Germany) were striped as test and control lines, respectively, using the S3 FlexArrayer (Scienion, Berlin, Germany) at 0.5 µL/cm. The conjugate and absorbent pads were mounted on the backing card with the membrane.

2.2. *Manufacturing of the Tube Insert: Mini LFT Holder*

The tube inserts, featuring a fitted slit for integrating the mini LFT, were designed to fit tightly into the lower portion of a standard 0.5 mL vial without coming into contact with the liquid reaction mix at the bottom of the tube. They were fabricated using a 3D-printed mold treated with a mold release agent (Figure S1). A two-component silicone was mixed in a 1:1 ratio, poured into the mold, and removed after curing for 30–60 min at room temperature.

2.3. *Enumeration of Campylobacter in the Samples via Microbiological Culture*

The microbiological detection of *Campylobacter* was based on the guidelines from EN ISO 10272-1 [16]. Chicken skin samples were trimmed; ~10 g of the sample was rinsed with 9 weight parts of buffered peptone water, stomached for 1 min, and settled for 5 min. Tenfold serial dilutions were made in peptone water; 0.1 mL of the dilutions was plated onto a 9 cm modified Charcoal-Cefoperazone-Deoxycholate (mCCD) agar plate. Characteristic colonies were counted after ~48 h of incubation in microaerophilic conditions at 41.5 °C.

2.4. *Sample Preparation for LAMP and PCR*

Sample preparation was based on rapid DNA extraction with the StarPrep® Three Kit (Hygiena Diagnostics), with additional pre-treatment steps with Reagent D (Hygiena Diagnostics) to detect only viable *Campylobacter* spp. This light-sensitive reagent penetrates dead cell membranes and binds to DNA upon light exposure, blocking the PCR amplification of DNA from dead cells [17].

Then, 800 µL of peptone-rinsed sample, as prepared for microbiological culture, was mixed with 100 µL of Reagent D and incubated in the dark for 10 min, followed by 5 min light exposure in the foodproof D-Light instrument (Hygiena Diagnostics). The sample was centrifuged for 10 min at 5400× *g*, and the supernatant was discarded. The pellet was resuspended in 200 µL of StarPrep® Three Lysis buffer containing 1.25% *v*/*v* BAX proteinase (Hygiena Diagnostics). The sample was incubated for 10 min at 37 °C and subsequently inactivated at 100 °C for 10 min. After cooling to room temperature, the sample was spun down for 5 min at 5400× *g*. Twenty-five µL of the supernatant was used for qPCR or the CampyTube test.

Alternatively, a treatment without the addition of proteinase was also investigated in paired samples. The treatment steps remained unchanged, except for the omission of the enzyme. A schematic overview of the sample preparation is shown in Figure 1.

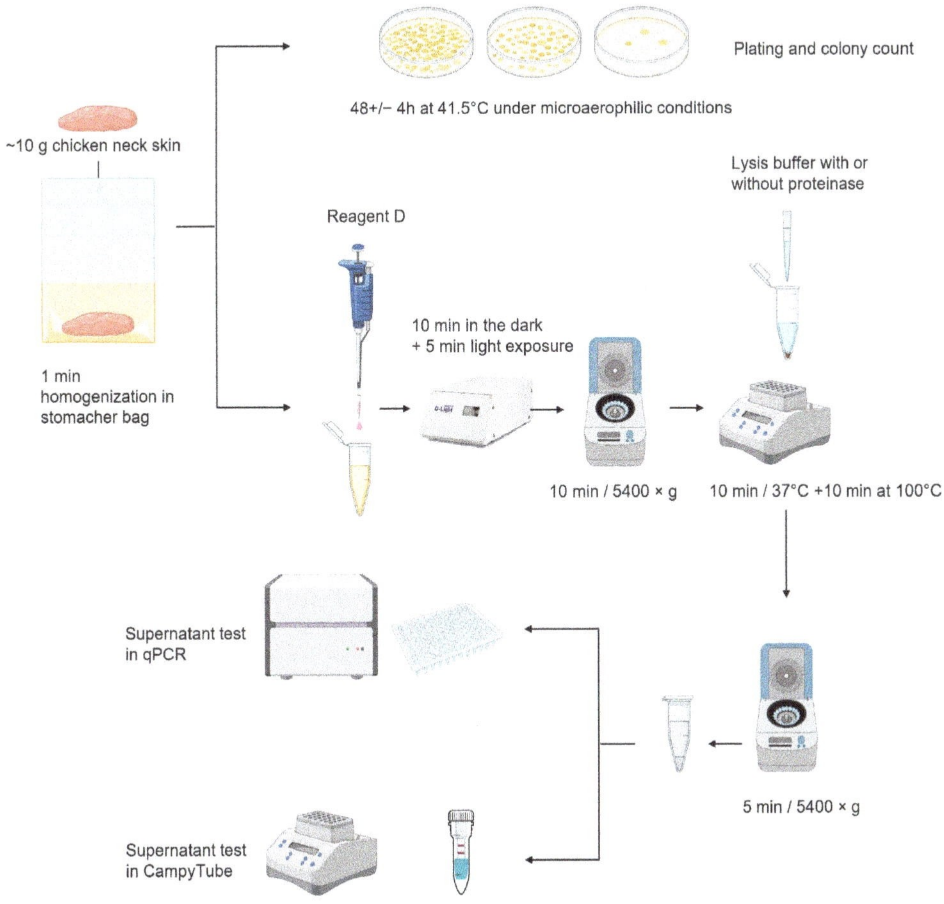

Figure 1. Schematic overview of the sample processing: description in text [18].

2.5. LAMP Reaction with Lateral Flow Detection

The LAMP assay targeted a region of the thermotolerant *Campylobacter* spp. (Accession number Y19244). The primer sequences published by Romero et al. were used in this study [19].

The reaction mix contained 20 mM of Tris-HCl pH 8.8, 50 mM of KCl, 8 mM of magnesium sulfate, 1.4 mM of each deoxynucleotide (dNTPs), 0.25 mg/mL of BSA, 6%

(*w*/*v*) trehalose, 0.2 µM each of the primers B3 and F3, 1.6 µM each of the primers BIP and FIP, 0.4 µM each of the labeled loop primers 5′-Biotin-LF and 5′-FAM-LB, and 320 U/mL of Bst 2.0 polymerase. The primer sequences were designed and published by Romero et al. [19] and are listed in Table S1. The reagents for the LAMP assay were freeze-dried in 0.5 mL screw cap vials in portions for 25 µL of reaction volume in an Epsilon 2–6 Freeze Dryer (Christ, Osterode am Harz, Germany).

For the initial characterization of the test performance, lateral flow detection was performed in a non-integrated format, either using the in-house produced mini LFT or commercially available LFT Amodia DetectLine Basic (Amodia, Braunschweig, Germany). The detection was performed in a separate closed area to prevent carry-over contamination with the products of DNA amplification. The mini LFTs (25 × 3 mm) were placed in the reaction vial with the 25 µL LAMP product; the bands were visible after ~1 min. The commercial 60 × 4 mm LFT required a larger detection volume, so 100 µL of chromatographic buffer 01 (Amodia, Germany) was added to the reaction vial with the 25 µL of LAMP product. The LFT was placed in this sample and the result was registered after ~10 min, when the bands had distinctly formed and the background had cleared.

2.6. *CampyTube Assembly and Application*

Twenty-five µL of the sample was pipetted into the reaction vial with the freeze-dried LAMP mix. For the determination of the limit of detection, 24 µL of water and 1 µL of purified *Campylobacter jejuni* DNA or 25 µL of water as a negative control were used. A mini LFT was placed into the slit of the vial insert, and this complex was embedded into the vial to fit tightly. The vial was closed with the screw cap and incubated in a heating block HB-2 (Wealtec, Taiwan) for 40 min at 65 °C.

After incubation, the vials were removed from the heating block, flipped over, and lightly snapped with the finger to force the liquid flow towards the mini LFT. The vials were left in the upside-down position for ~1 min until the test and/or control bands became visible. The results were interpreted qualitatively by the naked eye; pictures of the LFT results were taken for documentation.

2.7. *qPCR*

The qPCR was performed using a commercial kit prototype (Hygiena Diagnostics). The kit targets sequences specific to the entire group of thermotolerant *Campylobacter* species (FAM), as well as the individual species *C. jejuni* (HEX), *C. coli* (ROX), and *C. lari* (melt curve, FAM). The kit additionally contains an internal control amplification (Cy5) to monitor PCR inhibition. The qPCR mix in the kit is provided in lyophilized form for 25 µL of reaction volume in individual wells on a PCR plate. Furthermore, 25 µL of the sample was added to the reaction well. PCR was performed in a LightCycler 480 II (Roche, Mannheim, Germany). The reaction started with 4 min pre-incubation at 37 °C for uracil-DNA Glycosylase (UNG)-mediated decontamination, followed by initial denaturation for 5 min at 95 °C and 50 cycles of two-step amplification (5 s at 95° + 60 s at 60 °C). Fluorescence detection was performed at the end of the annealing step. A melting curve acquisition from 40 to 65 °C was added after the run to specifically detect the presence of *C. lari*. A digital PCR-quantified DNA standard containing the thermotolerant *Campylobacter* target sequence in four decadic dilutions from 10^1 to 10^4 genomic copies per reaction was included in every run. FAM-channel data from these standards were used to construct a standard curve for the quantification of the total thermotolerant *Campylobacter* in field samples. Quantitative interpretation of the results was performed for the samples ≥ 5 copies per reaction, which had been verified as a suitable limit of quantification with an acceptable $E_{LQ} \leq 0.25$.

3. Results

3.1. Development of Mini LFT

The development of the closed-tube mini LFT was driven by three requirements for the fully integrated format: (1) its dimensions must allow for fitting into a standard 0.5 mL reaction vial, (2) its activation must require $\leq$ 50 µL liquid, and (3) its sensitivity must not be compromised by the dimensions. The developed test has the dimensions of 25 mm $\times$ 3 mm, which was enabled not only by shrinking the sizes of the LFT elements, but also by sacrificing the sample pad, i.e., the conjugate pad in the test serves for the sample uptake. Twenty-five µL of liquid, the entire volume of the LAMP reaction, was sufficient for the effective detection of the labeled LAMP product: distinct test (T) and control (C) bands were visible, and the background was cleared without the addition of any chromatographic buffer.

The efficacy of the mini LFT was investigated in comparison with a conventional commercially available test strip that had dimensions of 60 mm $\times$ 4 mm and included all standard components of an LFT, i.e., a sample pad, a conjugate pad, a backed nitrocellulose membrane, and a wicking pad. The use of the commercial LFT required the addition of the chromatographic buffer to the LAMP product. The mini LFT demonstrated a similar performance to the commercial LFT: as little as 0.1 pg (60 genome copies) of *Campylobacter* DNA amplified in LAMP was stably detected, resulting in bright T-bands on both tests, while lower serial DNA dilution already lacked detection reproducibility, also with both tests (exemplary demonstration in Figure 2). It was noteworthy that the time-to-readout (i.e., the time from the LFT contact with the LAMP product to the complete clearing of the GNP background on the membrane and optimal visibility of the bands) was significantly shorter on the mini LFT, namely, ~1 min vs. ~10 min on the commercial test strip.

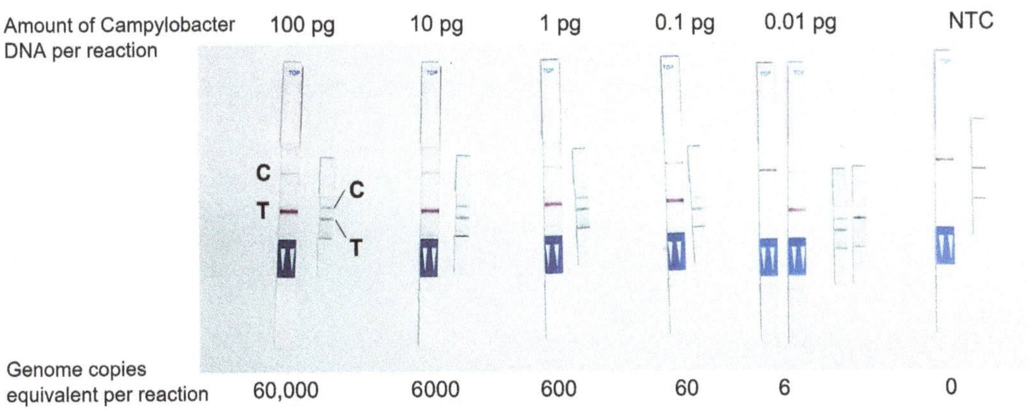

Figure 2. Detection of *Campylobacter* DNA amplified in LAMP with a commercial LFT and newly developed mini LFT. C, control band, T, test band.

3.2. Integration of the Test Components and Processes in Fully Closed Campytube Format

CampyTube is a fully closed test format for the molecular amplification and detection of *Campylobacter* nucleic acids that consists of three functional elements: a mini LFT, a vial insert (a mini LFT holder), and lyophilized LAMP reagents, integrated in a 0.5 mL reaction vial (Figure 3). The vial is a standard laboratory disposable item that enables the use of the CampyTube with common equipment and heating blocks and also obviates the need for any customized design and manufacturing of the test housing.

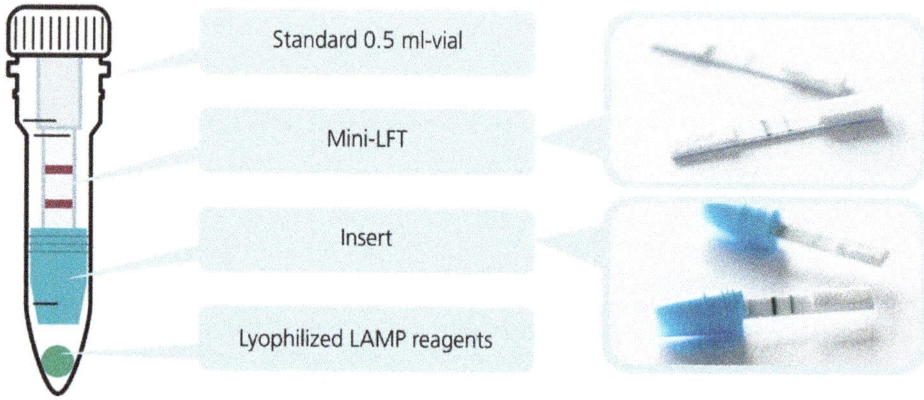

Figure 3. Structure of the CampyTube: Schematic assembly and its separate elements, a mini LFT and a mini LFT placed into a vial insert.

The vial insert was designed to fulfill several key functions in the CampyTube. First, it serves for the optimal positioning of the mini LFT in the vial, providing the necessary distance between the liquid reaction and the test strip. Second, it enables using a low reaction volume (here: 25 µL) without its evaporation during the heating step (65 °C for 40 min) via the tight sealing of the bottom part of the vial. Thus, although the vial itself is relatively large and could potentially lead to vapor condensation and the loss of the effective LAMP volume, the reaction runs within a small segregated compartment that eliminates these problems. We did not notice any significant condensation of the liquid during the tests, and no pre-term wetting of the mini LFT occurred. Third, the bottom part of the insert has a funnel shape that efficiently directs all the sample volume towards the mini LFT. For the targeted LFT activation after nucleic acid amplification, the vial must be flipped over and lightly snapped to let the liquid sample run into the insert towards the mini LFT (Figure 4). The "Flip & snap" mechanism activates the LFT only in the upside-down vial position and ensures that no spontaneous activation takes place during the handling or eventual dropping of the vials. This handling security is naturally maintained via liquid tension; only the intentional snapping of the vial (not dropping, shaking, or flipping alone) breaks the tension and allows the liquid to flow directly toward the LFT. The shape of the insert prevents any by-pass of the liquid and ensures reproducible and effective detection with a low sample volume.

The limit of *Campylobacter* DNA detection in the integrated CampyTube format remained the same as in the bench protocol (see Figure 2): sixty genome copies were successfully detectable (Figure 5).

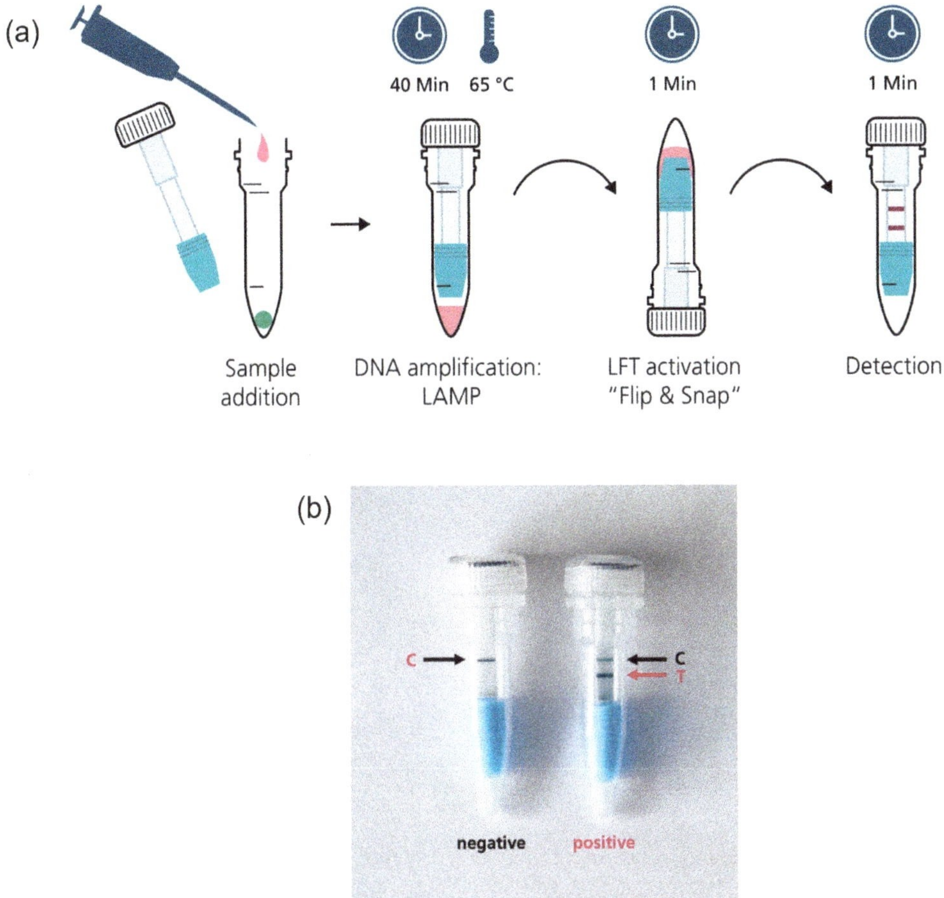

Figure 4. (**a**) CampyTube workflow; (**b**) Detection of the test result on the integrated mini LFT. T, test band, C, control band.

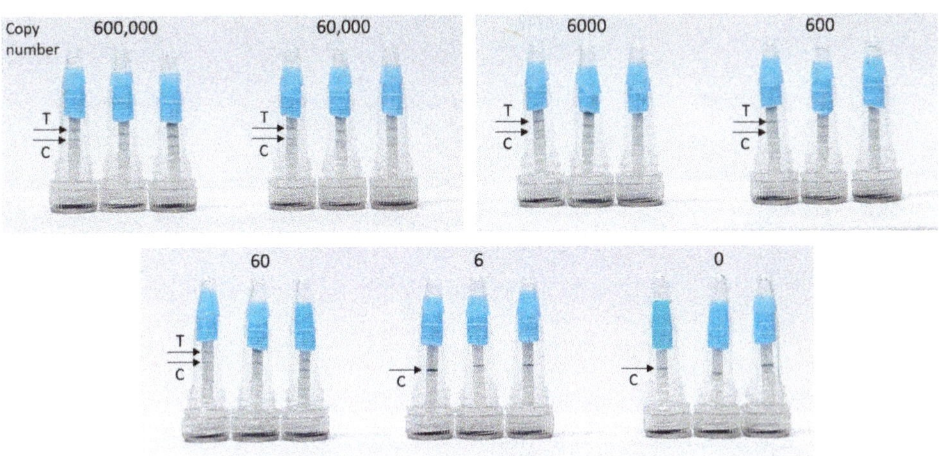

Figure 5. Detection of *C. jejuni* DNA in the integrated CampyTube. T, test band, C, control band.

3.3. *Detection of Campylobacter in Chicken Skin Samples with CampyTube and qPCR*

Sixty rinsed chicken skin samples prepared according to the specifications of Commission Regulation (EU) 2017/1495 were tested for the presence of *Campylobacter* in CampyTube using qPCR as the reference method. Two sample preparation protocols were compared for their impact on the detection efficiency: (1) without proteinase treatment and (2) with proteinase treatment. The first, simplified version of the protocol with omitted proteinase treatment was evaluated as a minimalistic time-saving procedure. However, its performance proved to be insufficient: only 34 out of 60 samples were detected as positive, resulting in a sensitivity of 61.8% and specificity of 88.5%.

In contrast, when proteinase treatment was included in the sample preparation, the detection efficiency improved significantly. Among the 45 out of the 60 samples that tested positive via qPCR (containing 1.4×10^2 to 2.5×10^6 genome copies/g meat), the sensitivity increased to 71.1% while maintaining 100% specificity. The sensitivity reached 100% for all the samples (17/17) containing > 1000 CFU/g, confirmed microbiologically (a hygienically relevant pathogen load in poultry meat proposed by the Commission Regulation (EU) 2017/1495) [10]. In the same subgroup of samples without proteinase, the sensitivity remained low and reached only 76.5%, thus demonstrating the clear advantage of using proteinase treatment for improved test performance.

The positive impact of the proteinase treatment on the detectability of *Campylobacter* in individual rinsed meat samples is visualized in the plot in Figure 6. It demonstrates that in all cases, the efficiency of qPCR detection was always higher with proteinase (lower Cq values in paired samples except for one sample that remained negative). Moreover, without the proteinase treatment, some samples tested negative in PCR and/or CampyTube despite a high pathogen load in the specimen. In most cases, a clearly positive result was obtained for these samples after the proteinase treatment.

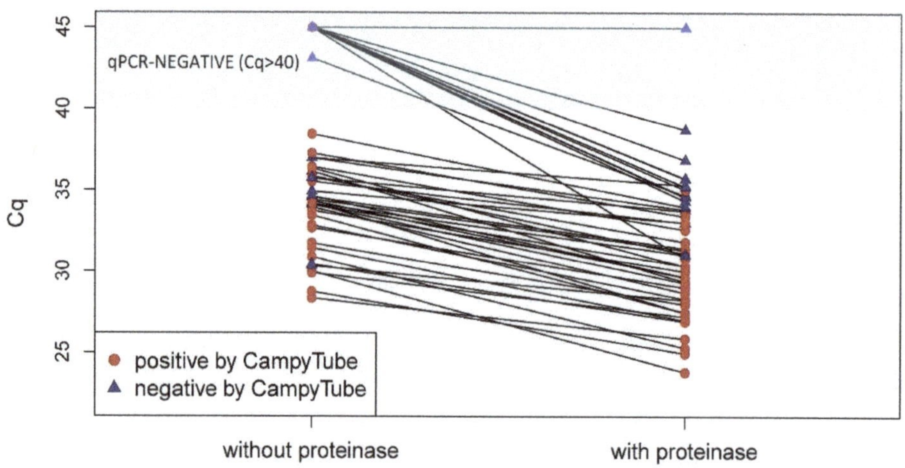

Figure 6. qPCR and CampyTube results in paired samples with and without proteinase treatment. A cut-off at Cq = 40 was applied to qPCR results.

4. Discussion

According to the Agricultural Outlook 2023–2032, the global meat supply will continue to expand over the next 10 years, reaching 382 million tons by 2032 [20,21]. In the upper middle-income countries, the production growth will increase by +14%. Poultry meat will account for half of this expansion in the next decade. The growing world population combined with the improvements in animal breeding and meat production technologies are the main drivers of this growth. [21]. The high density of poultry production bears high risks of increasing the prevalence of associated infections, such as Campylobacteriosis. Biosecurity measures and hygiene control at all stages of the meat production chain are therefore essential for the prevention of public health risks.

We have developed a simple and cost-effective format for the molecular detection of pathogens at the point of need. CampyTube is a first-of-its-kind seamless integration format of an LFT with molecular detection in a completely closed single vial. This is realized with two innovative components: a specially designed vial insert and a miniaturized LFT. The insert and the mini LFT enable integrated DNA amplification and its immediate subsequent detection in only 25 µL of liquid, without any additional buffer or further manual steps. All test components, from the dry amplification reagents to the mini LFT, are incorporated in a single vial, which is closed after the addition of liquid sample and never has to be opened again. This ensures the absolute prevention of carry-over contamination and makes the system very safe and simple to use in point-of-care settings. The readout of the results is performed by the naked eye; the test and control bands are visible through the transparent wall of the vial.

CampyTube was developed to speed up hygiene control in poultry production facilities. Conventional methods for *Campylobacter* detection include the microbiological culture of the pooled meat samples according to EN ISO 10272-2 [22]. This procedure is labor intensive and takes 48–72 h, while the processing of meat, its packaging, and delivery to consumers is already executed during this time. Hence, microbiological methods cannot enable a fast response to an eventual significant contamination in meat production facilities, so rapid detection methods are sought. The latest amendment to the EN ISO 10272-2 (A1:2023) refers to the use of molecular methods, particularly PCR, for the detection of *Campylobacter* spp. in the food chain. While being a faster method (2–3 h), PCR requires specially equipped molecular diagnostic laboratory and skilled personnel at the point of need; otherwise, the time-to-result is extended by the time of the sample transportation to an external laboratory.

Multiple reports address the development of LFT-based assays for the detection of foodborne bacteria in various sample matrices, from feces to milk products, including fresh, frozen, and processed meat. The most recent reviews describe the broad span of the achievable limits of detection, from 1 CFU/mL to ~10^{10} CFU/mL [23–25]. The limits of detection <100 CFU/mL mostly required the combination of a lateral flow assay either with signal enhancement, like fluorescent nanoparticles, nanozymes, quantum dots, magnetically, or electrochemically active labels, among others, and were coupled with an instrumental readout, and/or were combined with nucleic acid amplification like our method, yet lacked a fully integrated format. Ji et al. underline that such a kind of detection of bacterial nucleic acids on LFT "greatly improves the detection sensitivity and specificity, but the detection process is more complicated and susceptible to false positives" [23]. The latter challenge has motivated us in the development of the fully closed system for the detection of amplified nucleic acids in an instrument-free manner.

CampyTube can be used with a simple heating block as the only required instrument (the use of a centrifuge during sample preparation is considered to be optional for CampyTube; its use here was primarily driven by the needs of qPCR). The entire duration of the detection takes ~60–75 min; of this, 20–35 min is sample preparation (depending on the optional use of Reagent D), 40 min amplification, and ~1 min readout of the result. Sample preparation turned out to be a significant factor for the effective detection of *Campylobacter* both via qPCR and CampyTube. First, the sensitivity of *Campylobacter* detection in rinsed meat samples after their treatment with proteinase was higher than without the enzyme (Table 1). Beyond this fact, the absence of a correlation between the pathogen load and a negative result in PCR and/or CampyTube without the proteinase treatment suggests that the negative results are caused by the inhibition of DNA amplification in individual specimens. The successful neutralization of the inhibitors via the proteinase treatment allows them to assume their protein nature—either a total amount of protein or most likely nucleases may play a crucial role in the inhibition of DNA amplification. No detailed investigation of these aspects was performed in our study, yet we assume that the assay sensitivity may have been negatively affected by some false-negative results in CampyTube due to the inhibition of amplification. Besides the already mentioned types of inhibitors, other substances like fat content or blood traces may play a significant role. The eventual inhibition of the reaction in some samples implies the need for the introduction of an internal amplification control in the CampyTube. This feature is easily implementable via the addition of a synthetic target and primers to the LAMP mix and the introduction of an additional internal control line on the test strip. In order to enable good differentiation by the naked eye on the mini LFT, a different color or shape of this line may be used. We aim to add this feature to upcoming versions of the test format.

Table 1. Sensitivity and specificity of CampyTube test in rinsed chicken meat samples. PPV, Positive Predictive Value, NPV, Negative Predictive Value, and CI, confidence interval.

Without Proteinase Treatment		
Number of Samples	**Positive via qPCR**	**Negative via qPCR**
Positive via CampyTube	21 (of those, 13 with >1000 CFU/g)	3
Negative via CampyTube	13	23
Total:	34	26
	60	
Diagnostic performance	Value (95% CI)	
Sensitivity	61.76% (43.6–77.8%)	
Sensitivity in samples with >1000 CFU/g	76.5% (50.1–93.2%)	
Specificity	88.5% (69.9–97.6%)	
PPV	87.5% (67.6–97.3%)	
NPV	63.89% (46.2–79.2%)	
With Proteinase Treatment		
Number of samples	Positive via qPCR	Negative via qPCR
Positive via CampyTube	32 (of those, 17 with >1000 CFU/g)	0
Negative via CampyTube	13	15
Total:	45	15
	60	
Diagnostic performance	Value (95% CI)	
Sensitivity	71.1% (55.7–83.6%)	
Sensitivity in samples with >1000 CFU/g	100% (85–100%)	
Specificity	100% (78.2–100%)	
PPV	100% (89.11–100%)	
NPV	53.6% (33.9–72.5%)	

Worth mentioning is the option of speeding up the amplification time in CampyTube in the outlook. Novel bioengineered polymerases have become available recently and should enable a potential reduction in the amplification time to ~20 min, according to the information from the enzyme manufacturers (to be evaluated).

Further adjustment of the test can be implemented for the complete alignment of its sensitivity, with cut-off values specified in regulatory standards. The European Food Safety Authority (EFSA) proposes quantitative hygiene criteria that would reduce the public health risks associated with *Campylobacter*. According to EU legislation [10], adhering to a *Campylobacter* limit of 1000 CFU/g in chicken carcasses could reduce the public health risks associated with poultry consumption by over 50%. In our small-scale evaluation with a limited number of samples, CampyTube has shown 100% sensitivity for samples with >1000 CFU/g of *Campylobacter*. In order to use CampyTube for testing the compliance to this hygienic limit, lower concentrations should not be detectable by the test. While in the current study this cut-off was not integrated yet for the sake of a comprehensive test evaluation, it is technically possible to adjust the sensitivity to the desired limit of detection (LOD). For instance, this could be approached via a reduction in the duration of DNA amplification (see Supplementary Information, Table S2) or an adjustment of the LAMP primer concentration.

In our study, qPCR could detect more *Campylobacter*-positive samples than CampyTube, and enabled the precise quantification of viable *Campylobacter* (genome copies) in the samples, but it lacked the technical simplicity necessary for application at the point and time of need. Some available alternatives to PCR among the *Campylobacter* tests are also based on isothermal amplification, and therefore require less complex instrumentation in comparison to real-time PCR devices. Lin et al. presented a prototype of a system for smartphone-based *Campylobacter* detection in LAMP [26]. However, it requires a specialized device for the integration of heating and readout. Moreover, the use of a smartphone for the quality control, certification, and validation of a diagnostic test remains problematic due to the varying technical characteristics of different smartphone models. The Neogen® Molecular Detection Assay (Neogen Corporation, Lansing, MI, USA) is a commercial LAMP assay that can be run in a compact instrument, yet it requires the pre-enrichment of bacteria in a nutritional broth [27], as other proposed LAMP methods do [28–30]. At the same time, Romero and Cook (2018) have demonstrated that the detection of *Campylobacter* in chicken carcasses via LAMP is also possible without enrichment, which aligns well with our results [31]. The authors also proposed a convenient sampling method by swabbing meat and reached a detection limit of 800 CFU/swab. While it is not possible to assess this LOD in relation to the recommended limit of 1000 CFU/g due to the unstandardized size of the swabbed chicken carcasses, this method can be an attractive sampling technique to be combined with CampyTube, since it obviates more laborious rinsing and very likely reduces the proportion of the inhibitors in the sample. A 5 min heating at 85 °C was sufficient to inactivate the presumably present inhibitors in this study. A quantitative characterization of the combination of swab sampling and CampyTube could potentially enable even more rapid point-of-care testing. Feng et al. have also confirmed that cloth sampling yields identical results to the rinse method in the detection of Salmonella in meat samples, thus highlighting more convenient alternative sampling strategies [32].

Yet none of the isothermal amplification methods described above enabled a truly integrated simple point-of-care format: The reported developments are rather based on typical laboratory procedures with the manual setup of the reactions. The development of CampyTube was driven by the motivation to provide the simplest and most convenient possible molecular test format for non-proficient users or settings with limited technical and personnel resources. To our knowledge, CampyTube is the first-of-its-kind system that seamlessly combines molecular nucleic acid amplification and lateral flow detection in one standard laboratory vial in a single-step, fully integrated test. While lateral flow detection offers straightforward instrument-free detection by the naked eye, its application for DNA amplicons bears a high risk of carry-over contamination between the tests. Typical LFTs require the addition of a detection buffer or operate with a large (≥100 µL) total sample volume. The only reported system aiming to address the challenge of sealed lateral flow detection of amplified products is a UStar Nucleic Acid Probes (fragments) Detection Device (Ustar Biotechnologies, Hangzhou, China) [33]. However, it comprises a multi-component plastic cassette that needs to be pre-loaded with the vial of the detection buffer and serves only for the detection of the products amplified externally elsewhere, while being rather bulky and thus also expensive, and producing a considerable amount of plastic waste. Since CampyTube uses a common amplification vial as housing for closed lateral flow detection, our system is a sustainable option for a variety of point-of-care applications. The amount of plastic in such a vial is only 15% of the amount used in typical LFT cassettes (~0.95 g vs. ~6.35 g, compared to an exemplary test cassette (DCN, USA)). Remarkably, the production of the mini LFT requires ~3× less nitrocellulose than an average LFT, making CampyTube a cheaper and more sustainable rapid test.

Another product that combines integrated DNA amplification and lateral flow detection is Multitest COVID-19 (Selfdiagnostics, Leipzig, Germany) [34]. Both processes occur within a sealed microfluidic chip. The activation of the LFT in the Multitest requires the addition of the detection buffer, which is preloaded in a blister compartment on the cartridge, and its release is triggered by applying pressure to the blister. However, the cartridge is relatively large (approximately hand-sized) and contains disposable heating elements, as the amplification process operates at ~65 °C. Due to its size and design, regular use in point-of-care (PoC) applications would not be sustainable. Additionally, the complexity of the system results in high production costs, thus making the device unsuitable for high-throughput areas like hygiene control. We believe that the users of the CampyTube format can especially benefit from its integration into a standard disposable 0.5 mL lab vial due to its cost-saving manufacturing and compatibility with any standard heating blocks or racks. This format is especially favorable for application areas with high throughput or low resources. In the current study, we demonstrated a functional prototype of the system where the insert was produced manually; Figure S2 provides an idea of a single-piece insert merged with the cap. Upon need, the system allows for integration in various reaction vessels and volumes with minimal design adjustments.

Among the few non-integrated and thus cheap variants of LAMP-based tests amenable for detection by the naked eye are colorimetric LAMP tests [35,36]. Commercially available colorimetric LAMP assays utilize pH-sensitive dyes to visually indicate DNA amplification through color changes. Yet the weak buffering capacity of pH-based colorimetric LAMP master mixes can limit sample compatibility, as highly buffered or acidic samples may impact the color change, leading to unspecific results [37]. Visual interpretation of the indistinct color change by the operator can be incorrect and depends on the human factor. Furthermore, colored samples like blood can mask the color change completely. CampyTube enables working with such samples: as long as the LAMP itself is not inhibited, detection remains unaffected and unbiased. Our ongoing tests (unpublished) have demonstrated the direct compatibility of CampyTube with up to 8% v/v blood in LAMP reactions. Thus, CampyTube offers advantages over the available simple colorimetric LAMP tests, especially considering the potential of CampyTube for multiplexing (detection via multiple bands on the LFT), which is not possible in colorimetric reactions.

5. Conclusions

CampyTube enables the seamless integration of isothermal DNA amplification with instrument-free qualitative detection via a mini LFT placed inside a conventional 0.5 mL reaction vial. The main advantage of this system is its secure and simple use without the risk of carry-over contamination. The compact format is material-saving and convenient for logistics and application. The system is versatile and can be adapted for other pathogens. The novel test has demonstrated 100% sensitivity in chicken skin samples, exceeding the regulatory limit of 1000 CFU/g, with 71.1% overall sensitivity and 100% specificity. The current limit of detection is 60 genome copies/reaction. In the outlook, it would be of importance to integrate an internal amplification control into the test, as well as to evaluate the system with larger sample quantities and other sample matrices. We believe that this test format can enable safe and reliable molecular diagnostics in many point-of-care application areas.

6. Patents

German patent DE102024103902 [38]; international patent applications are filed.

Supplementary Materials: The following supporting information can be downloaded at: https://www.mdpi.com/article/10.3390/bios15080497/s1. Figure S1: Design of the tube insert and the negative mold for its manufacturing; Figure S2: An option of the insert design and its integration as a single piece merged with the vial cap; Table S1: Primer sequences used in this work; Table S2: LAMP detection time for different concentrations of *Campylobacter* DNA.

Author Contributions: Conceptualization, N.S., A.K., K.M. and D.K.; Methodology, A.K., K.M., V.V. and F.P.; Formal analysis, N.S. and V.V.; Investigation, N.S., A.K., V.V. and F.P.; Resources, V.V. and F.P.; Writing—original draft, N.S.; Writing—review and editing, N.S., V.V., A.K. and F.P.; Visualization, N.S. and K.M.; Supervision, D.K.; Project administration, N.S. and F.P.; Funding acquisition, N.S., D.K. and F.P. All authors have read and agreed to the published version of the manuscript.

Funding: This research was funded by the Federal Ministry of Food and Agriculture within the BMEL-Program to promote innovation; grant numbers 281C103A18 and 281C103D18.

Institutional Review Board Statement: Not applicable.

Data Availability Statement: The original contributions presented in this study are included in the article and Supplementary Material. Further inquiries can be directed to the corresponding author.

Acknowledgments: The authors would like to thank Plukon Food Group, especially the team of Plukon Vertrieb GmbH for the constructive consulting on the industry needs and for providing the meat samples for the test evaluation.

Conflicts of Interest: Authors Vanessa Vater and Florian Priller were employed by the diagnostic company Hygiena Diagnostics. Author D.K. is a co-founder of the diagnostic company INTU Diagnostics. The remaining authors declare that the research was conducted in the absence of any commercial or financial relationships that could be construed as a potential conflict of interest.

Abbreviations

The following abbreviations are used in this manuscript:

CFU	Colony-forming unit
LAMP	Loop-mediated isothermal amplification
LFT	Lateral flow test
LOD	Limit of detection

References

1. Tack, D.M.; Ray, L.; Griffin, P.M.; Cieslak, P.R.; Dunn, J.; Rissman, T.; Jervis, R.; Lathrop, S.; Muse, A.; Duwell, M.; et al. Preliminary Incidence and Trends of Infections with Pathogens Transmitted Commonly Through Food—Foodborne Diseases Active Surveillance Network, 10 U.S. Sites, 2016–2019. *MMWR Morb. Mortal. Wkly. Rep.* **2020**, *69*, 509–514. [CrossRef]
2. Havelaar, A.H.; Kirk, M.D.; Torgerson, P.R.; Gibb, H.J.; Hald, T.; Lake, R.J.; Praet, N.; Bellinger, D.C.; de Silva, N.R.; Gargouri, N.; et al. World Health Organization Global Estimates and Regional Comparisons of the Burden of Foodborne Disease in 2010. *PLoS Med.* **2015**, *12*, e1001923. [CrossRef]
3. Olsen, A.; Bonardi, S.; Barco, L.; Sandberg, M.; Langkabel, N.; Roasto, M.; Majewski, M.; Brugger, B.; Kautto, A.H.; Blagojevic, B.; et al. A comparison of European surveillance programs for *Campylobacter* in broilers. *Food Control* **2024**, *155*, 110059. [CrossRef]
4. Sadek, S.A.S.; Shaapan, R.M.; Barakat, A.M.A. Campylobacteriosis in Poultry: A Review. *J. World's Poult. Res.* **2023**, *13*, 168–179. [CrossRef]
5. European Food Safety Authority (EFSA). The European Union One Health 2020 Zoonoses Report. *EFSA J.* **2021**, *19*, e06971. [CrossRef] [PubMed]
6. Andritsos, N.D.; Tzimotoudis, N.; Mataragas, M. Prevalence and Distribution of Thermotolerant *Campylobacter* Species in Poultry: A Comprehensive Review with a Focus on the Factors Affecting the Detection and Enumeration of *Campylobacter jejuni* and *Campylobacter coli* in Chicken Meat. *Appl. Sci.* **2023**, *13*, 8079. [CrossRef]
7. Skarp, C.P.A.; Hänninen, M.-L.; Rautelin, H.I.K. Campylobacteriosis: The role of poultry meat. *Clin. Microbiol. Infect.* **2016**, *22*, 103–109. [CrossRef]
8. Hermans, D.; Pasmans, F.; Messens, W.; Martel, A.; van Immerseel, F.; Rasschaert, G.; Heyndrickx, M.; van Deun, K.; Haesebrouck, F. Poultry as a host for the zoonotic pathogen *Campylobacter jejuni*. *Vector Borne Zoonotic Dis.* **2012**, *12*, 89–98. [CrossRef]
9. Hakeem, M.J.; Lu, X. Survival and Control of *Campylobacter* in Poultry Production Environment. *Front. Cell. Infect. Microbiol.* **2020**, *10*, 615049. [CrossRef] [PubMed]
10. European Commission. Commission Regulation (EU) 2017/1495 of 23 August 2017 Amending Regulation (EC) No 2073/2005 as Regards *Campylobacter* in Broiler Carcasses; Official Journal of the European Union, L 218, 1–6, 2017 ((EU) 2017/1495). Available online: https://eur-lex.europa.eu/legal-content/EN/TXT/PDF/?uri=OJ:L:2017:218:FULL (accessed on 14 February 2025).
11. He, D.; Du, Z.; Wang, Y.; Xu, E.; Jin, Z.; Wu, Z. Quantitative detection of *Campylobacter jejuni* with a core-satellite assemblies-based dual-modular aptasensor. *Food Control* **2022**, *135*, 108828. [CrossRef]
12. Li, C.; Chen, X.; Wen, R.; Ma, P.; Gu, K.; Li, C.; Zhou, C.; Lei, C.; Tang, Y.; Wang, H. Immunocapture Magnetic Beads Enhanced the LAMP-CRISPR/Cas12a Method for the Sensitive, Specific, and Visual Detection of *Campylobacter jejuni*. *Biosensors* **2022**, *12*, 154. [CrossRef]
13. Asakura, H.; Yamamoto, S.; Yamada, K.; Kawase, J.; Nakamura, H.; Abe, K.-I.; Sasaki, Y.; Ikeda, T.; Nomoto, R. Quantitative detection and genetic characterization of thermotolerant *Campylobacter* spp. in fresh chicken meats at retail in Japan. *Front. Microbiol.* **2022**, *13*, 1014212. [CrossRef] [PubMed]
14. Joseph, L.A.; Griswold, T.; Vidyaprakash, E.; Im, S.B.; Williams, G.M.; Pouseele, H.A.; Hise, K.B.; Carleton, H.A. Evaluation of core genome and whole genome multilocus sequence typing schemes for *Campylobacter jejuni* and *Campylobacter coli* outbreak detection in the USA. *Microb. Genom.* **2023**, *9*, 001012. [CrossRef] [PubMed]
15. Pierce EDC, No-Weigh Format. Instruction to product #A35391; Pierce Biotechnology, Ed. Available online: https://assets.thermofisher.com/TFS-Assets/LSG/manuals/MAN0017099_2162629_PierceEDC_UG.pdf (accessed on 27 July 2025).
16. *ISO 10272-1:2017*; Microbiology of the Food Chain—Horizontal Method for Detection and Enumeration of *Campylobacter* spp. Part 1: Detection Method. International Organization for Standardization: Geneva, Switzerland, 2017.
17. Hygiena Diagnostics GmbH. Reagent D for the Elimination of DNA from Dead Cells to Avoid False-Positive PCR Results: Product Instruction. Available online: https://www.hygiena.com/documents/71126/reagent-d-product-instructions.pdf?_gl=1*jl8i7a*_up*MQ.*_ga*MTU1OTg3NTQ3OS4xNzQwNTYxNjcy*_ga_N1ZFCTH78J*MTc0MDU2MTY3MS4xLjAuMTc0MDU2MTY3MS4wLjAuMTgzMTUzNjEwMA (accessed on 26 February 2025).
18. Sandetskaya, N. Schematic Overview of the Sample Processing: Created in BioRender. Available online: https://BioRender.com/gz3mzl8 (accessed on 21 July 2025).
19. Romero, M.R.; D'Agostino, M.; Arias, A.P.; Robles, S.; Casado, C.F.; Iturbe, L.O.; Lerma, O.G.; Andreou, M.; Cook, N. An immunomagnetic separation/loop-mediated isothermal amplification method for rapid direct detection of thermotolerant *Campylobacter* spp. during poultry production. *J. Appl. Microbiol.* **2016**, *120*, 469–477. [CrossRef]
20. World Health Organization. *Measures for the Control of Campylobacter spp. in Chicken Meat: Meeting Report*; Microbiological Risk Assessment Series 46; World Health Organization & Food and Agriculture Organization of the United Nations: Geneva, Switzerland, 2024; ISBN 978-92-4-008808-5.
21. OECD. *OECD-FAO Agricultural Outlook 2023–2032*; OECD: Paris, France, 2023; ISBN 9789264619333.
22. *ISO 10272-2:2017*; Microbiology of the Food Chain—Horizontal Method for Detection and Enumeration of *Campylobacter* spp. Part 2: Colony-Count Technique. International Organization for Standardization: Geneva, Switzerland, 2017.

23. Ji, Y.; Huang, Y.; Cheng, Z.; Hao, W.; Liu, G.; Liu, Y.; Zhang, X. Lateral Flow Strip Biosensors for Foodborne Pathogenic Bacteria via Direct and Indirect Sensing Strategies: A Review. *J. Agric. Food Chem.* **2023**, *71*, 10250–10268. [CrossRef] [PubMed]
24. Wei, C.; Wu, A.; Xu, L.; Xu, C.; Liu, L.; Kuang, H.; Xu, X. Recent progress on lateral flow immunoassays in foodborne pathogen detection. *Food Biosci.* **2023**, *52*, 102475. [CrossRef]
25. Zhao, X.; Bhat, A.; O'Connor, C.; Curtin, J.; Singh, B.; Tian, F. Review of Detection Limits for Various Techniques for Bacterial Detection in Food Samples. *Nanomaterials* **2024**, *14*, 855. [CrossRef]
26. Lin, Q.; Jia, K.; Gou, H.; He, H.; Wen, J.; Shen, H.; Chen, K.; Wu, Y.; Lu, B.; Liao, M.; et al. A smartphone-assisted high-throughput integrated color-sensing platform for the rapid detection of *Campylobacter coli*. *LWT* **2022**, *167*, 113790. [CrossRef]
27. Neogen Corporation. Neogen® Molecular Detection Assay 2, Campylobacter. Available online: https://www.neogen.com/de/categories/pathogens/molecular-detection-assay-2-campylobacter/ (accessed on 14 February 2025).
28. Yamazaki, W.; Taguchi, M.; Kawai, T.; Kawatsu, K.; Sakata, J.; Inoue, K.; Misawa, N. Comparison of loop-mediated isothermal amplification assay and conventional culture methods for detection of *Campylobacter jejuni* and *Campylobacter coli* in naturally contaminated chicken meat samples. *Appl. Environ. Microbiol.* **2009**, *75*, 1597–1603. [CrossRef]
29. Kreitlow, A.; Becker, A.; Ahmed, M.F.E.; Kittler, S.; Schotte, U.; Plötz, M.; Abdulmawjood, A. Combined Loop-Mediated Isothermal Amplification Assays for Rapid Detection and One-Step Differentiation of *Campylobacter jejuni* and *Campylobacter coli* in Meat Products. *Front. Microbiol.* **2021**, *12*, 668824. [CrossRef]
30. Rajagopal, R.; Barnes, C.A.; David, J.M.; Goseland, J. Evaluation of a commercial loop-mediated isothermal amplification assay, 3M™ Molecular Detection Assay 2-*Campylobacter*, for the detection of *Campylobacter* from poultry matrices. *Br. Poult. Sci.* **2021**, *62*, 404–413. [CrossRef] [PubMed]
31. Romero, M.R.; Cook, N. A Rapid LAMP-Based Method for Screening Poultry Samples for *Campylobacter* Without Enrichment. *Front. Microbiol.* **2018**, *9*, 2401. [CrossRef] [PubMed]
32. Feng, Y.; Talukder, S.; Bolkenov, B.; Duarte, T.; Yang, X. Comparative Effectiveness of Cloth Sampling to Rinse Sampling on Microbial Recovery and Salmonella Detection in Poultry Meats. *J. Food Prot.* **2025**, *88*, 100425. [CrossRef]
33. Ustar Biotechnologies (Hangzhou) Ltd. Nucleic Acid Probes (fragments) Detection Device. Available online: https://en.bioustar.com/product/93.html (accessed on 14 February 2025).
34. Selfdiagnostics Deutschland GmbH. MULTITEST COVID-19-Selfdiagnostics. Available online: https://selfdiagnostics.eu/ (accessed on 14 February 2025).
35. Saifuddin, S.A.; Rashid, R.; Nor Azmi, N.J.; Mohamad, S. Colorimetric strategies applicable for loop-mediated isothermal amplification. *J. Microbiol. Methods* **2024**, *223*, 106981. [CrossRef]
36. De Oliveira Coelho, B.; Sanchuki, H.B.S.; Zanette, D.L.; Nardin, J.M.; Morales, H.M.P.; Fornazari, B.; Aoki, M.N.; Blanes, L. Essential properties and pitfalls of colorimetric Reverse Transcription Loop-mediated Isothermal Amplification as a point-of-care test for SARS-CoV-2 diagnosis. *Mol. Med.* **2021**, *27*, 30. [CrossRef] [PubMed]
37. SARS-CoV-2 Rapid Colorimetric LAMP Assay Kit: Instruction Manual to the Product #NEB #E2019S; New England Biolabs, Ed. Available online: https://www.neb.com/en/-/media/nebus/files/manuals/manuale2019.pdf?rev=640485747af74e879b323c9464380faf&hash=3B6B3A346323CFBA2EBA47965C6BCD23 (accessed on 27 July 2025).
38. Sandetskaya, N.; Kölsch, A.; Mattern, K.; Kuhlmeier, D. Nachweiskit für den Nachweis von Biomolekülen in Einem Reaktionsgefäß Mit-Tels Integration von Molekularbiologischer Amplifikation von Nukleinsäure und Deren Nachweis Mittels Lateral Flow Detektion. DE102024103902, 13 February 2025.

biosensors

MDPI

Article

Multiplexed CRISPR Assay for Amplification-Free Detection of miRNAs

P. I. Thilini De Silva [1,†], Keshani Hiniduma [1,†], Rachelle Canete [1], Ketki S. Bhalerao [1], Sherif M. Shawky [1], Hansana Gunathilaka [1], Jessica L. Rouge [1], Islam M. Mosa [1], David C. Steffens [2], Kevin Manning [2], Breno S. Diniz [2,3] and James F. Rusling [1,4,5,6,*]

1 Department of Chemistry, University of Connecticut, Storrs, CT 06269, USA; ketki.bhalerao@uconn.edu (K.S.B.); sherif_mohamed.shawky_abdou@uconn.edu (S.M.S.); jessica.rouge@uconn.edu (J.L.R.); islam.mosa@uconn.edu (I.M.M.)
2 Department of Psychiatry, UConn Health, Farmington, CT 06030, USA; steffens@uchc.edu (D.C.S.); diniz@uchc.edu (B.S.D.)
3 UConn Center on Aging, UConn Health, Farmington, CT 06030, USA
4 School of Chemistry, National University of Ireland at Galway, H91 TK33 Galway, Ireland
5 Department of Surgery and Neag Cancer Center, UConn Health, Farmington, CT 06030, USA
6 Institute of Material Science, University of Connecticut, Storrs, CT 06269, USA
* Correspondence: james.rusling@uconn.edu
† These authors contributed equally to this work.

Abstract: CRISPR-Cas proteins from bacteria are powerful tools for gene editing and molecular diagnostics. Expanding capacity of CRISPR to low cost, multiplexed assays of biomarkers is a key to future disease diagnostics, since multiple biomarker detection is essential for reliable diagnostics. Herein we describe a multiplexed assay in a 3D-printed 96-well plate with CRISPR-Cas13a immobilized in each well to target three circulating blood biomarker microRNAs (miRNAs 34c-5p, 200c-3p, and 30e-5p) for Alzheimer's disease (ALZ). Immobilized Cas13a is equipped with different crRNAs complementary to each miRNA target. MiRNA binding to crRNA complements activates the collateral RNase activity of Cas13a, cleaving a quenched fluorescent reporter (RNaseAlert) with fluorophore and quencher connected by an RNA oligonucleotide to enable fluorescence measurements. We achieved ultralow limits of detection (LOD) of 0.74 fg/mL for miRNA 34c-5p, 0.70 fg/mL for miRNA 30e-5p, and 7.4 fg/mL for miRNA 200c-3p, with dynamic ranges from LODs up to about 1800 pg/mL. The accuracy of the assay was validated by spike-recovery studies and good correlation of levels of patient plasma samples vs. a referee method. This new approach provides selective, sensitive multiplex miRNA biosensing, and simultaneously accommodates analysis of standards and controls.

Keywords: CRISPR; Alzheimer biomarkers; miRNAs; multiplexed assay; fluorescence

1. Introduction

Molecular biomarkers, including proteins and nucleic acids [1,2], play a vital role in disease diagnostics and monitoring patient health and therapeutic outcomes. Accurate analyses of multiple biomarkers, as opposed to single biomarkers, offer objective measurements of physiological and pathological processes. However, biomarker assays are currently underutilized due to the unavailability of assays in clinics and hospitals, partly due to the cost and complexity of commercial options [3]. Disease detection utilizing CT, MRI, and PET imaging can be used, but high cost and low sensitivity for early detection limit their effectiveness, e.g., for cancers and Alzheimer's disease [4–6].

Nevertheless, huge advances in molecular diagnostic test research have been made over the past 10–15 years, propelled by breakthroughs in microfluidics, biotechnology, and molecular biology [7–9]. For example, simple, low-cost microfluidic arrays [10–12] can achieve limits of detections below 1 fg/mL for multiple proteins in fast, semi-automated immunoassays [13,14]. New methodologies involving CRISPR-based assays [15], surface plasmon resonance (SPR) [16,17], and methylation signatures of blood-circulating cancer cells [18] also provide high sensitivity and specificity for biomarker detection. Such developments provide the power of early clinical detection of illnesses that can lead to faster and more effective, less invasive treatments [19]. Cutting-edge technologies facilitated by low-cost point-of-care procedures in a hospital, clinic, or physician's office have the potential to transform disease management and enhance positive patient outcomes by enabling early detection [20,21].

CRISPR-Cas was first identified as a bacterial defense mechanism against viruses. CRISPR (Clustered Regularly Interspaced Short Palindromic Repeats) and Cas (CRISPR-associated) proteins work together to recognize and cleave specific nucleic acid regions [22]. CRISPR-Cas systems were identified and developed in the early 2010s, and have found wide use in science, agriculture, and medicine [21,23]. In diagnostic tests, CRISPR offers simpler and more direct analyses than PCR for nucleic acids, and has better sensitivity [24,25].

Bioanalytical CRISPR methods such as SHERLOCK (Specific High-sensitivity Enzymatic Reporter unLOCKing) and DETECTR (DNA Endonuclease Targeted CRISPR Trans Reporter) have been developed to accurately identify bacterial and viral infections [26], dengue virus [27], zika virus [28], and SARS-CoV-2 [29–31]. However, current drawbacks include the requirement for amplification of analytes to achieve the necessary sensitivity, and the lack of a general approach to multiple analyte detection, or multiplexing.

There is a large diagnostic advantage to measuring multiple biomarkers in a single experiment [20,32]. For example, the measurement of four prostate-cancer biomarkers in patient serum provided greatly improved diagnostic performance in predicting whether a patient needs a biopsy compared to the clinically used single protein prostate specific antigen [13]. Multiplex assays can also detect co-infections and track multiple pathogens in infectious diseases [33–35]. For example, a multiplexed CRISPR assay can detect a panel of 21 respiratory viruses, including SARS-CoV-2 variants, thereby providing a general test for virus identification [36]. Multiplexed CRISPR technology has also been used to integrate comprehensive disease profiles in a single test [37–39].

Herein we describe a simple mix-and-read, high-sensitivity, high-specificity, CRISPR/Cas miRNA assay system that features a general approach to multiplexing and is illustrated by the detection of a miRNA panel for Alzheimer's disease (ALZ) detection (Figure 1). The platform is a 3D-printed 96-microtiter well plate in which Cas31a protein and different CRISPR RNAs (crRNA/Cas13a nucleoprotein complex) complementary to target miRNAs are chemically immobilized onto a hydrogel in specific wells in a generally applicable approach to multiplexing. These immobilized CRISPR plates are storable for at least one week. For the assay, samples or standards containing miRNAs mixed with fluorescent reporter are then added to each well to produce individual signals for each target. We illustrate assay performance by detecting three plasma miRNAs (miR-34c-5p, miR-30e-5p, and miR-200c-3p) previously linked to early-stage ALZ [40–42]. When the CRISPR-cas13a's bind their specific targets in individual wells, the resulting non-discriminate cas13a RNAse activity cleaves poly-RNA links in the reporter between fluorophore and quencher to produce fluorescence in each well measured by a plate reader. The RNA reporter is a short, single-stranded RNA molecule labeled with a fluorescent dye at one end and a quencher at the other. In its intact state, the close proximity of the fluorophore and quencher enables Fluorescence Resonance Energy Transfer (FRET), effectively suppressing fluorescence emission. Upon the specific binding of the CRISPR RNA (crRNA) to its complementary miRNA target, Cas13a is activated and exhibits collateral RNase activity, randomly cleaving RNA molecules in the solution, including the reporter. This cleavage separates the fluorophore from the quencher, restoring fluorescence that can be detected upon excitation at the dye's characteristic wavelength. The resulting fluorescence intensity is directly proportional to the concentration of the target miRNA, providing a quantitative, real-time readout of its presence. Calibration standards can be measured in the same plate at the same time as samples. This approach provides highly sensitive and selective assays for multiple miRNAs in plasma and other fluids without amplification.

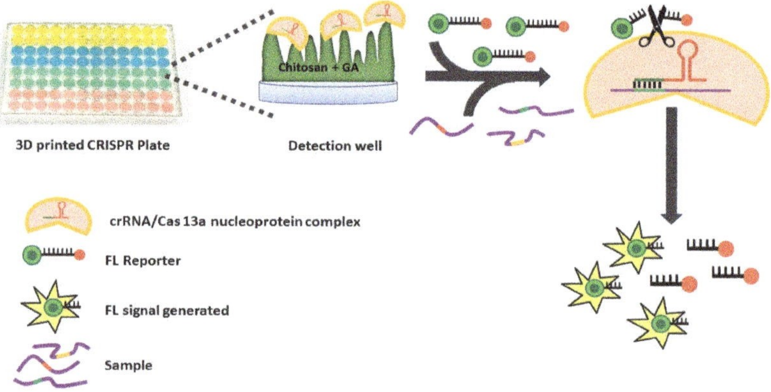

Figure 1. Schematic showing steps for the CRISPR assay in the 3D-printed well plate.

2. Materials and Methods

All reagents and chemicals were analytical grade. Chitosan (MW 448869), and glutaraldehyde, glycerol, and dithiothreitol solution (DTT) 1M were from Sigma Aldrich (St. Louis, MO, USA). RNaseAlert™ Lab Test Kit v2, SUPERase•In™ RNase Inhibitor (20 U/µL) UltraPure™ 1 M Tris-HCI Buffer pH 7.5, UltraPure™ water, was from Thermo Fisher (Waltham, MA, USA). Cas13a 1mg/mL (CAS 13a-200) was from MCLAB. miR-34c-5p, miR-200c-3p, miR-30e-5p, crRNA-miR-34c-5p, crRNA-miR-200c-3p, crRNA-miR-30e-5p were from Integrated DNA Technologies (Coralville, IA, USA). Pooled human plasma was from Innovative™ Research (Novi, MI, USA). The storage buffer was Tris-HCl (pH 7.5, 1 M), NaCl (5 M), glycerol, $MgCl_2$ (0.25 M), and DTT (1 M), in ultrapure water. Fluorescence was measured using a BioTek® Multi-Mode Microplate Reader (Agilent, Santa Clara, CA, USA) and Gen5™ software, version 2.0. A Form Labs Form 3B+ 3D printer and Clear Resin (RS-F2-GPCL-04) were used for 3D printing. Alzheimer's patient plasma samples were from UCONN Health Center (Farmington, CT, USA), (IRB approval 19-214S-1).

Well plates were printed using predetermined optimal 50 mL volume on a FormLabs Form 3B stereolithographic 3D printer (FormLabs, Somerville, MA, USA). The polyacrylate polymer used strongly chemisorbs chitosan [20] (see below) and was critical to assay sensitivity. Printed arrays were cleaned by sonicating and washing in isopropanol for 10–15 min to remove uncured resin, followed by air drying and curing at 60 °C for 30 min.

As shown in Figure 2, to each microwell 50 µL 0.5 mg/mL chitosan in 0.05 M HCl was added and incubated at room temperature for 3 h. Solution was removed by tapping the plate upside down, then it was left to dry overnight in a vacuum to provide hydrogel coats in each microwell. Then, wells were filled with 3% glutaraldehyde and incubated 1 h. to form Schiff's bases between amines on chitosan and glutaraldehyde [43]. Excess solution was drained by inverting the plate and tapping, followed by vacuum drying 1 h. Separate microwells were then reacted with crRNA/Cas13a nucleoprotein complex of miRNAs 30e-

5p, miRNA34c-5p, and miRNA 200c-3p. Cas13a/crRNA/nucleoprotein complex mixtures were prepared by mixing 400 mM Tris-HCl buffer (5 µL), ultrapure water (35 µL), Cas13a (63.3 µg/mL) (5 µL), SUPERase•In RNase Inhibitor (2.5 µL), and the appropriate CRISPR RNA (crRNA) (2.5 µg/mL) (2.5 µL), and incubating for 12 min. at 37 °C. The required Cas 13a/crRNA (50 µL) were then added to their respective wells and incubated overnight at 4 °C. This process facilitates a reaction between free glutaraldehyde and amines on the Cas13a protein in the mixture with the chitosan hydrogel already chemisorbed in the wells. The quantification of immobilized Cas13a in the microwells was carried out using the Bicinchoninic Acid (BCA) total protein assay [44], revealing 37×10^{11} Cas13a molecules/well. Excess crRNA/Cas13a was removed by tap drying, and washed with pH 7.5 (400 mM) Tris-HCl buffer. The reaction mixture contained 400 mM Tris-HCl (5 µL), RNase-free H_2O (38.25 µL), SUPERase•In RNase Inhibitor (1.25 µL), RNaseAlert reporter (2 µM) (2.5 µL), pooled human plasma (0.5 µL), and the respective target miRNA sample (2.5 µL) (total 50 µL per well based on the optimum volume for best assay performance, as shown in Figure S1), were kept in microcentrifuge tubes in an ice bath while preparing to add them to wells.

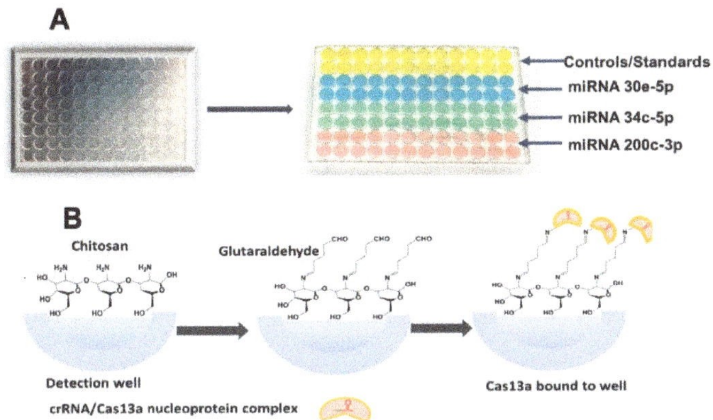

Figure 2. Three-dimensional-printed polyacrylate plates for multiplexed CRISPR miRNA assay: (**A**) 96-well microfluidic CRISPR plate with 50 uL vol. microwells (on left), and analytical CRISPR plate after immobilizing the desired crRNA/Cas13a's in the wells. (**B**) Binding of Cas13 into wells by chemical attachment to chemisorbed chitosan hydrogel.

In the assay procedure, optimized amounts of Cas13a and crRNA are added to 400 mM TRIS-HCl buffer, RNase-free water, Cas13a, inhibitor, and crRNA, and incubated at 37 °C for 12 min. This mixture was added to respective wells in the well-plate pre-functionalized with chitosan and glutaraldehyde. For crRNA optimization, the reaction mixture was prepared with different crRNA concentrations (1.25 to 10 µg/mL) with a 'maximum Cas13a concentration' (127 ug/mL) to ensure a maximum dynamic range of analyte concentrations was captured. Then, the assay was conducted across different analyte miRNA concentrations within the desired dynamic range [miRNA30e-5p] = (50, 500 and 1000 fg/mL). As mentioned above, the crRNA concentration that provided the largest signal separation in triplicate, and between low and high miRNA levels in the correct level of low to medium to high, was used as the optimized crRNA concentration (Figure 3a). Using this optimized crRNA concentration, the optimize dCas13a concentration was used as the largest signal separation between low and high miRNA (Figure 3b).

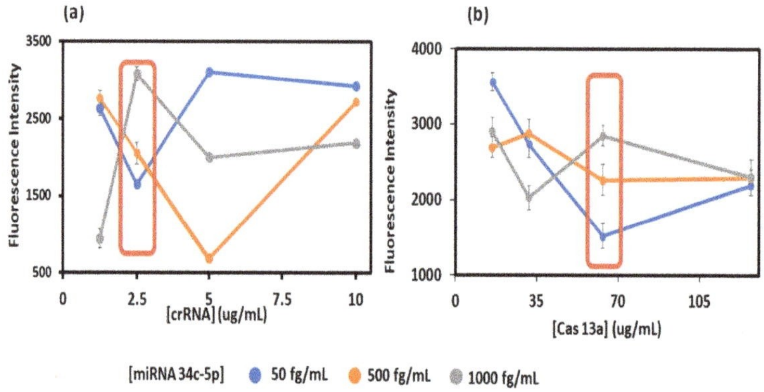

Figure 3. Optimization plots for (**a**) crRNA and (**b**) Cas13a with respect to the miRNA 34c-5p biomarker. The red boxes indicates the optimal concentrations, where the signal increase per unit concentration is the largest.

3. Results

3.1. Assay Optimization

The assay was designed to be conducted in 3D-printed 96-well plates with individual crRNA/Cas13a's chemically bound to chemisorbed chitosan in each well. CrRNA optimization was carried out by testing concentrations of the miRNA in the range of 1–10 µg/mL with maximum Cas13a (127 µg/mL) to maximize signal separation, e.g., shown for miRNA30e-5p detection (Figure 3a) [20,41]. The best crRNA was selected for each of the analyte miRNAs, as the value that provided increased fluorescence from low to high values in the correct order. This crRNA level was then used to optimize Cas13a (range 16–127 µg/mL) using the same criteria (Figure 3b).

3.2. Calibration and Validation

For each miRNA, calibrations were carried out using the optimized parameters described above over the dynamic range of the three miRNAs in solutions of 0.7 fg/mL to 1756 pg/mL miRNA, complementary to the immobilized crRNA/Cas-13a in the well, along with 2.5 uL of fluorescence reporter in 0.1% pooled human plasma for all samples. Figure 4 shows raw data from a typical calibration experiment showing fluorescent output from the plate reader using 490 and 520 nm as excitation and emission wavelengths, respectively, and nm incident light and measured fluorescence output at 520 nm vs. time after the start of the CRISPR Cas13a RNA cleavage activity upon the addition of an miRNA analyte. At the beginning of the experiment, fluorescence decays a bit, attributed to the addition of the miRNA reaction mixture that had been stored in an ice bath to stabilize the miRNAs. The small decrease is related to the time required for temperature equilibration in the well, and when the solution is equilibrated the reported cleavage rection that releases florescence increases until most of it is cleaved.

While the fluorescent traces are somewhat noisy, differences between the fluorescence rise at 140–150 min after subtracting the plateau at 90–100 min were quite reproducible and were used to prepare calibration curves. Differences between the 250 min and 90–100 min signals gave qualitatively similar calibrations, but added 100 min to the assay time. Fluorescence (FL) intensity differences at 140–150 minus 90–100 min as indicated were plotted against the concentration (Figure 5) to determine the linear dynamic range and the limit of detection (LOD), as 3× standard deviation (SD) of the blank above the blank signal for each miRNA.

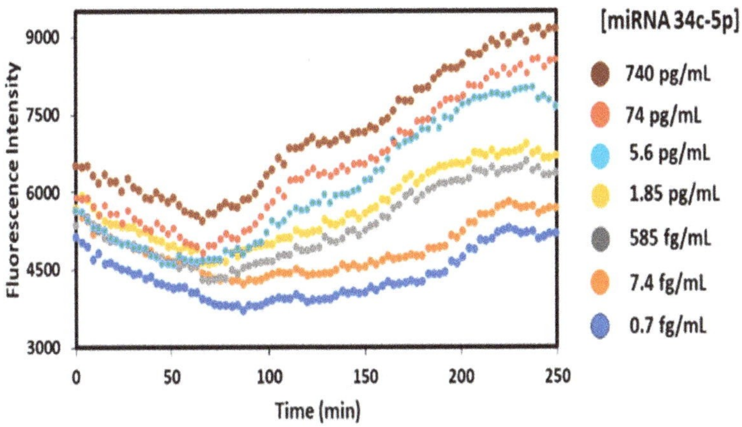

Figure 4. Fluorescence plate reader measurement using 490 and 520 nm as excitation and emission wavelengths for different concentrations of miRNA 34c-5p at 37 °C. For calibrations and sample analyses, emission intensity was measured after subtracting the plateau readings (90–100 min) from exponential phase points (140–150 min). All concentration-dependent signals were baseline-corrected by subtracting the corresponding control signal.

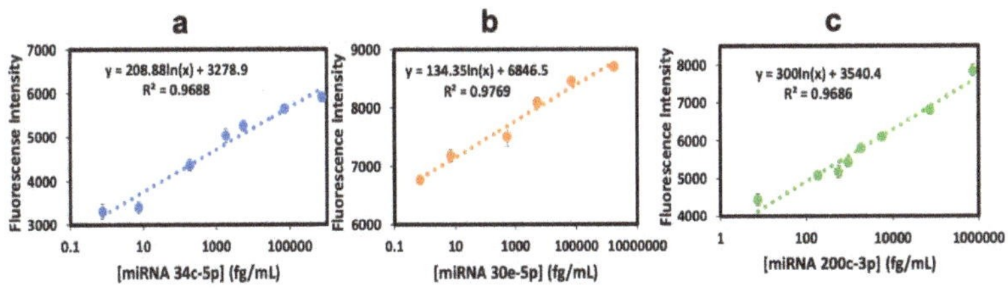

Figure 5. Calibration plots in 0.1% pooled human plasma using the fluorescence plate for simultaneous fluorescent detection of (**a**) miRNA 34c-5p, (**b**) miRNA 30e-5p, and (**c**) miRNA 200c-3p (error bars: SD for n = 3).

Calibration graphs in Figure 5 show good linearity with the corresponding equations, and (r^2) correlation coefficient values of 0.9688 for miRNA-34c-5p, 0.9769 for miRNA-30e-5p, and 0.9686 for miRNA-200c-3p. Calibration data show small standard deviations for each miRNA. Linear dynamic ranges are 0.74 fg/ mL–740 pg/mL (0.1 fM–100 pM) for miRNA 34c-5p, 0.7 fg/mL–1756 pg/mL (0.1 fM–250 pM) for miRNA 30e-5p, and 7.4 fg/mL–74.1 pg/mL (1 fM–10 pM) for miRNA 200c-3p, with the average relative standard deviation of ±5% plate to plate (n = 6). LODs obtained were 0.74 fg/mL for miRNA 34c-5p, 0.7 fg/mL for miRNA 30e-5p, and 7.4 fg/mL for miRNA 200c-3p. Reproducible signal intensity was found for the three microwells allotted to each miRNA within a single plate (Figure 5), giving an average standard deviation of ±3% well-to-well. Spike recovery tests (Table 1) gave values that were all within the acceptable analytical range of 100 ± 20% for biomedical assays [45]. A cross-reactivity assay involved testing one miRNA against the crRNA of the other two biomarkers and analyzing the resulting signals. Cross-reactivity was below 5% in all cases (Table 2, grey areas for no cross-reactivity possible).

Table 1. Spike recovery data for miRNAs from well-plate assay in 0.1% pooled human plasma (n = 3).

Spiked Conc, (fg/mL)	miRNA 30e-5p		miRNA 200c-3p		miRNA 34c-5p	
	Found (fg/mL)	Recovery %	Found (fg/mL)	Recovery %	Found (fg/mL)	Recovery %
70	83	118 ± 12	60	86 ± 8	79	112 ± 11
850	808	95 ± 9	902	106 ± 10	703	83 ± 7
3500	4224	120 ± 14	4100	117 ± 13	3380	96 ± 7
35,000	40,254	115 ± 11	40,790	116 ± 12	39,900	114 ± 7

Table 2. Cross-reactivity between different miRNAs in pooled human plasma miRNA.

miRNA	% Cross-Reactivity with CRISPR RNA		
	30e-5p	34c-5p	200c-3p
30e-5p		3.3	3.3
34c-5p	4.1		4.7
200c-3p	2.5	3.6	

Grey means no possibility of cross reactivity.

Spike recovery studies were conducted by spiking pooled human plasma with known concentrations (70 fg/mL, 850 fg/mL, 0.35 pg/mL, and 35 pg/mL) of each analyte miRNA and measuring blank-subtracted fluorescence (FL) intensities. Calibrations in pooled human plasma were used to determine the concentration of each analyte. Spike recovery % was established by comparing the known spike concentrations with the concentration found from the assay. Cross-reactivity between different miRNAs in pooled human plasma was established by testing one miRNA against the crRNA of the remaining two markers under study.

Elevated levels of miRNAs 30e, 34c, and 200c have been reported in human blood and brain in ALZ patients, and which have been shown to be highly correlated with early and later stages of the disease [38–40]. Preliminary studies showed that it was necessary to extract the RNAs from plasma to remove interferences in the assay, so all results below involving patient samples are shown for RNA extracted from plasma.

Samples from eight Alzheimer's patients were analyzed in a blind study. miRNAs were extracted using the mirVana™ PARIS™ RNA and Native Protein Purification Kit (Thermo Fisher), and the optimized fluorescence-based assay was performed in the 96-well plate to detect three miRNA analytes.

Of eight patient samples analyzed, six were diagnosed as AD-positive, while the remaining two negative samples exhibited very low concentrations of all three miRNAs (Table 3). To further validate the assay, results were compared with our developed ECL-based CRISPR assay for the eight patient samples displayed in the Supplementary Materials, Figure S2, showing slopes near 1.0 and intercepts close to zero for each miRNA, indicating excellent agreement between the two methods. Additionally, R^2 values close to 1.0 further confirm an excellent correlation.

Table 3. Sample results.

Patient	34c-5p (ng/mL)	30e-5p (ng/mL)	200c-3p (ng/mL)	ALZ Diagnosis
1	0.000028	0.000023	0.000030	Negative
2	0.000041	0.000033	0.000044	Negative
3	0.655	0.0001834	0.965	Positive
4	2.16	3.11	1.02	Positive
5	2.74	6.52	1.21	Positive
6	3.74	21.2	1.69	Positive
7	9.12	86.4	3.11	Positive
8	24.2	349	3.01	Positive

4. Discussion

The mix-and-read CRISPR assay described above features immobilized Cas13a protein loaded with corresponding crRNAs in a 96-well, 50 mL plate designed to detect multiple, unamplified miRNAs in plasma or serum samples. It is easy-to-use, low-cost (~USD 0.70 per printed plate and USD 2.80 for reagents per plate), specific, and highly sensitive using fluorescence. The new features for this CRISPR assay that is applicable to a collection of miRNAs are (1) the stable binding of Cas13a into the 50-mL wells of the 3D-printed polyacrylate 96-well plate, allowing the pre-storage of loaded plates for one week in a refrigerator, (2) a reaction mixture including the sample and fluorescent reporter that enables a one-step addition for the detection measurements, and (3) very high sensitivity that can directly address miRNAs in clinical plasma and serum samples without preamplification.

Excellent assay calibrations (Figure 4), spike recoveries (Table 1), and low cross reactivities (Table 2) were demonstrated for the detection of three known miRNA biomarkers for Alzheimer's disease (ALZ) in plasma. This is important since blood miRNAs are being intensely investigated as likely biomarkers for AD, since about 70% of the currently known miRNAs are expressed in the brain and involved in neurological processes [46–49]. Our new multiplexed CRISPR assay can be readily applied to any set of RNA molecules. Our assay offers potential benefits and distinct advantages in sensitivity and multiplexity over established miRNA detection techniques, including genome sequencing and PCR methods [50,51]. While PCR tests are regarded as gold standards for RNA detection, they do not directly detect miRNA concentrations, and require specialized instrumentation as well as amplification of the targets for use in plasma and serum. Also, PCR applications at low-resource sites, such as third-world countries, is restricted due to costly equipment and the need for specialized expertise [51,52]. The earlier SHERLOCK method utilizing CRISPR requires amplified RNA or DNA by recombinase polymerase amplification (RPA). Assay time is about 1–2 h, and does not feature multiplexing [15,51]. miSHERLOCK and CARMEN are multiplexing CRISPR systems, which require comparably large amounts of sample and additional amplification steps [15,38,52].

In the new CRISPR assay procedure described above, crRNA/Cas13a is chemically bound in fully active form to chitosan hydrogel in microwells of a 96-well plate, representing to our knowledge the first immobilized CRISPR platform. Previous work [41] in our lab showed that chitosan is strongly chemisorbed onto the 3D-printed polyacrylate polymer used. The film contains 98% water, resulting in a highly porous hydrogel that provides a high surface area for binding the crRNA/Cas13a complex, and a porous, water-rich network that allows the efficient internal mass transportation of water soluble molecules in the wells. Well plates with immobilized cas13a can be refrigerated at 4 °C for up to five days and used when needed. However, it is crucial to eliminate or inactivate any RNases present that can cleave reporter molecules. In our method, we recommend using an RNAse inhibitor during sample preparation to deactivate the RNases that may be present.

The new CRISPR miRNA assay as presented here achieves extremely low limits of detection (LODs) for multiple analytes without RNA amplification. Detection limits were 0.74 fg/mL for miRNA 34c-5p, 0.7 fg/mL for miRNA 30e-5p, and 7.4 fg/mL for miRNA 200c-3p. The assay can be applied to many more miRNAs than in this first example. Due to this high sensitivity, plasma samples need to be diluted so that concentration levels of target miRNAs fall within the assay's dynamic range, allowing the detection of miRNAs with plasma sample volumes as small as ~2.5 µL. Large dilutions are beneficial to significantly reduce concentrations of interfering substances in biomedical samples such as blood plasma and serum, and thus minimize cross-reactivity (Table 2), resulting in much lower background signals. The accuracy of the assay was validated through spike-recovery rates (Table 1) and low cross reactivity (Table 2), and confirmed by excellent correlations of the found concentrations of each miRNA from the FL assay with an ECL referee method (Supplementary Materials, Figure S2). Additionally, results on a small cohort of patient samples illustrate the capability to differentiate between Alzheimer-positive and Alzheimer-negative patients based on levels of analyzed miRNA biomarker from the assay.

In summary, we report above a new approach to develop an accurate, ultrasensitive, multiplexed, mix-and-read 3D-printed CRISPR well-plate assay using the first example of immobilized crRNA/CAS protein complexes to detect three ALZ-specific miRNAs in diluted plasma samples without amplification. This new 96-well-plate assay can measure standards and multiple samples in the same plate. Future goals include incorporating additional miRNA biomarkers related to ALZ and exploring novel multiple miRNA panels for cancer and cancer metastasis biomarkers, as well as developing modifications designed to measure both protein and miRNA biomarkers in the same CRISPR assay.

Supplementary Materials: The following supporting information can be downloaded at: https://www.mdpi.com/article/10.3390/bios15060346/s1, Figure S1: Well size optimization; Figure S2: Correlation plots with alternative method; Table S1: crRNA sequences. The document provides details on miRNA sequences and assay optimization related to Alzheimer's disease (AD). List of target miRNAs (34c-5p, 30e-5p, 200c-3p, and a control) along with their associated crRNA sequences, except for the control, which lacks a crRNA. The development of a miRNA fluorescence assay focused on optimizing CRISPR RNA and Cas13a. For CRISPR well-plate optimization, the assay was tested using miRNA34c-5p across four well volumes. This finding supports the use of 50-µL wells as a cost-effective and resource-efficient option, reducing the required sample and reagent volumes without compromising performance. Reference [53] is cited in Supplementary Materials.

Author Contributions: P.I.T.D.S. and K.H. contributed equally to the design of the assay, experimental work, and writing of the paper. J.F.R., I.M.M., B.S.D. and J.L.R. developed the initial concepts of the assay, and J.F.R. supervised the overall project. J.L.R. planned aspects of maintaining RNA stability and assay reliability, and verified the robustness of the results. S.M.S. supervised the initial and intermediate studies, and developed and refined experimental procedures. R.C. and H.G. performed auxiliary miRNA assays, ran miRNAs purity assays, necessary reagents, and controls. K.S.B. contributed to the initial assay design optimization. D.C.S., K.M. and B.S.D. discussed and planned all clinical testing aspects of miRNA ALZ biomarkers, verified data interpretation, obtained and provided ALZ samples from Uconn Health patients, and applied for and secured UConn Health ethical approval Internal Review Board (IRB) 19-214S-1. All authors have read and agreed to the published version of the manuscript.

Funding: This work was supported financially by the U. S. National Science Foundation, Award Number: 2231490, and by the University of Connecticut Research Excellence Program.

Institutional Review Board Statement: Samples were provided by the Uconn Health Center, and obtained with Institutional Review Board approval for the analysis of Alzheimer's patient samples

17. under Uconn Health Center ethical approval IRB 19-214S-1. 16 March 2023 for the analysis of the human samples. Prior informed consent was obtained from all subjects in the sample cohort.

Informed Consent Statement: Prior informed consent was obtained from all subjects in the sample cohort and samples were obtained by Uconn Health with no possibility of identifying their identities.

Data Availability Statement: All relevant data are presented in this paper or Supplementary Materials.

Acknowledgments: The authors thank Ahmed Farrag for exploratory experimental assistance, Oscar Clement and Palash Vora for assistance with the robustness studies, and Sherif M. Shawky thanks the Biochemistry Department, Faculty of Pharmacy, Misr University for Science and Technology, Giza, Egypt, for approving leave and support for his participation in this project at the University of Connecticut. The authors also than Diniz, Steffans and Manning for obtaining pateint samples and securing Uconn Health Center ethical approval under IRB 19-214S-1. 16 March 2023 for analysis of the human samples.

Conflicts of Interest: The authors declare no conflicts of interest.

References

1. Rusling, J.F.; Kumar, C.V.; Gutkind, J.S.; Patel, V. Measurement of Biomarker Proteins for Point-of-Care Early Detection and Monitoring of Cancer. *Analyst* **2010**, *135*, 2496–2511. [CrossRef] [PubMed]
2. Ahmad, A.; Imran, M.; Ahsan, H. Biomarkers as Biomedical Bioindicators: Approaches and Techniques for the Detection, Analysis, and Validation of Novel Biomarkers of Diseases. *Pharmaceutics* **2023**, *15*, 1630. [CrossRef]
3. Song, J.G.; Barol, K.C.; Kim, G.-L.; Park, J.-W.; Seo, S.-H.; Kim, D.-H.; Jung, D.H.; Ifekpolugo, N.L.; Han, H.-K. Quantitative analysis of therapeutic proteins in biological fluids: Recent advancement in analytical techniques. *Drug Deliv.* **2023**, *30*, 2183816. [CrossRef] [PubMed]
4. Kaminski, M.M.; Abudayyeh, O.O.; Gootenberg, J.S.; Zhang, F.; Collins, J.J. CRISPR-based diagnostics. *Nat. Biomed. Eng.* **2021**, *5*, 643–656. [CrossRef]
5. Costelloe, C.M.; Rohren, E.M.; Madewell, J.E.; Hamaoka, T.; Theriault, R.L.; Yu, T.K.; Lewis, V.O.; Ma, J.; Stafford, R.J.; Tari, A.M.; et al. Imaging Bone Metastasis in Breast Cancer: Techniques and recommendations for diagnosis. *Lancet* **2009**, *10*, 606–614. [CrossRef]
6. Tucaszewski, B.; Nazar, J.; Goch, M. Diagnostic Methods for Detection of bone metastasis. *Contemp. Oncol.* **2017**, *21*, 98–103. [CrossRef]
7. Self, W.K.; Holzman, D.M. Emerging diagnostics and therapeutics for Alzheimer disease. *Nat. Med.* **2023**, *29*, 2187–2199. [CrossRef]
8. Jiang, L.; Lin, X.; Chen, F.; Qin, X.; Yan, Y.; Ren, L.; Yu, H.; Chang, L.; Wang, Y. Current research status of tumor cell biomarker detection. *Microsyst. Nanoeng.* **2023**, *9*, 123. [CrossRef]
9. Rusling, J.F.; Forster, R.J. Biosensors Designed for Clinical Applications. *Biomedicines* **2021**, *9*, 702. [CrossRef]
10. Sharafeldin, M.; Kadimisetty, K.; Bhalero, K.R.; Chen, T.; Rusling, J.F. 3D-printed Immunosensor arrays for cancer diagnostics. *Sensors* **2020**, *20*, 4514. [CrossRef]
11. Dhanapala, L.; Krause, C.E.; Jones, A.L.; Rusling, J.F. Printed electrodes in microfluidic arrays for cancer biomarker protein detection. *Biosensors* **2020**, *10*, 115. [CrossRef] [PubMed]
12. Jones, A.; Dhanapala, L.; Kankanamage, R.N.T.; Kumar, C.V.; Rusling, J.F. Multiplexed Immunosensors and Immunoarrays. *Anal. Chem.* **2020**, *92*, 345–362. [CrossRef] [PubMed]
13. Jones, A.; Dhanapala, L.; Baldo, T.A.; Sharafeldin, M.; Krause, C.; Shen, M.; Faria, R.C.; Dey, D.; Moghaddam, S.; Watson, R.W.; et al. Prostate Cancer Diagnosis in the Clinic Using an 8-Protein Biomarker Panel. *Anal. Chem.* **2021**, *93*, 1059–1067. [CrossRef]
14. Dhanapala, L.; Jones, A.L.; Czarnecki, P.; Rusling, J.F. Sub-Zeptomole Detection of Biomarker Proteins Using a Microfluidic Immunoarray with Nanostructured Sensors. *Anal. Chem.* **2020**, *92*, 8021–8025. [CrossRef]
15. de Puig, H.; Lee, R.A.; Najjar, D.; Tan, X.; Soenksen, L.R.; Angenent-Mari, N.M.; Donghia, N.M.; Weckman, N.E.; Ory, A.; Ng, C.F.; et al. Minimally Instrumented SHERLOCK (miSHERLOCK) for CRISPR-Based Point-of-Care Diagnosis of SARS-CoV-2 and Emerging Variants. *Sci. Adv.* **2021**, *7*, eabh2944. [CrossRef]
16. Krishnan, S.; Mani, V.; Wasalathanthri, D.; Kumar, C.V.; Rusling, J.F. Attomolar Detection of a Cancer Biomarker Protein in Serum by Surface Plasmon Resonance Using Superparamagnetic Particle Labels. *Angew. Chem. Int. Ed.* **2011**, *50*, 1175–1178. [CrossRef]
17. Das, S.; Devireddy, R.; Gartia, M.R. Surface plasmon resonance (SPR) sensor for cancer biomarker detection. *Biosensors* **2023**, *13*, 396. [CrossRef] [PubMed]

18. Chen, X.; Gole, J.; Gore, A.; He, Q.; Lu, M.; Min, J.; Yuan, Z.; Yang, X.; Jiang, Y.; Zhang, T.; et al. Non-invasive early detection of cancer four years before conventional diagnosis using a blood test. *Nat. Commun.* **2020**, *11*, 3475. [CrossRef]
19. Shen, Y.; Yang, W.; Liu, J.; Zhang, Y. Minimally invasive approaches for the early detection of endometrial cancer. *Mol. Cancer* **2023**, *22*, 53. [CrossRef]
20. Hiniduma, K.; Bhalerao, K.S.; De Silva, P.I.T.; Chen, T.; Rusling, J.F. Design and Fabrication of a 3D-Printed Microfluidic Immunoarray for Ultrasensitive Multiplexed Protein Detection. *Micromachines* **2023**, *14*, 2187. [CrossRef]
21. Ghorbani, A.; Hadifar, S.; Salari, R.; Izadpanah, K.; Burmistrz, M.; Afsharifar, A.; Eskandari, M.H.; Niazi, A.; Denes, C.E.; Neely, G.G. A short overview of CRISPR-Cas technology and its application in viral disease control. *Transgen. Res.* **2021**, *30*, 221–238. [CrossRef] [PubMed]
22. Horvath, P.; Barrangou, R. CRISPR/CAS, the immune system of bacteria and archaea. *Science* **2010**, *327*, 167–170. [CrossRef]
23. Lau, A.; Ren, C.; Lee, L.P. Critical review on where CRISPR meets molecular diagnostics. *Prog. Biomed. Eng.* **2020**, *3*, 012001. [CrossRef]
24. Kellner, M.J.; Koob, J.G.; Gootenberg, J.S.; Abudayyeh, O.O.; Zhang, F. SHERLOCK: Nucleic acid detection with CRISPR nucleases. *Nat. Prot.* **2019**, *14*, 2986–3012. [CrossRef]
25. Green, C.M.; Spangler, J.; Susumu, K.; Stenger, D.A.; Medintz, I.L.; Díaz, S.A. Quantum Dot-Based Molecular Beacons for Quantitative Detection of Nucleic Acids with CRISPR/Cas(N) Nucleases. *ACS Nano* **2022**, *16*, 20693–20704. [CrossRef] [PubMed]
26. Xu, J.; Bai, X.; Zhang, X.; Yuan, B.; Lin, L.; Guo, Y.; Cui, Y.; Liu, J.; Cui, H.; Ren, X.; et al. Development and application of DETECTR-based rapid detection for pathogenic Bacillus anthracis. *Anal. Chim. Acta* **2023**, *1247*, 340891. [CrossRef] [PubMed]
27. Wang, J.; Xia, Q.; Wu, J.; Lin, Y.; Ju, H. A sensitive electrochemical method for rapid detection of dengue virus by CRISPR/Cas13a-assisted catalytic hairpin assembly. *Anal. Chim. Acta* **2021**, *1187*, 339131. [CrossRef]
28. Dukhovny, A.; Lamkiewicz, K.; Chen, Q.; Fricke, M.; Jabrane-Ferrat, N.; Marz, M.; Jung, J.U.; Sklan, E.H. A CRISPR Activation Screen Identifies Genes That Protect against Zika Virus Infection. *J. Virol.* **2019**, *93*, e00211-19. [CrossRef]
29. Ding, X.; Yin, K.; Li, Z.; Lalla, R.V.; Ballesteros, E.; Sfeir, M.M.; Liu, C. Ultrasensitive and Visual Detection of SARS-CoV-2 Using All-in-One Dual CRISPR-Cas12a Assay. *Nat. Commun.* **2020**, *11*, 4711. [CrossRef]
30. Jiang, Y.; Hu, M.; Liu, A.-A.; Lin, Y.; Liu, L.; Yu, B.; Zhou, X.; Pang, D.-W. Detection of SARS-COV-2 by CRISPR/CAS12A-Enhanced Colorimetry. *ACS Sens.* **2021**, *6*, 1086–1093. [CrossRef]
31. Palaz, F.; Kalkan, A.K.; Can, Ö.; Demir, A.N.; Tozluyurt, A.; Özcan, A.; Ozsoz, M. CRISPR-CAS13 system as a promising and versatile tool for cancer diagnosis, therapy, and research. *ACS Synth. Biol.* **2021**, *10*, 1245–1267. [CrossRef] [PubMed]
32. Kadimisetty, K.; Malla, S.; Bhalerao, K.S.; Mosa, I.M.; Bhakta, S.; Lee, N.H.; Rusling, J.F. Automated 3D-Printed Microfluidic Array for Rapid Nanomaterial-enhanced (ECL) Detection of Multiple Proteins. *Anal. Chem.* **2018**, *90*, 7569–7577. [CrossRef]
33. Corstjens, P.L.M.; Chen, Z.; Zuiderwijk, M.; Bau, H.H.; Abrams, W.R.; Malamud, D.; Niedbala, R.S.; Tanke, H.J. Rapid assay format for multiplex detection of humoral immune responses to infectious disease pathogens (HIV, HCV, and TB). *Ann. N. Y. Acad. Sci.* **2007**, *1098*, 437–445. [CrossRef] [PubMed]
34. Engin, E.D. The use of multiplexing technology in the immunodiagnosis of infectious agents. *J. Immunoass. Immunochem.* **2019**, *40*, 109–122. [CrossRef]
35. Hao, L.; Zhao, R.T.; Welch, N.L.; Tan, E.K.W.; Zhong, Q.; Harzallah, N.S.; Ngambenjawong, C.; Ko, H.; Fleming, H.E.; Sabeti, P.C.; et al. CRISPR-Cas-amplified urinary biomarkers for multiplexed and portable cancer diagnostics. *Nat. Nanotechnol.* **2023**, *18*, 798–807. [CrossRef]
36. Welch, N.L.; Zhu, M.; Hua, C.; Weller, J.; Mirhashemi, M.E.; Nguyen, T.G.; Mantena, S.; Bauer, M.R.; Shaw, B.M.; Ackerman, C.M.; et al. Multiplexed CRISPR-Based Microfluidic Platform for Clinical Testing of Respiratory Viruses and Identification of SARS-CoV-2 Variants. *Nat. Med.* **2022**, *28*, 1083–1094. [CrossRef] [PubMed]
37. Zhang, C.; Huang, Z.; Li, Z.; Hu, J.; Liu, R.; Lv, Y. CRISPR-Associated "Genetic Scissors" for Multiplexing Analysis. *Trends Anal. Chem.* **2024**, *170*, 117431. [CrossRef]
38. Zhao, L.; Qiu, M.; Li, X.; Yang, J.; Li, J. CRISPR-Cas13a system: A novel tool for molecular diagnostics. *Front. Microbiol.* **2022**, *13*, 1060947. [CrossRef]
39. Swarbrick, S.; Wragg, N.; Ghosh, S.; Stolzing, A. Systematic review of MIRNA as biomarkers in Alzheimer's Disease. *Mol. Neurobiol.* **2019**, *56*, 6156–6167. [CrossRef]
40. Cogswell, J.P.; Ward, J.; Taylor, I.A.; Waters, M.; Shi, Y.; Cannon, B.; Kelnar, K.; Kemppainen, J.; Brown, D.; Chen, C.; et al. Identification of miRNA Changes in Alzheimer's Disease Brain and CSF Yields Putative Biomarkers and Insights into Disease Pathways. *J. Alzheimer's Dis.* **2008**, *14*, 27–41. [CrossRef]
41. Müller, M.; Kuiperij, H.B.; Claassen, J.A.; Küsters, B.; Verbeek, M.M. MicroRNAs in Alzheimer's disease: Differential expression in hippocampus and cell-free cerebrospinal fluid. *Neurobiol. Aging* **2014**, *35*, 152–158. [CrossRef] [PubMed]
42. Sharafeldin, M.; Chen, T.; Ozkaya, G.U.; Choudhary, D.; Molinolo, A.A.; Gutkind, J.S.; Rusling, J.F. Detecting Cancer Metastasis and Accompanying Protein Biomarkers at Single Cell Levels Using a 3D-Printed Microfluidic Immunoarray. *Biosens. Bioelectron.* **2021**, *171*, 112681. [CrossRef]

43. Sharafeldin, M.; Kadimisetty, K.; Bhalerao, K.R.; Bist, I.; Jones, A.; Chen, T.; Lee, N.H.; Rusling, J.F. Accessible Telemedicine Diagnostics with ELISA in a 3D Printed Pipette Tip. *Anal. Chem.* **2019**, *91*, 7394–7402. [CrossRef] [PubMed]
44. Smith, P.K.; Krohn, R.I.; Hermanson, G.T.; Mallia, A.K.; Gartner, F.H.; Provenzano, M.; Fujimoto, E.K.; Goeke, N.M.; Olson, B.J.; Klenk, D.C. Measurement of Protein Using Bicinchoninic Acid. *Anal. Biochem.* **1985**, *150*, 76–85. [CrossRef]
45. Shah, V.P.; Midha, K.K.; Findlay, J.W.; Hill, H.M.; Hulse, J.D.; McGilveray, I.J.; McKay, G.; Miller, K.J.; Yacobi, A. Bioanalytical Method Validation—A Revisit with a Decade of Progress. *Pharm. Res.* **2000**, *17*, 1551–1557. [CrossRef] [PubMed]
46. Lv, J.; Hao, Y.-N.; Wang, X.-P.; Lu, W.-H.; Xie, L.-Y.; Niu, D. Bone marrow mesenchymal stem cell-derived exosomal miR-30e-5p ameliorates high-glucose induced renal proximal tubular cell pyroptosis by inhibiting ELAVL1. *Ren. Fail.* **2023**, *45*, 2177082. [CrossRef]
47. Tu, Y.; Hu, Y. MiRNA-34c-5p protects against cerebral ischemia/reperfusion injury: Involvement of anti-apoptotic and anti-inflammatory activities. *Metabol. Brain Dis.* **2021**, *36*, 1341–1351. [CrossRef]
48. Li, W.; Yang, S.; Chen, G.; He, S. MiR-200c-3p regulates pyroptosis by targeting SLC30A7 in diabetic retinopathy. *Hum. Exp. Toxicol.* **2022**, *41*, 096032712210995. [CrossRef]
49. Zahra, A.; Shahid, A.; Shamim, A.; Khan, S.H.; Arshad, M.I. The SHERLOCK Platform: An Insight into Advances in Viral Disease Diagnosis. *Mol. Biotechnol.* **2022**, *65*, 699–714. [CrossRef]
50. Yin, L.; Man, S.; Ye, S.; Liu, G.; Ma, L. CRISPR-Cas based virus detection: Recent advances and perspectives. *Biosens. Bioelectron.* **2021**, *193*, 113541. [CrossRef]
51. Kevadiya, B.D.; Machhi, J.; Herskovitz, J.; Oleynikov, M.D.; Blomberg, W.R.; Bajwa, N.; Soni, D.; Das, S.; Hasan, M.; Patel, M.; et al. Diagnostics for SARS-CoV-2 infections. *Nat. Mater.* **2021**, *20*, 593–605. [CrossRef] [PubMed]
52. Ackerman, C.M.; Myhrvold, C.; Thakku, S.G.; Freije, C.A.; Metsky, H.C.; Yang, D.K.; Ye, S.H.; Boehm, C.K.; Kosoko-Thoroddsen, T.-S.F.; Kehe, J.; et al. Massively multiplexed nucleic acid detection with Cas13. *Nature* **2020**, *582*, 277–282. [CrossRef] [PubMed]
53. Hiniduma, K.; De Silva, P.I.T.; Canete, R.; Gunathillaka, H.; Clement, O.; Vora, P.; Shawky, S.M.; Rouge, J.L.; Mosa, I.; Breno, D.; et al. ECL-CRISPR Array for Multiplexed Detection of miRNAs. *ACS Sens.* 2025; *submitted*.

biosensors

MDPI

Article

Strain Plethysmography Using a Hermetically Sealed MEMS Strain Sensor

Xinyu Jiang [1], Brian Sang [1], Haoran Wen [2], Gregory Junek [2], Jin-Woo Park [2] and Farrokh Ayazi [1,2,*]

[1] School of Electrical and Computer Engineering, Georgia Institute of Technology, Atlanta, GA 30332, USA; xjiang362@gatech.edu (X.J.); bsang3@gatech.edu (B.S.)
[2] StethX Microsystems Inc., Atlanta, GA 30308, USA; haoran@stethxmicro.com (H.W.); greg@stethxmicro.com (G.J.); jin@stethx.com (J.-W.P.)
* Correspondence: ayazi@gatech.edu

Abstract: We present a hermetically sealed capacitive microelectromechanical system (MEMS) strain sensor designed for arterial pulse waveform extraction using the strain plethysmography (SPG) modality. The MEMS strain sensor features a small form factor of 3.3 mm × 3.3 mm × 1 mm, leverages a nano-gap fabrication process to improve the sensitivity, and uses a differential sensing mechanism to improve the linearity and remove the common mode drift. The MEMS strain sensor is interfaced with an application-specific integrated circuit (ASIC) to form a compact strain sensing system. This system exhibits a high strain sensitivity of 316 aF/$\mu\varepsilon$, a gauge factor (GF) of 35, and a strain sensing resolution of 1.26 $\mu\varepsilon$, while maintaining a linear range exceeding 700 $\mu\varepsilon$. SPG signals have been reliably captured at both the fingertip and wrist using the MEMS strain sensor with high signal quality, preserving various photoplethysmography (PPG) features. Experimental results demonstrate that heart rate (HR) and heart rate variability (HRV) can be estimated from the SPG signal collected at the fingertip and wrist using the sensor with an accuracy of over 99%. Pulse arrival time (PAT) and pulse transit time (PTT) have been successfully extracted using the sensor together with a MEMS seismometer, showcasing its potential for ambulatory BP monitoring (ABPM) application.

Keywords: arterial pulse waveform; Blood Pressure; cardiovascular disease; MEMS; pulse transit time; strain sensors

1. Introduction

Wearable devices capture various physiological signals generated during cardiovascular activities to extract features related to diseases. Common modalities include electrocardiography (ECG), which detects electrical activity from ion flux [1,2]; photoplethysmography (PPG), which measures blood volume changes in peripheral arteries [3]; and seismocardiography (SCG), which senses heart-induced vibrations in the chest region [4].

PPG is a widely used non-invasive modality in clinical settings and wearable devices due to its simplicity and ability to monitor continuously. It collects the arterial pulse waveform (APW) using a light-based sensor. This APW has been extensively studied, for example, for blood pressure (BP) estimation [5] and cardiac output (CO) estimation [6]. Strain plethysmography (SPG) is another modality that extracts the arterial pulse waveform and shares the same morphological features as PPG [7]. It measures the strain variation on peripheral limbs to determine changes in blood flow volume. Initially, strain plethysmography was conducted using a cumbersome strain gauge consisting of a rubber cuff with mercury filling [8]. Recent studies have developed novel strain sensors and successfully extracted the SPG signal from the fingertip using a Fiber-Bragg Grating strain sensor [9] and from the wrist using a customized strain-gauge-based strain sensor [10]. However, further investigation is needed to reduce the form factor and optimize the sensor readout schemes to enable the practical use of these sensors in ambulatory monitoring settings. Additionally, despite recent advancements in high-sensitivity stretchable strain sensors for physiological signal monitoring [11], their limited lifespan due to polymer degradation hinders their suitability for long-term wearable sensing applications.

Microelectromechanical systems (MEMS) sensors exhibit great potential in wearable sensing applications due to their small size, low weight, low power consumption, and low cost (SWaP-C). For example, MEMS accelerometers have been used as contact microphones (ACM) to sense mechano-acoustic cardiopulmonary signals [12], and MEMS gyroscopes have been used in gyrocardiography (GCG) for myocardial deformation monitoring [13]. Piezoresistive MEMS pressure sensors have been reported for pulse waveform extractions as well; however, the pulse waveform collected from the pressure sensor was distorted and caused an error in pulse wave velocity calculation [14].

Here, we introduce a wafer-level packaged (WLP) capacitive MEMS strain sensor featuring a compact form factor of 3.3 mm × 3.3 mm × 1 mm for arterial pulse waveform extractions. This sensor has a large linear range and high sensitivity, designed for physiological strain sensing, which offers a robust and energy-efficient hardware-level solution to the skin-tone artifact issue commonly encountered in PPG-based wearable sensors for arterial pulse waveform extraction. As the artery dilation generates three-dimensional strain, the out-of-plane transduction mechanism of our sensor allows it to capture the SPG signal with greater fidelity compared to previously proposed MEMS-based strain sensors [15]. The wafer-level packaging enables hermetic encapsulation of the sensor, shielding it from corrosives like sweat, hence enhancing its repeatability and lifetime.

A highly integrated strain sensing system is built for signal sampling and transmission, including the MEMS strain sensor, an application-specific integrated circuit (ASIC) sensor interface, a thin printed circuit board (PCB), and a top metal cap to protect the wire bonding. The small form of the sensing system allows it to be easily placed on various pulse sites such as the fingertip and the wrist. It is demonstrated that the SPG signals collected closely track the PPG signals and preserve various PPG features.

The potential integration of the sensor into daily objects to facilitate wearable data collection is explored, proposing a novel method for collecting pulse signals by simply holding a smartphone with a strain sensor attached to its side. This innovative approach leverages everyday interactions with mobile devices, enabling pulse monitoring without needing dedicated equipment, making it an accessible solution for continuous health tracking.

We also investigate the application and relevance of the SPG signal in cardiovascular parameter extraction and estimation. Heart rate (HR) and heart rate variability (HRV) are derived from SPG signals collected consistently from users' fingertips and wrists. Additionally, we demonstrate the mounting of the SPG sensor on the side of a smartphone and the acquisition of HR and HRV while being held by a user. The strain sensor is also used synchronously with a MEMS seismometer [12] to form a multimodality sensing platform with a 40 μs timing resolution and a synchronous output. Using this platform, we successfully extracted pulse arrival time (PAT) and pulse transit time (PTT) to demonstrate the potential of the sensor for continuous cuffless BP monitoring [16,17].

2. Materials and Methods

2.1. Strain Sensor Design

Figure 1a shows the scanning electron microscopy (SEM) image of the MEMS strain sensor built on a silicon-on-insulator (SOI) substrate [18], which consists of a suspended silicon plate loosely coupled to the Si handle layer through two folded beam springs, where the ends of the springs are anchored to the handle layer via the buried oxide layer of an SOI substrate. The sensor resolves the input strain through axial compression and vertical deflection. The design utilizes four variable capacitors for transduction with a nominal transduction gap d_0 of 200 nm. The variable capacitors are configured to form a differential sensing pair, C+ and C−. For example, in Figure 1b, when the strain is compressive, the anchors move upwards with the handle layer due to vertical deflection while the suspended plate remains flat. This causes a gap change between the poly-Si electrodes and the Si device layer. The gap change causes the capacitance output C+ to increase and C− to decrease. This differential sensing mechanism improves the linearity of the strain sensor by removing the second-order non-linear component:

$$\Delta C_{out} = \frac{\epsilon_0 A}{d_0 - \delta d} - \frac{\epsilon_0 A}{d_0 + \delta d} = \frac{2\epsilon_0 A}{d_0^2}\delta d + O\left(\delta d^3\right), \tag{1}$$

where ϵ_0 is the electric permittivity of the air, d_0 is the nominal gap size of the variable capacitors with zero input strain, δd is the capacitor gap change due to the input strain, and A is the area of the variable capacitor. The differential sensing mechanism also removes common-mode drifts.

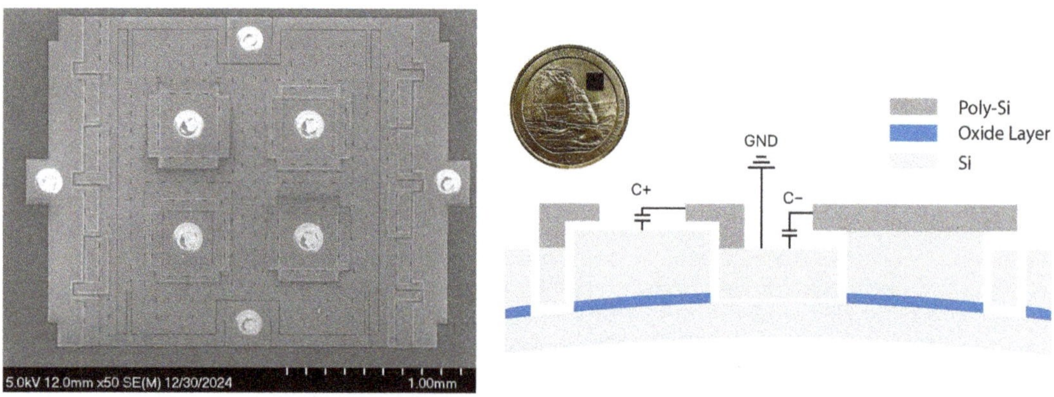

Figure 1. Capacitive MEMS strain sensor: (**a**) SEM image of the MEMS strain sensor. (**b**) Cross-section of the MEMS sensor showing the differential capacitive sensing pair. C+ and C− share a common terminal connected to the ground. The inset shows the wafer-level-packaged MEMS strain sensor die placed on a US quarter.

Figure 2 shows the process flow of the capacitive MEMS strain sensor. First, a silicon-on-insulator (SOI) wafer is etched using deep reactive ion etching (DRIE) to create trenches defining the structure. The trenches are subsequently filled with low-pressure chemical vapor deposition (LPCVD) tetraethyl orthosilicate (TEOS), as shown in Figure 2a. The oxide on the top surface is then patterned to define a 200 nm air gap for the variable capacitor (Figure 2b). Then, a layer of LPCVD poly-Si is deposited on top of the oxide and patterned. After that, a vapor hydrogen fluoride (VHF) etching of sacrificial SiO2 is performed to release the suspended plate (Figure 2d). Figure 2e shows the cap wafer prepared through oxidation, patterning, and metal electroplating (Figure 2e). Then, the cap wafer and the device wafer are bonded together using an Au-Si eutectic bond process. The handle layer of the wafer is ground and is metalized for electric connection. Finally, stealth dicing reveals the wire bonding pad and singulates the dies (Figure 2f).

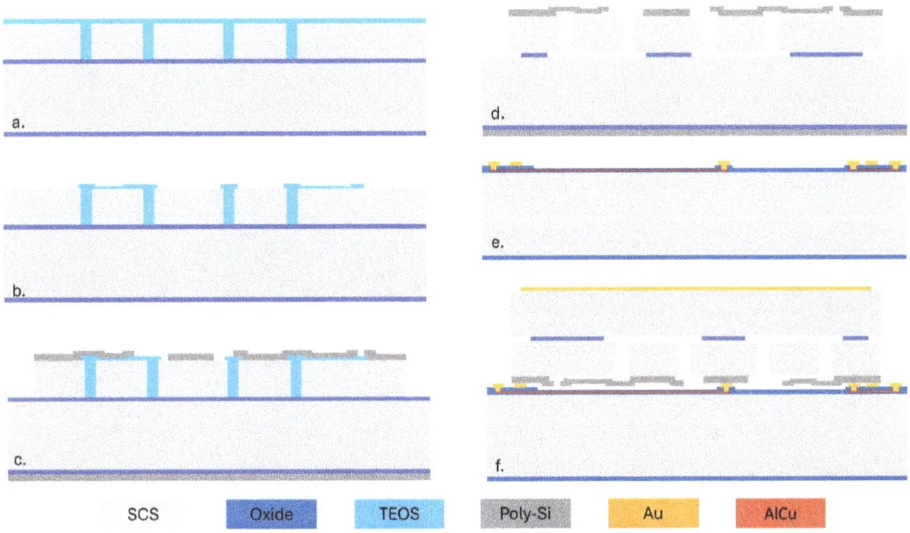

Figure 2. Fabrication process flow for the capacitive MEMS strain sensor. (**a**) The SOI device layer is patterned and filled with TEOS. (**b**) Pattern TEOS. (**c**) Deposit and pattern poly-Si. (**d**) Oxide etching to release the suspended plate. (**e**) Prepare the cap wafer. (**f**) Wafer bonding and stealth dicing.

We developed a theoretical model for analyzing the strain sensitivity of the MEMS sensor based on this cross-sectional view to help the sensor design, as shown in Figure 3. In the model, the sensor is approximated as a prismatic beam for simplicity. We can view the sensor to be symmetric about the x–z plane. Under this assumption, there will be no displacement along the z-axis at the symmetric plane. We can impose the first boundary condition on the system by choosing the cross-section of the beam on the symmetric plane as a fixed plane.

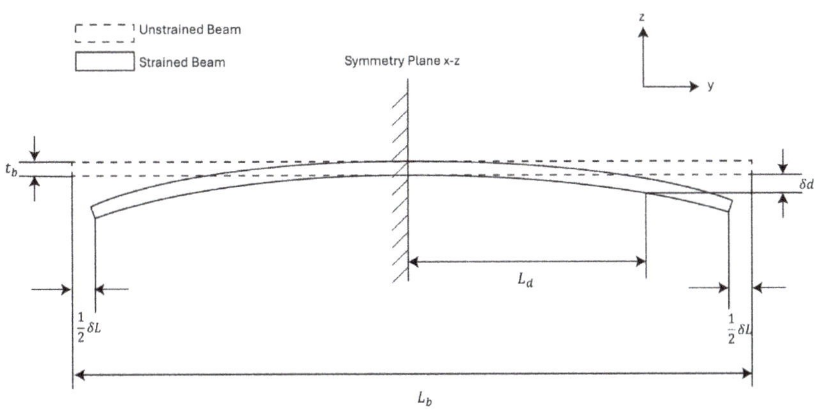

Figure 3. Analytical model for strain sensors. The strain sensor is treated as a prismatic beam with a central symmetric plane.

In Figure 3, the thickness of the beam t_b is the sum of the handle layer thickness of the sensor and the device layer thickness of the sensor. The length of the beam L_b is the length of the MEMS die. The terminals of the variable capacitors are positioned at a distance L_b from the symmetry plane, and the vertical displacement at the terminals is denoted as δd. The external strain is represented by prescribing a $\frac{\delta L}{2}$ displacement to the bottom edge of both sides of the beam. This gives us the second boundary condition of the system. The input strain ϵ to the beam is then represented as:

$$\epsilon = \frac{\delta L}{L_b}. \quad (2)$$

2.2. Strain Sensing System

A strain sensing system is developed to read the capacitance output of the MEMS strain sensor, as shown in Figure 4a. The strain sensor is interfaced with an ASIC with a 14-bit digital output. The ASIC and the strain sensor are placed on a thin PCB and wire-bonded together. The small form of the total assembly enables it to be easily placed at the fingertips and wrist for pulse signal collection (Figure 4b). A customized cable connected to the through-hole pins is used to transmit the ASIC data to an external signal processing hub. The signal processing hub then transmits the data to a personal computer for further signal processing. The power consumption for the entire sensing system is ~200 µW.

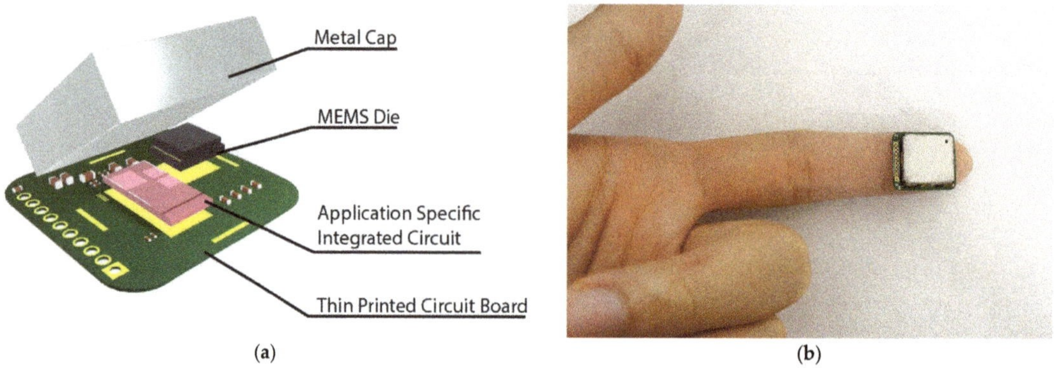

Figure 4. Strain sensing system: (**a**) The strain sensing system includes the MEMS die, an application-specific integrated circuit, a thin printed circuit board, and a metal cap. The complete strain sensing system features a form factor of 15 mm × 17 mm × 3 mm. (**b**) An image showing the small form sensor placed on the fingertip.

3. Results and Discussion

3.1. Strain Sensor Finite Element Simulation

Finite element simulations (FEM) using COMSOL 6.1 are performed to evaluate the strain sensitivity of the MEMS sensor. The simulation utilized the solid mechanics physics interface coupled with the electrostatic physics interface to model the capacitance change induced by input strain. The simulation setup follows the analytical model discussed in the previous section, where strain is applied by prescribing a displacement to the bottom edges of the sensor, and the symmetry plane is treated as a fixed boundary condition, as shown in Figure 3. The capacitance change ΔC is simulated using the electrostatic physics interface embedded in COMSOL.

Figure 5a illustrates that the strain is primarily concentrated in the handle layer of the sensor, while the suspended plate experiences minimal strain. Figure 5b shows the relationship between capacitance output and input strain for the MEMS sensor, with a linear fit applied to estimate the strain sensitivity within the sensor's linear operating range. This die-level simulation shows that the MEMS strain sensor exhibits a strain sensitivity of 2.34 fF/με and a linear range of 800 με. The strain sensitivity S of the sensor is calculated using:

$$S = \frac{\Delta C}{\epsilon}. \quad (3)$$

where ΔC is the capacitance change and ε is the input strain. The simulation also shows the sensor has a nominal capacitance C_0 of 8.8 pF. The gauge factor (GF) of the sensor is calculated to be 265 using:

$$\mathrm{GF} = \frac{\Delta C}{\epsilon C_0}. \quad (4)$$

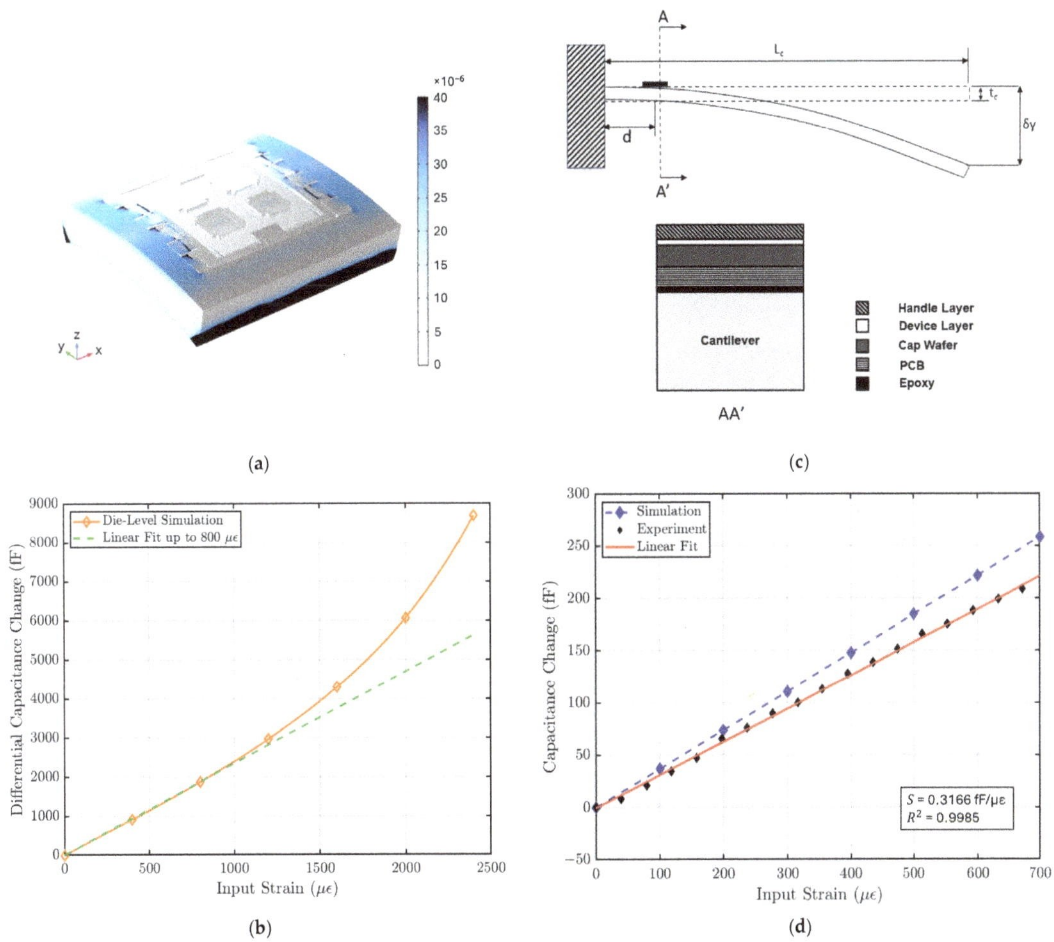

Figure 5. (**a**) COMSOL simulation of strain distribution in the strain sensor. (**b**) Die-level FEM simulation result shows a strain sensitivity of 2.34 fF/με with a linear range of 800 με. (**c**) Strain sensitivity characterization test setup. The strain sensor is placed near the fixed end of a long cantilever, and the input strain is controlled by the displacement δy at the free end of the cantilever using a micrometer. The cross-section AA' is shown on the right [18]. (**d**) The characterization result is compared with FEM simulation result using the cantilever setup specifications. The strain sensor system shows a sensitivity of 316 aF/με, a linear range of 700 με, and a linearity of 99.85%.

The impact of design parameters on the strain sensitivity was also studied parametrically. The prescribed displacement at the edge of the MEMS die shown in Figure 3 is accounted for by a combination of lateral compression and bending of the sensor. For a simple beam, the lateral spring constant is proportional to the beam thickness t_b, and the bending spring constant is proportional to t_b^3 [19]. The strain sensitivity can be improved substantially by reducing the handle layer thickness of the MEMS die. Figure A1 shows the FEM parametric simulation of the handle layer and the device layer thickness. In this work, the strain sensor was implemented in a SOI wafer with a 40 μm-thick device layer to be compatible with the process flow of an out-of-plane seismometer used for cardiovascular SCG monitoring as described later in this paper. A comparison between our design and other strain sensor designs is presented in Table 1.

Table 1. Comparison between this work and reported strain sensors.

	Mechanism	Material	Linear Range	Scale Factor	Gauge Factor
This work	Capacitive	Si	800 με	2.34 fF/με	293
[15]	Capacitive	Si	1000 με	265 aF/με	602
[20]	LC Resonant	Si	1000 με	34 kHz/με	430
[21]	Piezoresistive	SU-8-Graphene	50,000 με	NA	19.1
[22]	Piezoresistive	Silver NW on PDMS	20,000 με	NA	289
[23]	Piezoresistive	MoS_2	7000 με	NA	102

3.2. *Strain Sensing System Characterization*

3.2.1. Strain Sensitivity Characterization

The silicon MEMS die has a significantly higher Young's modulus compared to the PCB. This causes less strain to be coupled from the test surface to the MEMS sensor and reduces the strain sensitivity of the sensing system. This requires us to conduct a strain sensitivity characterization on the complete sensing system. As shown in Figure 5c, a cantilever beam setup was used to characterize the strain sensitivity. The strain sensing system was bonded with epoxy near the fixed end of the cantilever beam. The strain input to the sensing system was controlled by the displacement at the end of the cantilever, following the relationship [15]:

$$\epsilon = \frac{3t_c\delta y(L_c - d)}{2L_c^3}, \quad (5)$$

where L_c is the length of the cantilever, t_c is the thickness of the cantilever, δy is the displacement at the end, which is controlled by a micrometer, and d is the distance of the sensor to the end of the cantilever beam.

Since the strain sensing system loads the cantilever beam and causes deviations from theoretical behavior in Equation (5), a FEM simulation is performed to characterize the effect of this loading. In the simulation, the entire strain-sensing system is modeled and placed at the end of a cantilever, and the strain is introduced by the displacement at the other end, as shown in Figure 5c. It is found that strain sensitivity approaches an asymptote as cantilever beam thickness increases, showing that the effect of loading becomes less significant (Figure A2). To optimize the strain input error and the loading-effect error, a cantilever beam thickness t_c of 4 mm is used for characterization.

The characterization result is compared with FEM simulation result with the same cantilever setup. Figure 5d shows that the strain sensing system has a strain sensitivity of 316 aF/με and a gauge factor of 35 with a linear range of 700 με, where the nominal capacitance is measured to be ~2.2 pF for each variable capacitor using an Agilent E4980A

Precision LCR Meter (Keysight, Santa Rosa, CA, USA). The non-linear behavior during the characterization is not limited by the MEMS sensor but by the ASIC.

3.2.2. Response to Dynamic Loads

The response of the strain sensing system to dynamic loads is characterized using the strain induced by a rate table. The strain sensing system is first bonded to the cantilever beam, similar to the static characterization setup. Then, the end of the cantilever beam is attached to the edge of the rate table, clamped by two metal plates, as shown in Figure 6a. A sinusoidal function is commanded to the rate table, with an amplitude of 0.36 degrees and a frequency of 1 Hz. The strain input to the sensing system can be calculated from the displacement using (5), and the displacement at the end of the cantilever, δy, is the radius of the rate table multiplied by the angular position of the rate table. The angular position of the rate table is read by the rate table's embedded sensor. This corresponds to a sinusoid strain input with an amplitude of 78 $\mu\varepsilon$. We can then plot the input strain as a function of time. Figure 6b shows that the strain sensor output closely follows the strain input with high repeatability. The strain sensing system's sensitivity to the dynamic input strain is ~300 aF/$\mu\varepsilon$, which agrees with the static characterization result.

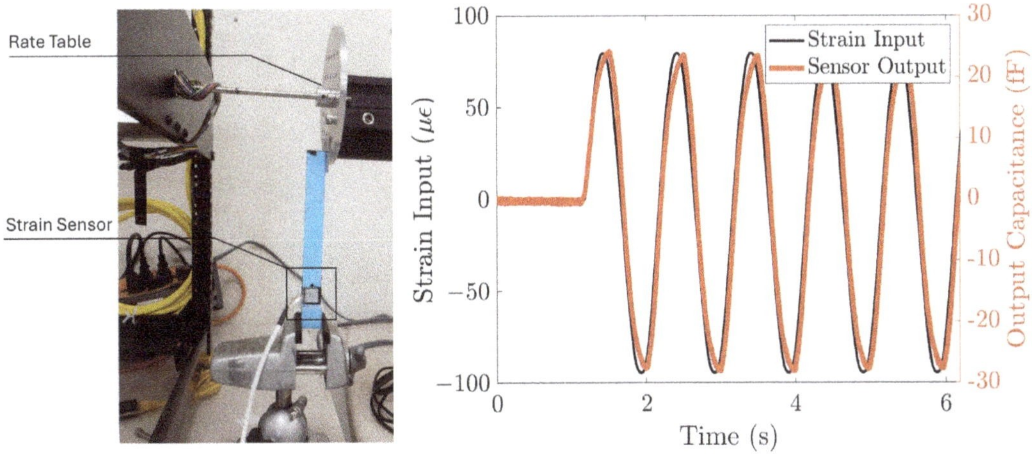

Figure 6. The strain sensing system responds to dynamic input strain. (**a**) Test setup. The strain input to the strain sensor is controlled using a rate table. (**b**) The sensor output closely follows the dynamic input strain with ~300 aF/$\mu\varepsilon$ sensitivity and high repeatability.

The sensor frequency response is also studied using COMSOL as shown in Figure 7, showing a flat gain bandwidth of greater than 10 kHz. A Q of 10 was assumed in the simulation, which causes peaking in the frequency response. The inset shows the first eigenmode of the strain sensor at 46.3 kHz.

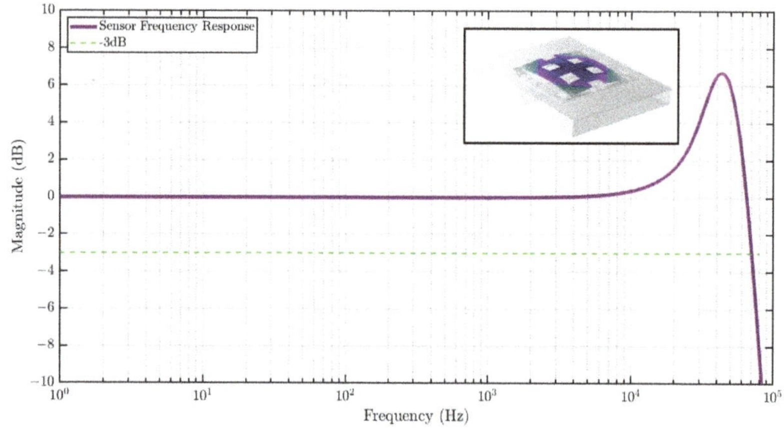

Figure 7. Frequency response of the MEMS strain sensor to strain input simulated using FEM analysis. The sensor exhibits a high bandwidth of 65 kHz. The inset shows the first eigenmode of the sensor at 46.3 kHz.

3.2.3. Resolution

Three major noise sources, the Brownian noise of the MEMS sensor, the electronic noise of the ASIC, and quantization noise, limit the resolution of the strain sensing system. To characterize the resolution of the strain sensing system, the sensing system output is recorded for 30 s using the data processing hub without signal input. The result in Figure 8 shows a sensor noise of ~400 aF. This corresponds to a strain sensing resolution of 1.26 $\mu\varepsilon$.

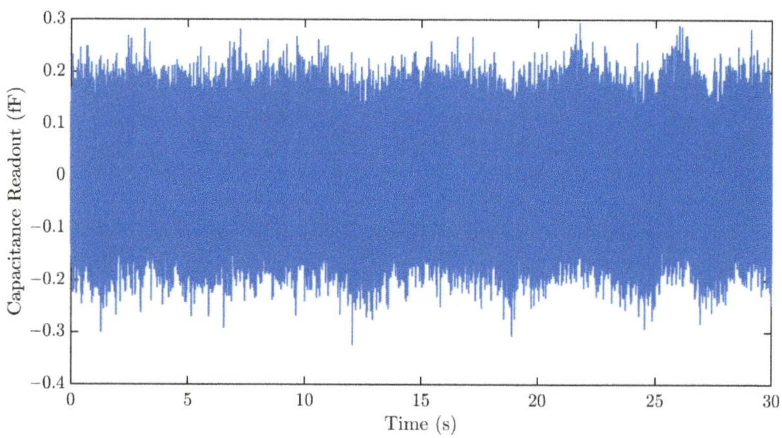

Figure 8. Time series noise output of the strain sensing system. The sensing system exhibits a noise of ~400 aF. This corresponds to a strain sensing resolution of 1.26 $\mu\varepsilon$.

3.3. Cardiovascular Parameter Extraction

3.3.1. Experimental Setup

We conducted four experiments to verify the MEMS strain sensor's reliability in collecting the SPG signals. Figure 9 depicts the general setup for all four experiments, where a MEMS seismometer is used together with the strain sensor. The MEMS seismometer was placed in the tricuspid region to collect the SCG signal for all four experiments, while the strain sensor placement varied across experiments for SPG signal collection.

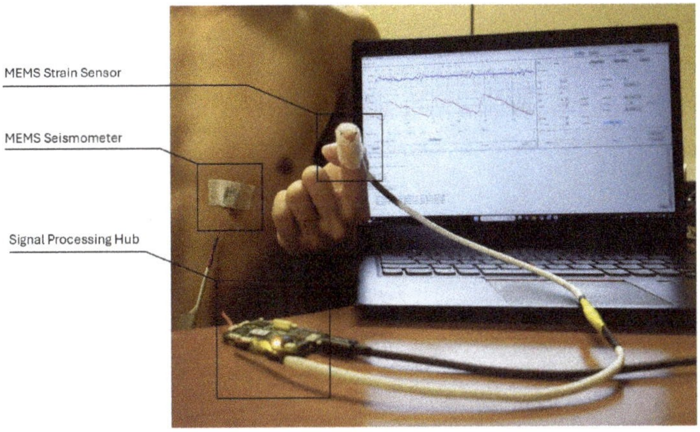

Figure 9. Experiment setup for cardiovascular parameter estimation. The SCG signal (blue) and SPG signal (red) collected from the sensors are sampled by the signal processing hub and displayed on the computer monitor.

- In experiments 1 and 4, the MEMS strain sensor is wrapped around the fingertip using medical tape as shown in Figure 10a.

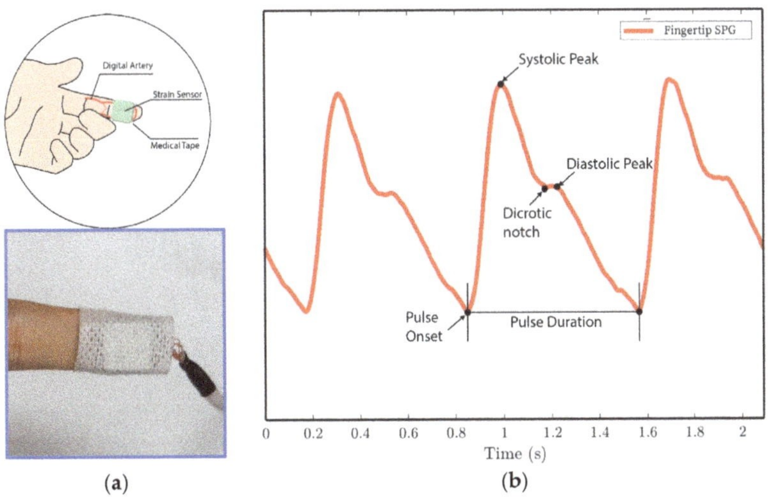

Figure 10. SPG signals are collected from the fingertip pulse location for experiments 1 and 4. (**a**) Illustration of sensor placement at the digital artery of the fingertip. Medical tape is used to attach the sensor to the fingertip. (**b**) SPG signal collected from the fingertip possesses the key features of the PPG signal.

- In experiment 2, the MEMS strain sensor is placed on the wrist and fixed with medical tape, as shown in Figure 11a.

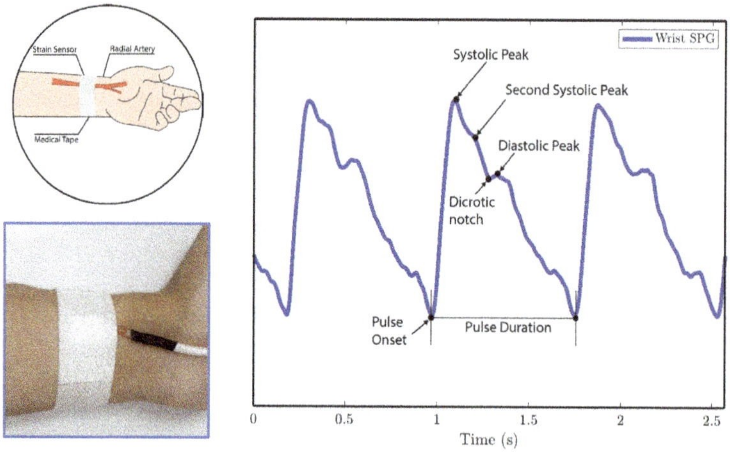

(a) (b)

Figure 11. SPG signals are collected from the wrist pulse location for experiment 2. (**a**) Illustration of sensor placement at the radial artery of the wrist. Medical tape is used to attach the sensor to the wrist. (**b**) SPG signal collected from the wrist possesses key features of the PPG signal.

- In experiment 3, the MEMS strain sensor is first attached to the side of a cell phone using double-sided tape. The subject then holds the phone to collect the SPG signal, as shown in Figure 12a.

(a)

(b)

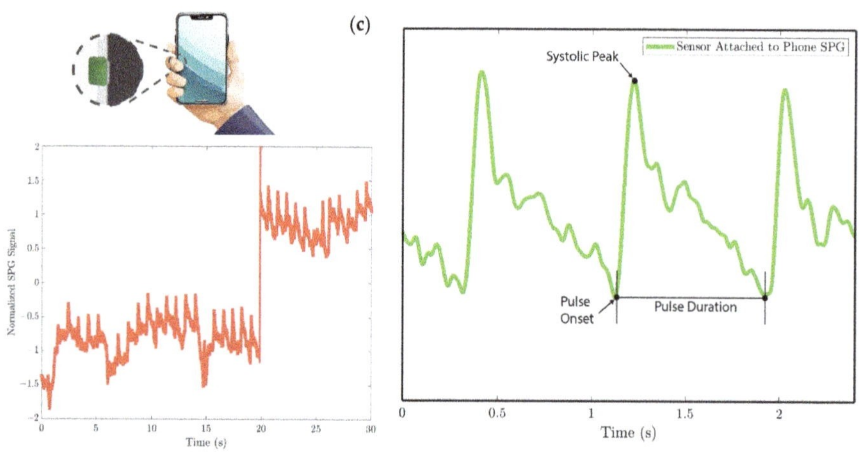

Figure 12. SPG signal is collected using the strain sensor attached to the cell phone for experiment 3. (**a**) Illustration of collecting the pulse signal from the digital artery at the fingertip by holding the cell phone with the strain sensor attached to the side. (**b**) Baseline drift due to the gripping force variation. (**c**) SPG signal collected preserves the shape and some features of the pulse waveform.

Throughout all experiments, both the SCG and SPG signals were recorded synchronously. The SCG signal and the SPG signal were displayed on a laptop screen during the tests, represented by the blue line and the red line in the screen monitor in Figure 9. The seismometer and the strain sensor are both sampled by the data processing hub with a sampling rate of 25 kHz.

Experiments 1, 2, and 3 were designed to evaluate the sensor's capability to measure the SPG signal at different pulse locations with high fidelity and accuracy. The fidelity of the

SPG signal was assessed by examining the preservation of the morphological features of the arterial pulse waveform. The collected signals are then processed and used to extract heart rate and heart rate variability. The temporal accuracy of the SPG signal was determined by comparing the HR and HRV extracted from the SPG modality with those derived from the SCG modality.

Experiment 4 aims to test the sensor's capability of measuring pulse arrival time and pulse transit time. This experiment involved comparing the PAT/PTT under normal and elevated blood pressure conditions. The subject first rested for 3 min to collect signals under the normal BP condition. Then, after performing a 3 min stair-climbing exercise to elevate blood pressure, a second set of signals was collected to represent the signals under elevated BP conditions.

3.3.2. Signal Processing

The collected SPG signals are first filtered using a second-order Butterworth filter with a pass band of 0.5–25 Hz. To remove the baseline drift, the first peak detection is performed to detect the pulse onset points of the SPG signal. A piecewise linear function is constructed by connecting the pulse onset points, which represent the sensor's baseline drift. This piecewise linear baseline is then subtracted from the filtered SPG signal to remove the baseline drift.

The SPG signal is then normalized, and a second peak detection is performed to identify the systolic peak of the SPG signal, which will be used for cardiovascular parameter estimation. Another modality, acceleration plethysmography (APG), is also generated from the normalized SPG signal by taking the first time derivative of the SPG signal. The APG signal is an indicator of vascular aging [9]. It is also useful for HR and HRV estimation when the SPG signal is corrupted with pressure artifacts in experiment 3, and the systolic peaks could not be accurately identified using a peak detection algorithm.

The collected SCG signals are first filtered using a second-order Butterworth filter with a pass band of 0.5–80 Hz and then normalized. As shown in Figure 13, the S1 peak of the SCG signal includes a mitral valve closure peak (MC), an aortic opening peak (AO), and an isovolumic moment point (IM). As MC and AO have similar amplitudes, it is difficult for the peak detection algorithm to distinguish between these two peaks. To resolve the issue, we first run a peak detection on IM peaks. The IM peaks sit between the MC peak and the AO peak for each S1 peak. We then segment the SCG signals into single heartbeat segments using IM peaks. For each segment, we lower the threshold of the peak detection so that it captures both MC and AO peaks. The first peak location for each segment is the AO peak of the previous heartbeat, and the last peak location for each segment is the MC peak for the current heartbeat.

Heart rates are extracted from the peak-to-peak interval (PPI) of the systolic peak for the SPG signal, the peak of the APG signal, and the AO peaks of the SCG signal. The performance of the HR estimation using the SPG signal and the APG signal is evaluated by using the SCG signal as the reference signal. We also use the APG signal as a reference to the SPG signal to assess the accuracy of the SPG signal, which provides insight into how accurately the SPG signal represents the true pulse waveform, as the max slope point of the SPG signal is the peak of the APG signal. Average HR (AVG), mean absolute error (MAE), mean error (ME), standard deviation of the error (SD), and root mean square error (RMSE) are calculated to evaluate the performance of the HR estimation. The accuracy (ACC) is calculated using normalized RMSE using:

$$\mathrm{ACC} = 100\frac{\mathrm{AVG} - \mathrm{RMSE}}{\mathrm{AVG}} \tag{6}$$

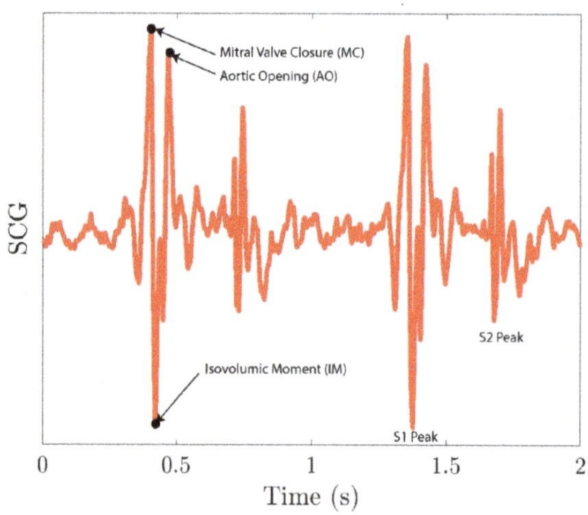

Figure 13. SCG signal collected using the MEMS seismometer.

Three different HRVs are then calculated using the time domain root mean square of successive differences (RMSSD) method from the estimated HR. The accuracy of HRV estimation is calculated using:

$$\mathrm{ACC} = 100(1 - 2\left|\frac{\mathrm{HRV}_1 - \mathrm{HRV}_2}{\mathrm{HRV}_1 + \mathrm{HRV}_2}\right|) \tag{7}$$

where HRV_1 is calculated from the test modality, and HRV_2 is calculated from the reference modality. PAT is extracted between the systolic peak of the SPG signal and the MC peak of the S1 peak of the SCG signal, and PTT is extracted between the systolic peak of the SPG signal and the AO peak of the S1 peak of the SCG signal [17]. The mean and standard deviation of the PAT and PTT are also reported.

3.3.3. Cardiovascular Parameter Extraction Results

We analyze the morphological features of the collected SPG signal compared to PPG signals. Key features of PPG signals include the pulse onset point, the systolic peak, the diastolic peak, the max slope point, and the dicrotic notch [24]. Due to wave reflection, PPG signals also possess features such as the second systolic peak [25]. Preserving these features with high fidelity plays a crucial role in machine learning (ML) based cardiovascular monitoring applications [26]. Figure 10b shows the SPG signal collected from the fingertip, and Figure 11b shows the SPG signal collected from the wrist. Both pulse waveforms preserve morphological features, demonstrating that the SPG signal can be collected using the MEMS strain sensor with high fidelity. Figure 12b shows the raw signal collected from experiment 3. The sensor's output is susceptible to pressure artifacts due to the gripping force variance. However, as shown in Figure 12c, the SPG signal collected from the sensor attached to a cell phone still preserves the distinct shape of the arterial pulse waveform.

The results of experiments 1, 2, and 3 are shown in Figure 14a–c, respectively. The SPG and SCG signals are plotted continuously for 30 s. For experiments 1 and 2, it is found that the HR extracted from three different modalities closely follow each other, where the mean absolute error (MAE) between SPG-APG, SPG-SCG, and APG-SCG is 0.117, 0.145, and 0.234 beats per minute (BPM) for experiment 1. The MAE between SPG-APG, SPG-SCG, and APG-SCG is 0.52, 0.653, and 0.31 BPM for experiment 2. In experiment 3, the SPG signal is distorted due to pressure artifacts. This hinders the identification of the systolic peaks from the SPG signal and reduces the accuracy of estimating HR directly from the SPG signal, with an MAE of 4.32 BPM between the SPG and SCG. However, the APG modality still has a high accuracy with an MAE of 1.35. This shows that while the systolic peaks are distorted, the SPG signal still reliably captures the maximum slope point of the arterial pulse waveform. The evaluation metrics for HR are reported in Table 2. HRV has also been estimated using the SPG with high accuracy, as reported in Table 3. These results demonstrate that SPG signals can be reliably collected from the fingertip (Experiment 1), the wrist (Experiment 2) using the MEMS strain sensor, and by holding a cell phone with the sensor attached (Experiment 3). The collected signals exhibit exceptional quality, high temporal accuracy, and precise cardiovascular information.

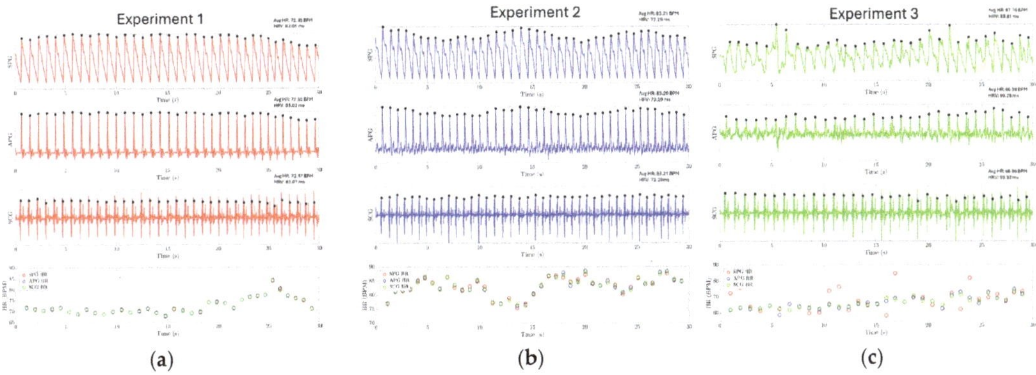

Figure 14. Experiment results for HR and HRV estimation. The HR and HRV for each experiment are reported on the top right of each subplot for each modality. (**a**) HR estimation using the SPG signal collected from the fingertip for experiment one. (**b**) HR estimation using the SPG signal collected from the wrist for experiment 2. (**c**) HR estimation using the SPG signal collected from the strain sensor attached to the cell phone.

Table 2. Performance Evaluation of Heart Rate Estimation.

	SPG-APG				SPG-SCG				APG-SCG			
	MAE	ME ± SD	RMSE	ACC%	MAE	ME ± SD	RMSE	ACC%	MAE	ME ± SD	RMSE	ACC%
Fingertip	0.117	−0.03 ± 0.14	0.145	99.80	0.194	0.016 ± 0.24	0.234	99.68	0.225	−0.047 ± 0.28	0.281	99.61
Wrist	0.520	−0.01 ± 0.66	0.653	99.22	0.533	−0.012 ± 0.63	0.626	99.25	0.31	0.004 ± 0.40	0.394	99.52
Phone	4.32	1.17 ± 6.41	6.4176	90.36	4.28	1.18 ± 6.37	6.38	90.41	0.99	−0.019 ± 1.38	1.35	97.96

The performance of HR is evaluated using a second modality as the ground truth, for example, SPG-SCG uses SCG as a reference to assess the accuracy of HR estimated by SPG. SPG: Strain Plethysmography APG: Accelerated Plethysmography SCG: Seismocardiography. Unit: BPM.

Table 3. Accuracy of Heart Rate Variability Estimation.

	SPG-APG	SPG-SCG	APG-SCG
Fingertip	99.95	99.98	99.93
Wrist	99.99	99.99	99.99
Phone	99.21	99.28	99.92

This table reports the accuracy of HRV estimation calculated using (7).

The result of experiment 4 is shown in Figure 15. For normal BP conditions, PAT has a mean value of 222.8 ms with a standard deviation of 2.7 ms, and PPT has a mean value of 152.4 ms with a standard deviation of 3.3 ms, as shown in Figure 15a. Under elevated BP conditions, PAT has a mean value of 187.9 ms with a standard deviation of 8.1 ms, and PTT has a mean value of 143.4 ms with a standard deviation of 5.7 ms, as shown in Figure 15b.

A significant difference has been found between the PAT/PTT values under the normal BP condition and the elevated BP condition. This result demonstrates the capability of using the MEMS strain sensor together with a MEMS seismometer for monitoring the PAT and PTT.

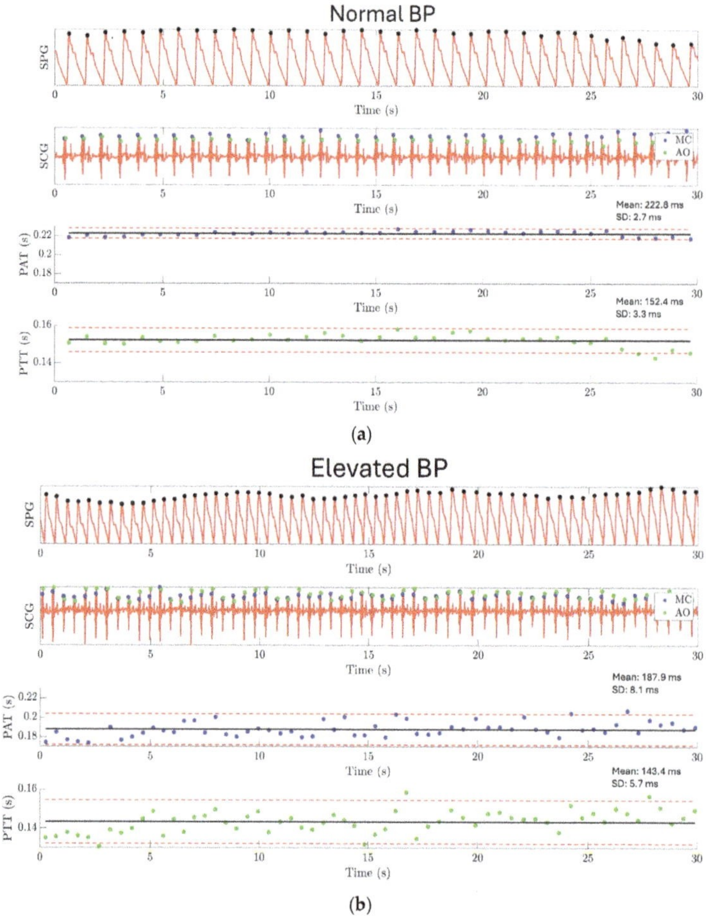

Figure 15. Experiment 4: Pulse arrival time and pulse transit time estimation results. (**a**) PAT and PTT are extracted under the normal BP condition. PAT has a mean value of 222.8 ms with a standard deviation of 2.7 ms, and PTT has a mean value of 152.4 ms with a standard deviation of 3.3 ms. (**b**) PAT and PTT are extracted under the elevated BP condition. PAT has a mean value of 187.9 ms with a standard deviation of 8.1 ms, and PTT has a mean value of 143.4 ms with a standard deviation of 5.7 ms. The result shows a significantly shorter PAT and PTT for the elevated BP condition compared to the normal BP condition.

4. Conclusions

We introduced a capacitive hermetically sealed MEMS strain sensor with high sensitivity, wide linear range, and high durability for continuous monitoring of cardiovascular parameters using the SPG modality. The arterial pulse waveform has been reliably extracted from both the fingertip and wrist using the SPG, demonstrating strong agreement with the signal extracted using PPG. Heart rate and heart rate variability are also accurately estimated from the arterial pulse waveform. This further proves the high temporal resolution and high fidelity of the SPG signal collected using the strain sensor.

The strain sensor is also used in conjunction with a MEMS seismometer to form a multimodality sensing platform that collects the SPG and the SCG signal synchronously. PAT and PTT are successfully extracted using this sensing platform. The synchronous acquisition of SPG and SCG signals enhances its capability for multimodal physiological monitoring. This could play a crucial role in cardiovascular signal acquisition under a wearable setup, supporting ambulatory BP monitoring and other continuous monitoring tasks, thereby aiding in the management of chronic conditions and cardiovascular diseases.

For future studies, we are exploring a new flexible PCB design that integrates a Bluetooth module, removing the need for a separate data processing hub. This eliminates cables and enables more seamless wearable data collection.

Author Contributions: Conceptualization: F.A., X.J. and H.W.; data curation: X.J.; formal analysis: X.J.; funding acquisition: F.A.; investigation: B.S.; methodology: X.J., B.S., H.W., G.J. and J.-W.P.; resources: H.W., G.J. and J.-W.P.; software: J.-W.P.; supervision: F.A.; writing—original draft: X.J.; writing—review and editing: F.A. and B.S. All authors have read and agreed to the published version of the manuscript.

Funding: This research received no external funding.

Institutional Review Board Statement: Not applicable.

Informed Consent Statement: Informed consent was obtained from all subjects involved in the study.

Data Availability Statement: The data and code are available on https://github.com/xjiang362/Strain-Sensor-Data (accessed on 1 August 2024).

Acknowledgments: The authors wish to thank StethX Microsystems Inc. for providing wafer-level-packaged MEMS die and ASIC used in this work.

Conflicts of Interest: Farrokh Ayazi is an inventor of the technology being studied, and the purpose of this project is to explore its commercialization. The terms of the arrangement have been reviewed and approved by the Georgia Institute of Technology in accordance with its conflict-of-interest policies. Authors Haoran Wen, Gregory Junek, and Jin-woo Park were employed by the company StethX Microsystems Inc. The remaining authors declare that the research was conducted in the absence of any commercial or financial relationships that could be construed as a potential conflict of interest.

Appendix A

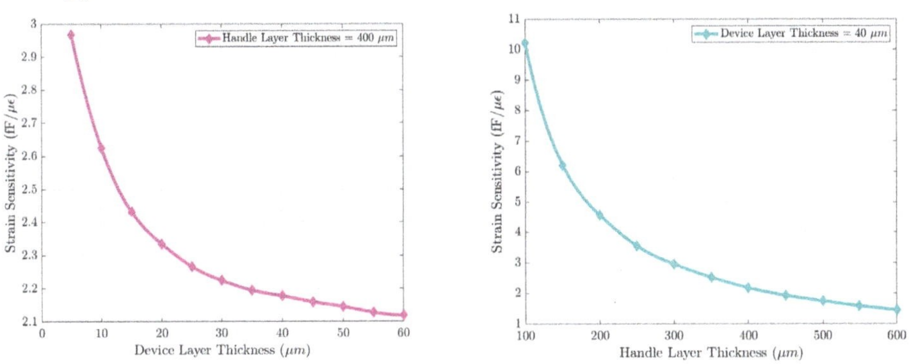

(**a**) (**b**)

Figure A1. Die level FEM parametric simulation describing effects of the device layer and the handle layer thickness on strain sensor sensitivity: (**a**) The strain sensor sensitivity improves slightly with a thinner device layer (5–60 μm). (**b**) The strain sensor sensitivity can be improved substantially by reducing the thickness of the handle layer (100–600 μm).

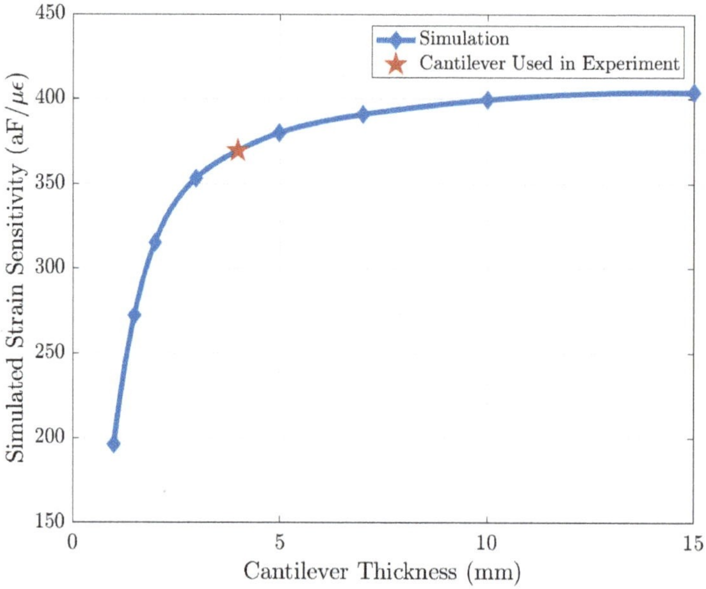

Figure A2. Simulation result of strain sensitivity with different cantilever thicknesses. The error is due to the loading of the strain sensor being less than 10% when the thickness of the cantilever beam is 4 mm.

References

1. Mirvis, D.M.; Goldberger, A.L. Electrocardiography. *Heart Dis.* **2001**, *1*, 82–128.
2. Feng, J.; Huang, Z.; Zhou, C.; Ye, X. Study of continuous blood pressure estimation based on pulse transit time, heart rate and photoplethysmography-derived hemodynamic covariates. *Australas. Phys. Eng. Sci. Med.* **2018**, *41*, 403–413. [CrossRef]
3. Slapničar, G.; Luštrek, M.; Marinko, M. Continuous Blood Pressure Estimation from PPG Signal. *Informatica* **2024**, *42*, 33–42.
4. Taebi, A.; Solar, B.E.; Bomar, A.J.; Sandler, R.H.; Mansy, H.A. Recent Advances in Seismocardiography. *Vibration* **2019**, *2*, 64–86. [CrossRef] [PubMed]
5. Liu, Z.; Zhang, Y.; Zhou, C. BiGRU-attention for Continuous blood pressure trends estimation through single channel PPG. *Comput. Biol. Med.* **2024**, *168*, 107795. [CrossRef]
6. Saugel, B.; Kouz, K.; Scheeren, T.W.; Greiwe, G.; Hoppe, P.; Romagnoli, S.; de Backer, D. Cardiac output estimation using pulse wave analysis—Physiology, algorithms, and technologies: A narrative review. *Br. J. Anaesth.* **2021**, *126*, 67–76. [CrossRef]
7. Shokouhmand, A.; Jiang, X.; Ayazi, F.; Ebadi, N. MEMS Fingertip Strain Plethysmography for Cuffless Estimation of Blood Pressure. *IEEE J. Biomed. Health Inform.* **2024**, *28*, 2699–2712. [CrossRef]
8. Whitney, R.J. The measurement of volume changes in human limbs. *J. Physiol.* **1953**, *121*, 1–27. [CrossRef] [PubMed]
9. Pant, S.; Umesh, S.; Asokan, S. A Novel Approach to Acquire the Arterial Pulse by Finger Plethysmography Using Fiber Bragg Grating Sensor. *IEEE Sens. J.* **2020**, *20*, 5921–5928. [CrossRef]
10. Wang, Y.-J.; Chen, C.-H.; Sue, C.-Y.; Lu, W.-H.; Chiou, Y.-H. Estimation of Blood Pressure in the Radial Artery Using Strain-Based Pulse Wave and Photoplethysmography Sensors. *Micromachines* **2018**, *9*, 556. [CrossRef]
11. Boutry, C.M.; Nguyen, A.; Lawal, Q.O.; Chortos, A.; Rondeau-Gagné, S.; Bao, Z. A Sensitive and Biodegradable Pressure Sensor Array for Cardiovascular Monitoring. *Adv. Mater.* **2015**, *27*, 6954–6961. [CrossRef] [PubMed]
12. Sang, B.; Wen, H.; Junek, G.; Neveu, W.; Di Francesco, L.; Ayazi, F. An Accelerometer-Based Wearable Patch for Robust Respiratory Rate and Wheeze Detection Using Deep Learning. *Biosensors* **2024**, *14*, 118. [CrossRef] [PubMed]
13. Tadi, M.J.; Lehtonen, E.; Pankäälä, M.; Saraste, A.; Vasankari, T.; Terás, M.; Koivisto, T. Gyrocardiography: A new non-invasive approach in the study of mechanical motions of the heart. Concept, method and initial observations. In Proceedings of the 2016 38th Annual International Conference of the IEEE Engineering in Medicine and Biology Society (EMBC), Lake Buena Vista, FL, USA, 16–20 August 2016; pp. 2034–2037. [CrossRef]
14. Sun, Y.; Dong, Y.; Gao, R.; Chu, Y.; Zhang, M.; Qian, X.; Wang, X. Wearable Pulse Wave Monitoring System Based on MEMS Sensors. *Micromachines* **2018**, *9*, 90. [CrossRef]

15. Suster, M.; Guo, J.; Chaimanonart, N.; Ko, W.H.; Young, D.J. A High-Performance MEMS Capacitive Strain Sensing System. *J. Microelectromech. Syst.* **2006**, *15*, 1069–1077. [CrossRef]
16. Poon, C.C.Y.; Zhang, Y.T. Cuff-less and Non-invasive Measurements of Arterial Blood Pressure by Pulse Transit Time. In Proceedings of the 2005 IEEE Engineering in Medicine and Biology 27th Annual Conference, Shanghai, China, 1 September–4 January 2005; pp. 5877–5880. [CrossRef]
17. Yang, C.; Tavassolian, N. Pulse Transit Time Measurement Using Seismocardiogram, Photoplethysmogram, and Acoustic Recordings: Evaluation and Comparison. *IEEE J. Biomed. Health Inform.* **2018**, *22*, 733–740. [CrossRef]
18. Jiang, X.; Shokouhmand, A.; Ebadi, N.; Ayazi, F. Fingertip Plethysmography Using a Sensitive Wafer-Level-Packaged Capacitive MEMS Strain Sensor. In Proceedings of the 2024 IEEE 37th International Conference on Micro Electro Mechanical Systems (MEMS), Austin, TX, USA, 21–25 January 2024; pp. 348–351. [CrossRef]
19. Senturia, S.D. (Ed.) Structures. In *Microsystem Design*; Springer US: Boston, MA, USA, 2001; pp. 201–238. [CrossRef]
20. Zhang, C.; Zhang, S.-Y.; Wang, L.-F. A Sawtooth MEMS Capacitive Strain Sensor for Passive Telemetry in Bearings. *IEEE Sens. J.* **2021**, *21*, 22527–22535. [CrossRef]
21. Beigh, F.T.; Beigh, N.T.; Mallick, D. Lithographically patternable SU-8/Graphene nanocomposite based strain sensors for soft-MEMS applications. *J. Micromech. Microeng.* **2024**, *34*, 095003. [CrossRef]
22. Wu, S.; Kim, D.; Tang, X.; King, M.W.; Zhu, Y. Encapsulated stretchable amphibious strain sensors. *Mater. Horiz.* **2024**, *11*, 5070–5080. [CrossRef]
23. Rana, V.; Gangwar, P.; Meena, J.S.; Ramesh, A.K.; Bhat, K.N.; Das, S.; Singh, P. A highly sensitive wearable flexible strain sensor based on polycrystalline MoS2 thin film. *Nanotechnology* **2020**, *31*, 385501. [CrossRef]
24. Hasanzadeh, N.; Ahmadi, M.M.; Mohammadzade, H. Blood Pressure Estimation Using Photoplethysmogram Signal and Its Morphological Features. *IEEE Sens. J.* **2020**, *20*, 4300–4310. [CrossRef]
25. Solà, J.; Delgado-Gonzalo, R. (Eds.) *The Handbook of Cuffless Blood Pressure Monitoring: A Practical Guide for Clinicians, Researchers, and Engineers*; Springer International Publishing: Cham, Switzerland, 2019. [CrossRef]
26. Ajmal; Boonya-Ananta, T.; Rodriguez, A.J.; Le, V.N.D.; Ramella-Roman, J.C. Monte Carlo analysis of optical heart rate sensors in commercial wearables: The effect of skin tone and obesity on the photoplethysmography (PPG) signal. *Biomed. Opt. Express* **2021**, *12*, 7445–7457. [CrossRef] [PubMed]

biosensors

MDPI

Article

Analysis and Tracking of Intra-Needle Ultrasound Pleural Signals for Improved Anesthetic Procedures in the Thoracic Region

Fu-Wei Su [1,2], Chia-Wei Yang [2], Ching-Fang Yang [2], Yi-En Tsai [2], Wei-Nung Teng [1,3] and Huihua Kenny Chiang [2,*]

1 Department of Anesthesiology, Taipei Veterans General Hospital, Taipei City 11217, Taiwan; fwsu@vghtpe.gov.tw (F.-W.S.); wnteng@vghtpe.gov.tw (W.-N.T.)
2 Department of Biomedical Engineering, National Yang-Ming Chiao-Tung University, Taipei City 112304, Taiwan; kevin_yang@cino.com.tw (C.-W.Y.); f910820@gm.ym.edu.tw (C.-F.Y.); ianye.tsai@garmin.com (Y.-E.T.)
3 School of Medicine, National Yang-Ming Chiao-Tung University, Taipei City 112304, Taiwan
* Correspondence: chiang@nycu.edu.tw; Tel.: +886-2-2826-7027; Fax: +886-2-2821-0847

Abstract: Background: Ultrasonography is commonly employed during thoracic regional anesthesia; however, its accuracy can be affected by factors such as obesity and poor penetration through the rib window. Needle-sized ultrasound transducers, known as intra-needle ultrasound (INUS) transducers, have been developed to detect the pleura and fascia using a one-dimensional radio frequency mode ultrasound signal. In this study, we aimed to use time-frequency analysis to characterize the pleural signal and develop an automated tool to identify the pleura during medical procedures. Methods: We developed an INUS system and investigated the pleural signal it measured by establishing a phantom study, and an in vivo animal study. Signals from the pleura, endothoracic fascia, and intercostal muscles were analyzed. Additionally, we conducted time- and frequency-domain analyses of the pleural and alveolar signals. Results: We identified the unique characteristics of the pleura, including a flickering phenomenon, speckle-like patterns, and highly variable multi-band spectra in the ultrasound signal during the breathing cycle. These characteristics are likely due to the multiple reflections from the sliding visceral pleura and alveoli. This automated identification of the pleura can enhance the safety for thoracic regional anesthesia, particularly in difficult cases. Conclusions: The unique flickering pleural signal based on INUS can be processed by time-frequency domain analysis and further tracked by an auto-identification algorithm. This technique has potential applications in thoracic regional anesthesia and other interventions. However, further studies are required to validate this hypothesis. Key Points Summary: Question: How can the ultrasound pleural signal be distinguished from other tissues during breathing? Findings: The frequency domain analysis of the pleural ultrasound signal showed fast variant and multi-band characteristics. We suggest this is due to ultrasound distortion caused by the interface of multiple moving alveoli. The multiple ultrasonic reflections from the sliding pleura and alveoli returned in variable and multi-banded frequency. Meaning: The distinguished pleural signal can be used for the auto-identification of the pleura for further clinical respiration monitoring and safety during regional anesthesia. Glossary of Terms: intra-needle ultrasound (INUS); radio frequency (RF); short-time Fourier transform (STFT); intercostal nerve block (ICNB); paravertebral block (PVB); pulse repetition frequency (PRF).

Keywords: ultrasound; pleura; signal analysis

 https://doi.org/10.3390/bios15040201

1. Introduction

During thoracic regional anesthesia, it is imperative to exercise utmost caution to avoid accidental puncture of the pleura, as the intended injection site is often close to this delicate membrane [1–5].

Ultrasonography is commonly used to identify the pleura and relevant anatomy and to guide needle insertion and injection in real time [1,5–7]. The pleural features appear as sliding shimmering lines on B-mode ultrasound when the visceral and parietal pleura slide against each other during the expansion and deflation of the lung during breathing cycles [8–10].

Similar to other miniaturized transducers used in cardiovascular catheterization for vascular imaging [11–14], tissue identification [15], and puncture safety [16], needle-sized transducers have been developed for use in anesthetic procedures [17–19]. Intra-needle ultrasound (INUS) transducers have been successfully used to identify the pleura during intercostal blocks [20]. The pleural signal and adjacent intercostal muscles can be visualized using INUS at the tip of the needle in real-time along the puncture using a one-dimensional radio frequency (RF) mode ultrasound signal, which presents as a unique flickering signal [20].

Previous studies have suggested that the ultrasonic features of the pleura originate from the acoustic interaction of the soft tissue and alveoli with different impedances [10]; however, the biological and physical principles of the RF-mode flickering signal of the pleura have not been fully investigated.

To differentiate the distinct patterns between ultrasound signals obtained from the pleural tissue and other tissues, we employed the short-time Fourier transform (STFT), which involves a sequence of Fourier transforms applied to a windowed signal. The STFT provides time-localized frequency information, making it suitable for situations in which the frequency components of a signal change over time [21–23].

The penetration and resolution of conventional ultrasound can be affected when the target is deep beneath the skin or the patient is obese [24,25]. Echogenic needles feature laser-etched marks on the needle shaft to enhance visibility. However, studies indicate that not all echogenic needle designs uniformly improve needle visualization, particularly when the needle follows a steep puncture trajectory [26]. Therefore, INUS could serve as a valuable tool to address this unmet medical need by providing image guidance directly at the needle tip, unaffected by the distance from the skin to the deep target.

We attempted to use time-frequency analysis to characterize the pleural signal and develop an automated tool that can identify the pleura during medical procedures. We used ex vivo and in vivo animal studies to investigate the pleural ultrasound signal in this study. Validation of the signal identification was also performed during clinical regional anesthesia.

2. Methods

2.1. Common Procedures in the Thoracic Region

Figure 1a shows an intercostal nerve block (ICNB) procedure with a needle inserted into the intercostal space. The tip was placed near the intercostal nerve, which lies between the innermost and internal intercostal muscles, as shown in Figure 1b,c. In the paravertebral block (PVB) procedure, the anesthesiologist inserts a puncture needle into the paravertebral space formed among the transverse process bone, superior costotransverse ligament, and the pleura, as shown in Figure 1d,e. Local anesthetics were administered to achieve regional anesthesia.

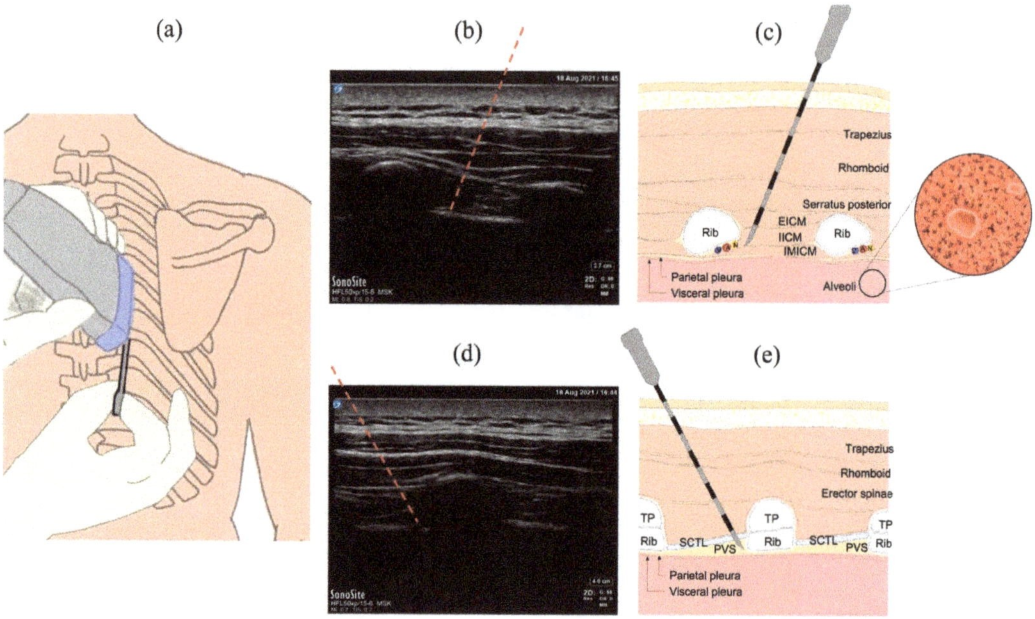

Figure 1. (**a**) Common procedures in thoracic anesthesia with risk of causing pleural injury and pneumothorax, (**b**) the ultrasound B-mode image, (**c**) the anatomy structure in the ICNB procedure, (**d**) the ultrasound B-mode image, and (**e**) the anatomy structure in the PVB procedure. Intercostal nerve block (ICNB); paravertebral block (PVB). The red dashed lines in (**b**,**c**) represent the proposed needle trajectories. Ultrasound images from piglets are shown in Figure 1b,d.

2.2. The INUS System

An INUS system was custom-developed to guide the induction of thoracic regional anesthesia (Figure 2a): ❶ indicates the portable ultrasound pulser/receiver, ❷ indicates the miniature INUS transducer, which was inserted into an 18-gauge puncture needle (inner diameter, 1.14 mm; outer diameter, 1.27 mm), and ❸ shows the RF-mode Display App on the Android smartphone system (Pixel 4 XL, Google Inc., Mountain View, CA, USA). The app receives and records the ultrasound signals and displays the RF mode signals on the screen of the Pixel 4 cell phone in real-time, with a pulse repetition frequency (PRF) of 30.

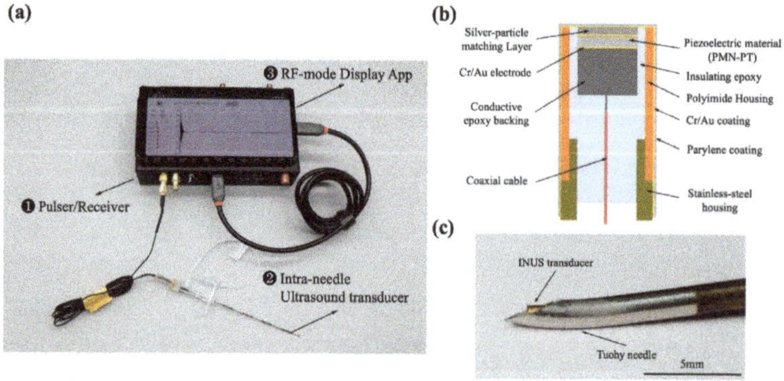

Figure 2. (**a**) Intra-needle ultrasound system, (**b**) the design cross-section of the needle ultrasound, (**c**) the needle transducer fit inside the Tuohy needle.

The needle transducer comprised a single-element lead magnesium niobate-lead titanate crystal (CTS Corp., Lisle, IL, USA). The crystal was polished to a thickness of 95 μm and cut into an area of 0.5 × 0.5 mm^2. The design cross-section of the needle transducer is illustrated in Figure 2b. The top segment of the transducer provides a 10–15° bending flexibility, enabling the tip of the transducer to fit well in the curve-shaped anesthesia Tuohy needle (Figure 2c).

The specifications of the portable INUS system are presented in Table 1.

Table 1. Specifications of the INUS system.

Type	Parameter	Value
Transducer	Type	Pulse-echo
	Number of elements	Single element
	Natural Focus	1.5 mm
Pulser	Frequency	20 MHz
	Pulse-repetition frequency	30 Hz
	Analog-to-Digital Converter	12 bits
	Sampling rate	125 MHz
	Pulse-type	Sine wave
	Excitation voltage	125 V
	Dimensions	163 mm × 90 mm × 42 mm
	Weight	670 g
	Display	Pixel 4 XL

Calibration testing was conducted using a steel block as a reflector to evaluate performance and technical specifications. The energy intensity of the acoustic field distribution was simulated using COMSOL Multiphysics (version 5.6, COMSOL Inc., Burlington, MA, USA) in Supplemental Figure S1, which yielded a natural focus at 1.5 mm. The time-domain pulse-echo waveform and frequency-domain spectrum of the INUS transducer measured in the natural focus are shown in Supplemental Figure S2. From the Figure, the center frequency of the INUS transducer was 21.5 MHz. The fractional bandwidth at −6 dB was approximately 33%. The maximum output voltage of the pulse-echo signal was 2.04 V using the INUS system with an excitation voltage of 125 V. In our previous proof-of-concept study [20], we found the INUS system could distinguish two interfaces separated by 0.15 mm (axial resolution) and penetrate the tissue to a depth of 5 mm, and also had the ability to identify the endothoracic fascia, parietal pleura, and visceral pleura.

2.3. Ex Vivo Animal Study

An ex vivo phantom study was designed to simulate breathing and lung sliding to investigate if the "flickering" RF-mode pleural signal could be reproduced. A motorized platform with a resected porcine lung was designed to move back and forth to simulate the pleural sliding movement during the breathing cycle (Supplemental Figure S3a). The needle transducer was inserted into a gel pad (Parker Laboratories Inc., Fairfield, NJ, USA) and placed on top of the visceral pleural (lung) surface (Supplemental Figure S3b). Both the needle transducer and the gel pad were fixed during the lung-sliding movement. The distance between the needle tip and the visceral pleura was approximately 1.5 mm. The raw RF mode data were recorded, the PRF was set at 30 Hz, and the lung slid along the gel pad surface to simulate breathing, with a moving speed of approximately 3 cm/s, driven by the motor platform. A study on healthy subjects reported that the lung expansion distance during inspiration is approximately 4.82 ± 1.84 cm. Given that the inspiration time ranges from 1.0 to 1.2 s, a motor speed of 3 cm/s is reasonable for the simulation [27]. The motor was stopped with 0–1 s and 4–4.5 s, and the M-mode image showed a static lung image.

This ex vivo animal study would also be valuable if the results demonstrate reproducibility with in vivo animal studies.

2.4. In Vivo Animal Study

This study was approved by the Institutional Animal Care and Use Committee of the Taipei Veterans General Hospital, Taiwan (IACUC No. 2018-96 and No. 2021-210). In this study, punctures and measurements were performed on 10 native Chinese pigs with an average weight of 25 kg [20].

After the induction of general anesthesia, a needle was inserted between the adjacent ribs, as shown in Figure 3a. Conventional surface ultrasound with a linear array probe was then used to confirm the insertion path and the distance from the skin to the pleura. After the skin puncture, the original stylet was replaced with an INUS transducer to guide the needle and inserted slowly toward the pleura until the pleural signal was detected 1.5 mm away from the needle tip. RF-mode ultrasound data were recorded using a 12-bit analog-to-digital converter with a sampling frequency of 125 MHz for post-analysis and M-mode image reconstruction.

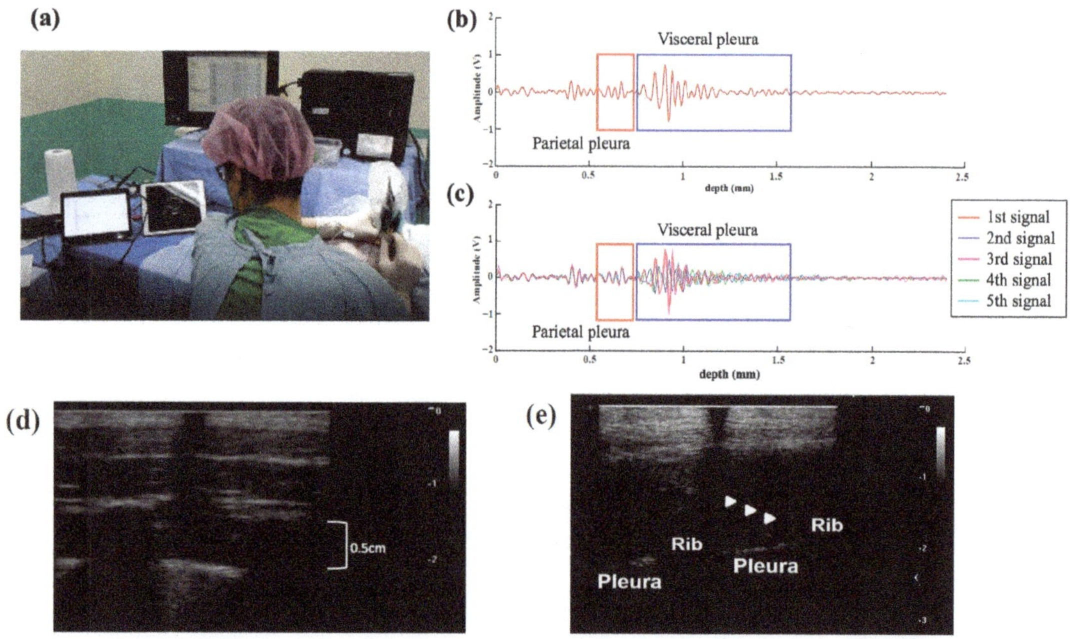

Figure 3. (**a**) In vivo animal study of thoracic regional anesthesia. (**b**) The RF-mode signal of the two-layered pleural membrane, the outer parietal pleura in the red box, and the inner visceral pleura in the blue box. (**c**) The overlapped plotting of five consecutive pleural signals showing the signals and flickering characteristics of the visceral pleura and alveoli during the breathing cycles. (**d**) The image obtained from surface ultrasound; the target area for injection is between ribs. (**e**) The image obtained from surface ultrasound with the needle (arrow head) in the target position of intercostal nerve block. Radiofrequency (RF).

The INUS transducer was excited and displayed using 30 PRF during the puncture process. Flickering pleural RF mode signals similar to pig breathing were observed on a Pixel 4 XL smartphone. Figure 3b presents the two-layered pleural signals with the parietal pleura in the red box and the visceral pleura and alveoli in the blue box. Figure 3c illustrates the overlapping RF mode signals of the pleura over five consecutive measurements during the breathing cycles. The parietal pleural layer was stationary, and the RF-mode signals consistently overlapped during the five cycles, whereas the RF-mode signals of the visceral pleura and alveoli flickered.

2.5. Time-Frequency Domain Analysis

An STFT analysis was used to analyze and compare the differences between the ultrasound signals of the pleural tissues and other tissues in the frequency domain. The STFT involved dividing the whole length of the acquired RF mode signals into several short depths. Then, we computed the Fourier transform for each short depth. The STFT of the ultrasound RF-mode signal $s(t)$ can be written as

$$S(\tau,\omega) = \int_{-\infty}^{\infty} s(t)\theta(t-\tau)e^{-i\omega t}\Big] dt \tag{1}$$

where θ $(t - \tau)$ is the window function, τ is the time resolution, and ω is the radial frequency. The transformation of the RF-mode signals was computed using MATLAB (R2020a, MathWorks Inc., Natick, MA, USA). The flow chart is presented in Supplemental Figure S4.

To quantify the flickering degree of the pleural ultrasound signal during breathing, we calculated the correlation coefficient (r) between every two consecutive pleural signals, including the RF-mode signal, M-mode image, and STFT spectrum. We also calculated the number of STFT spectral bands (−6 dB) of the RF-mode signals of the endothoracic fascia and the parietal and visceral pleura to quantify the multi-band characteristics of the ultrasonic signals from different tissues.

2.6. In Vivo Study of Pleura Tracking

The tracking algorithm analyzed the RF mode signal and utilized four distinct parameters to assess the region of interest (ROI) and determine its association with the pleura. These parameters are the average amplitude of the ROI, the average number of frequency bands per second, the standard deviation of the per-second frequency band counts, and the disparity between successive RF-mode data.

The signal was imported into MATLAB software and subjected to the tracking algorithm. The flowchart of pleura tracking is presented in Supplemental Figure S5.

3. Results

We observed the flickering phenomenon from the INUS-RF signals in the ex vivo and in vivo studies at the pleura during motor moving and breathing. The signals are presented in Figures 4 and 5 and Supplemental Video S1.

In this figure, we aim to demonstrate the multiple and random reflections from lung tissue. When scanning a moving but relatively flat surface, the signals collected over a short period should be able to be overlaid. This indicates that the ultrasound reflections from this specific tissue are stable and do not exhibit multiple random reflections, as observed in the endothoracic fascia or parietal pleura in the latter part of our study. However, when we attempted to overlay signals from moving lung tissue collected over a short period, the signals could not be aligned. We suggest that this is caused by the random and multiple reflections from the air within the lung.

Frequency domain analysis of the returned RF signals revealed that the frequency bands of visceral pleura signals were multiple and distorted in the −6 dB frequency bands and rapidly shifted. The frequency bands of the visceral pleura were random and of mixed frequency. We speculated that this character of ultrasound waves led to the flickering phenomena observed in the RF mode and the "shimmering phenomena" in the M mode.

Endothoracic fascia and parietal pleura showed stable frequency bands and fixed signals but no change in the frequency.

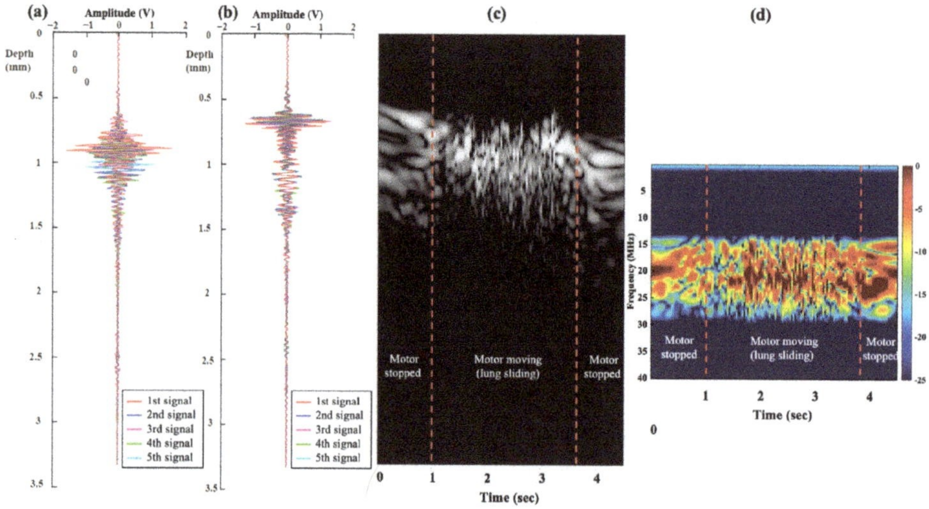

Figure 4. Overlapped five RF-mode signals of the ex vivo phantom when the motor was (**a**) moving (lung sliding and flickering signals) and (**b**) motor stopped (non-flickering, stable RF-signals). (**c**) M-mode image covering both the stopped and moving process. (**d**) The stable (motor stopped) and highly variable (motor moving) time-frequency spectral distribution. Radiofrequency (RF).

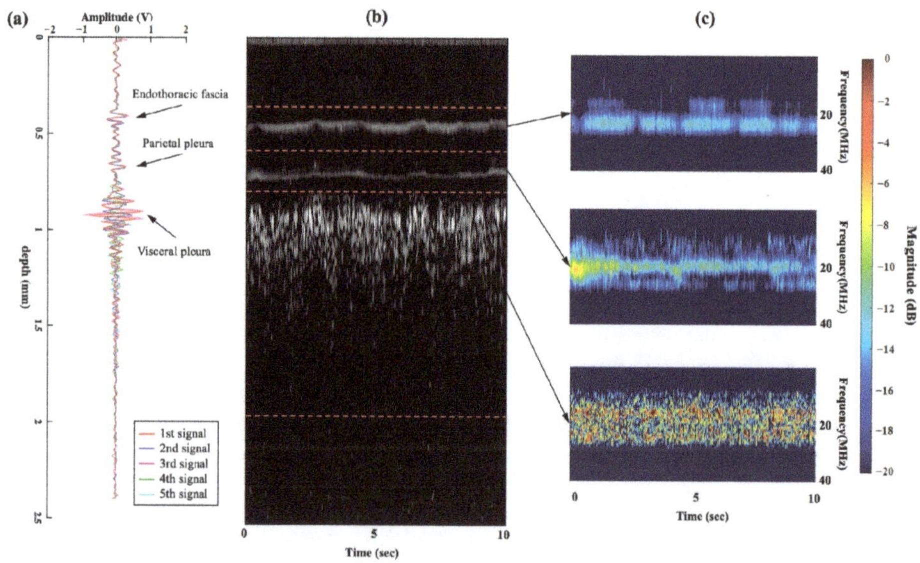

Figure 5. (**a**) Consecutive RF-mode signals of the endothoracic fascia and parietal pleura tissues were well overlapped, whereas those of the visceral pleura and alveoli tissues flickered in the time domain. (**b**) In five breathing cycles, the M-mode image of the endothoracic fascia and the parietal pleura was stable, whereas that of the visceral pleura and the alveoli had a speckle-like pattern. (**c**) The STFT spectra of the endothoracic fascia and parietal pleura had stable spectral bands, whereas that of the visceral pleura and alveoli were multi-banded and chaotic in the frequency domain. Radiofrequency (RF), Short-time Fourier Transform (STFT).

In the M-mode signal analysis, flickering characteristics of the visceral pleura and alveoli were also observed, with a much lower mean (m = 0.53) and higher standard deviation (σ = 0.25) of the correlation coefficient than that of the endothoracic fascia and parietal pleura tissues (Table 2). These results also validate the speckle-like characteristics in the M-mode images of the visceral pleura and alveolar tissues (Figure 5b).

Table 2. Quantitative analysis of the flickering effects of the fascia and pleura tissue signals (m $\pm$ σ).

	Endothoracic Fascia	Parietal Pleura	Visceral Pleura + Alveoli
Correlation coefficient of RF-mode signals	0.87 ± 0.13	0.80 ± 0.16	0.16 ± 0.33
Correlation coefficient of M-mode signals	0.94 ± 0.04	0.80 ± 0.15	0.53 ± 0.25
Correlation coefficient (STFT Spectra)	0.93 ± 0.07	0.85 ± 0.14	0.39 ± 0.25
Average number per second of STFT bands (m $\pm$ σ)	2.7 ± 0.84	2.7 ± 0.81	5.5 ± 2.13

Mean (m); standard deviation (σ); radiofrequency (RF); short-time Fourier transform (STFT).

Figure 5 shows the RF-mode signals, an M-mode image, and the time-frequency spectra information from the in vivo animal study. In Figure 5a, we overlapped five consecutively measured RF-mode signals from the in vivo animal study. We observed that the RF-mode signals of the endothoracic fascia and parietal pleura were stable, whereas that of the visceral pleura and alveoli flickered. In Figure 5b, we collected the M-mode image transformed from 10 s consecutively measured RF-mode signals in the in vivo animal study during five breathing cycles. The M-mode image shows two stable, bright lines representing the endothoracic fascia and parietal pleura, as well as a speckle-like region representing the visceral pleura and alveoli region. In Figure 5c, we analyzed the corresponding time-frequency spectral information covering three different depth regions. The spectral information in the 0.4–0.6 mm and 0.6–0.8 mm regions, endothoracic fascia, and parietal pleura region was stable and covered 2–3 narrower bands between 19–23 MHz. In the 0.8–2.0 mm region, the visceral pleura, and shallow alveoli region, the spectra were highly variable multi-bands and covered a significantly wider bandwidth between 12 and 28 MHz.

Table 2 shows the results of quantitative analysis with the correlation coefficient (r) analysis of fascia and pleura tissue during breathing. The M-Mode correlation coefficient for different layers is calculated by analyzing the M-Mode intensity over a continuous sequence of images within the same scanning region. With INUS set at a pulse repetition frequency of 30, there are 30 RF-mode images per second. Running STFT on these RF signals reveals multiple frequency bands in each tissue layer. We then calculate the average number of bands per second for each layer. Therefore, the higher average per second with higher standard deviation of STFT bands suggests "multiple, random, rapid changing reflections". Table 2 compares the correlation coefficients of the STFT spectra for these tissues. The mean correlation coefficient (m = 0.39) of the visceral pleura and alveoli was lower than that of the other tissues, whereas the standard deviation (σ = 0.25) was higher.

Table 2 lists the averaged number per second of spectral bands (m = 5.5) and standard deviation (σ = 2.13) in the STFT spectra of the visceral pleura and alveoli. The average number per second of STFT bands and standard deviation of the visceral pleura and alveoli tissues were higher than that of the endothoracic fascia and parietal pleura tissues.

These results validate the chaotic (random) and multi-banded characteristics of the STFT spectra of the visceral pleura and alveoli in the frequency-domain analysis (Figure 5c).

The time-frequency spectra results of the ex vivo and in vivo studies are shown in Figures 4d and 5c, respectively, revealing the multi-band and highly variable spectral characteristics of the visceral pleura and alveoli during sliding and breathing. In contrast, the spectra of the endothoracic fascia and parietal pleural signals were stable and invariant (Figure 5c), with a 0.4–0.8 mm depth region. Therefore, we can conclude that the highly variant spectral characteristics of the INUS signals were caused by lung sliding (visceral pleura and alveolar movement) during breathing.

In this study, we compared the flickering effects of visceral pleural ultrasound signals with those of the endothoracic fascia and parietal pleura (Table 2). The mean correlation coefficient of the consecutive visceral pleura and alveoli signals was significantly lower (m = 0.16) than that of the endothoracic fascia (m = 0.87) and parietal pleura (m = 0.80) signals in the RF mode. Additionally, the standard deviation of the visceral pleura and alveoli ($\sigma = 0.33$) was higher than that of the endothoracic fascia ($\sigma = 0.13$) and parietal pleura ($\sigma = 0.16$). These findings demonstrate the stable RF-mode signals of the endothoracic fascia and parietal pleura tissues and the flickering RF-mode signals of the visceral pleura and alveoli tissues (Figure 5a).

The multiband count within the spectrum held considerable relevance in identifying the pleura. The recorded signal was further processed with the tracking algorithm, and the pleura was traced successfully (Figure 6). The INUS system allows users to switch between RF-mode and M-mode, improving confidence in identifying the pleura's location during procedures.

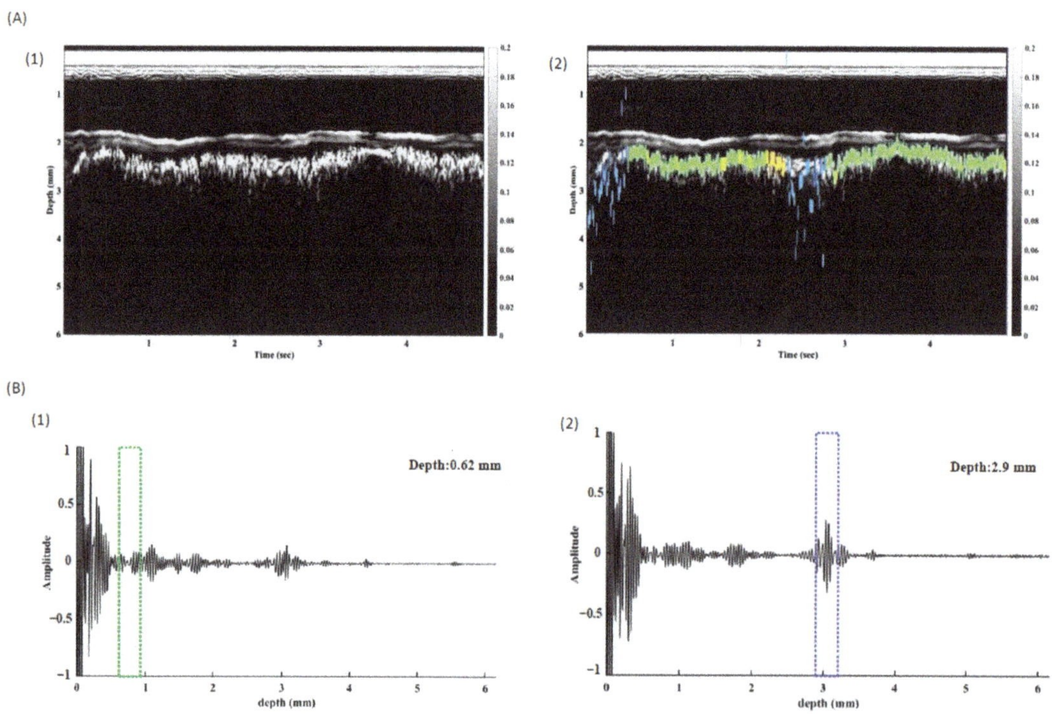

Figure 6. Tracking of pleura in the in vivo study. (**A**) The original (**1**) and the annotated (**2**) M-mode images from INUS. The pleura can be identified by the tracking algorithm. (**B**) A screenshot of the RF-mode video clip (Supplemental Video S2) shows the pleural auto-identification process. After applying the STFT algorithm, the green box represents a "guessed but not confirmed" status during the pleural location search. Furthermore, the blue box in the RF-mode diagram and the green line in the M-mode represent the signal that fits the pleural characteristics of our analysis and are visually highlighted for clinicians to avoid accidental puncture. Intra-needle ultrasound (INUS), Radiofrequency (RF).

4. Discussion

As our goal was to avoid pleural puncture, identifying the pleura from the needle tip, i.e., from INUS, could present a promising strategy when dealing with limited image quality from external ultrasound, such as in cases of deep punctures or in overweight patients [24]. We effectively discerned the distinct features of the pleural signal from INUS. Only the visceral pleura and alveoli showed high variability and flickering characteristics in the time-domain RF-mode signals and multibanded and chaotic characteristics in the frequency-domain spectra during respiration in the ICNB and PVB procedures. We suggest this is ultrasound distortion caused by the interface of multiple moving alveoli, and the signal returned in variable and multi-banded frequency. The RF-mode signals of the visceral pleura and alveoli showed high variability during the breathing cycle, causing a "flickering phenomenon" due to their fast changes and a frame rate of display of 30 Hz. This is due to the persistence of vision, which lasts for approximately 100 ms [28].

The "flickering phenomenon", the speckle-like M-mode image, and the shimmering B-mode image (lung sliding sign) disappear when the lung is not breathing. We suggest that these essentially originate from the same mechanism: visceral pleura and alveoli movement during lung sliding during breathing.

Some difficulties may be encountered when using ultrasound guidance to perform thoracic regional anesthesia in patients with obesity because of the presence of subcutaneous fat, which has strong attenuation and may reduce image quality, making it difficult to accurately discern the location of the needle tip relative to the pleura. Additionally, procedures near the spine, such as PVB, can make it more challenging for ultrasound to fully delineate the pleura, as it lies deeper in this region. Inaccurate identification of the pleura or the distance between the pleura and needle tip can result in complications, such as inadvertent pleural puncture, pneumothorax, and ineffective analgesia [5–7,9–11]. Although the quality of conventional surface ultrasonic images continues to improve, accurately advancing the needle to deeply located targets remains a challenge for anesthesiologists [8].

Pleura identification was successfully performed using the INUS transducer in the ex vivo and the in vivo study, and the characteristics were successfully integrated into an auto-identification algorithm for real-time monitoring and imaging. The M-mode image can be constructed in real-time from the digitized RF-mode signals to provide a detailed display of the forward view of the INUS to guide the needle insertion process regardless of the depth of the skin or the limitations of conventional surface ultrasound.

These distinctive and specific results may provide valuable information for our group to develop a visceral pleura auto-identification algorithm. In addition, the sliding of the pleura (lungs) during breathing may differ between animals and humans. Therefore, further clinical studies are warranted.

In conclusion, the unique flickering signal of the pleura from INUS can be processed by time-frequency domain analysis and further tracked by an auto-identification algorithm. This technique can be used for thoracic regional anesthesia and other interventions. However, further studies are required to confirm this hypothesis and validation of the auto-identification of the pleura will be performed.

Supplementary Materials: The following supporting information can be downloaded at: https://www.mdpi.com/article/10.3390/bios15040201/s1, Figure S1: Simulated Acoustic Field Energy Intensity; Figure S2: Time-Domain Pulse-Echo Waveform and Frequency-Domain Spectrum of the INUS Transducer; Figure S3: Platform of the Ex Vivo Animal Study; Figure S4: Flowchart of the STFT Analysis; Figure S5: Overview of the Pleural Tracking Algorithm Flowchart; Video S1: Pleural Flickering Signal Observed in INUS-RF Signals; Video S2: Video Demonstration of Needle Tip-to-Pleura Distance Estimation Using INUS-RF Signal Tracking.

Author Contributions: F.-W.S.: This author helped with designing and conducting the animal experiment, data acquisition, and drafting and revising the manuscript. C.-W.Y.: This author helped with designing and conducting the signal analysis, data acquisition, figure preparation, drafting and revising the manuscript. C.-F.Y.: This author helped with the statistical analysis and manuscript revision. Y.-E.T.: This author helped with designing and conducting the signal analysis, data acquisition, and figure preparation. W.-N.T.: This author helped with the statistical analysis and manuscript revision. H.K.C.: This author helped with drafting and revising the manuscript. All authors have read and agreed to the published version of the manuscript.

Funding: This work was supported by the National Science and Technology Council, Taiwan, under Grant No. NSTC 112-2622-B-A49-001.

Institutional Review Board Statement: This study does not involve human subjects. The animal study was approved by the Institutional Animal Care and Use Committee of Taipei Veterans General Hospital, Taiwan (IACUC No. 2018-96, approved on 5 July 2018, and No. 2021-210, approved on 11 November 2021).

Informed Consent Statement: Not applicable.

Data Availability Statement: Data supporting the findings of this study are available from the corresponding author HK Chiang on request.

Acknowledgments: The authors acknowledge the funding supplied by the National Science and Technology Council, Taiwan, under Grant No. NSTC 112-2622-B-A49-001.

Conflicts of Interest: Fu-Wei Su, Chia-Wei Yang, Ching-Fang Yang, and Hui-Hua Kenny Chiang are credited as inventors on patents linked to the research discussed in this article. Taipei Veterans General Hospital and National Yang-Ming Chiao-Tung University hold the patent rights. The patented technology involves the intra-needle ultrasound system and its method for analyzing, tracking, and displaying pleural data. Despite this potential conflict of interest, we affirm that it does not impact the objectivity, validity, or interpretation of the findings presented in this manuscript. Patent numbers: United States (US11786217), United Kingdom (GB2582201B), Taiwan (I720398), China (2019111430), and Japan (7057948).

References

1. Byas-Smith, M.G.; Gulati, A. Ultrasound-guided intercostal nerve cryoablation. *Anesth. Analg.* **2006**, *103*, 1033–1035. [CrossRef] [PubMed]
2. Matchett, G. Intercostal Nerve Block and Neurolysis for Intractable Cancer Pain. *J. Pain Palliat. Care Pharmacother.* **2016**, *30*, 114–117. [CrossRef] [PubMed]
3. Moore, D.C.; Bridenbaugh, L.D. Intercostal nerve block in 4333 patients: Indications, technique, and complications. *Anesth. Analg.* **1962**, *41*, 1–11. [PubMed]
4. Karmakar, M.K.; Ho, A.M. Acute pain management of patients with multiple fractured ribs. *J. Trauma* **2003**, *54*, 615–625. [CrossRef]
5. Karmakar, M.K. Thoracic paravertebral block. *Anesthesiology* **2001**, *95*, 771–780. [CrossRef]
6. Peng, P.W.; Narouze, S. Ultrasound-guided interventional procedures in pain medicine: A review of anatomy, sonoanatomy, and procedures: Part I: Nonaxial structures. *Reg. Anesth. Pain Med.* **2009**, *34*, 458–474. [CrossRef]
7. Elkhashab, Y.; Wang, D. A Review of Techniques of Intercostal Nerve Blocks. *Curr. Pain Headache Rep.* **2021**, *25*, 67. [CrossRef]
8. Marini, T.J.; Rubens, D.J.; Zhao, Y.T.; Weis, J.; O'connor, T.P.; Novak, W.H.; Kaproth-Joslin, K.A. Lung Ultrasound: The Essentials. *Radiol. Cardiothorac. Imaging* **2021**, *3*, e200564. [CrossRef]
9. Doerschug, K.C.; Schmidt, G.A. Intensive care ultrasound: III. Lung and pleural ultrasound for the intensivist. *Ann. Am. Thorac. Soc.* **2013**, *10*, 708–712. [CrossRef]
10. Oelze, M.L.; Miller, R.J.; Blue, J.P., Jr.; Zachary, J.F.; O'Brien, W.D., Jr. Estimation of the acoustic impedance of lung versus level of inflation for different species and ages of animals. *J. Acoust. Soc. Am.* **2008**, *124*, 2340–2352. [CrossRef]
11. Yoon, S.; Williams, J.; Kang, B.J.; Yoon, C.; Cabrera-Munoz, N.; Jeong, J.S.; Lee, S.G.; Shung, K.K.; Kim, H.H. Angled-focused 45 MHz PMN-PT single element transducer for intravascular ultrasound imaging. *Sens. Actuators A Phys.* **2015**, *228*, 16–22. [CrossRef] [PubMed]
12. Zhu, B.; Zhang, Z.; Ma, T.; Yang, X.; Li, Y.; Shung, K.K.; Zhou, Q. (100)-Textured KNN-based thick film with enhanced piezoelectric property for intravascular ultrasound imaging. *Appl. Phys. Lett.* **2015**, *106*, 173504. [CrossRef] [PubMed]

13. Qiu, W.; Wang, X.; Chen, Y.; Fu, Q.; Su, M.; Zhang, L.; Xia, J.; Dai, J.; Zhang, Y.; Zheng, H. Modulated Excitation Imaging System for Intravascular Ultrasound. *IEEE Trans. Biomed. Eng.* **2017**, *64*, 1935–1942. [CrossRef] [PubMed]
14. Yu, M.; Li, Y.; Ma, T.; Shung, K.K.; Zhou, Q. Intravascular Ultrasound Imaging with Virtual Source Synthetic Aperture Focusing and Coherence Factor Weighting. *IEEE Trans. Med. Imaging* **2017**, *36*, 2171–2178. [CrossRef]
15. Jiang, Y.; Qiu, Z.; McPhillips, R.; Meggs, C.; Mahboob, S.O.; Wang, H.; Duncan, R.; Rodriguez-Sanmartin, D.; Zhang, Y.; Schiavone, G.; et al. Dual Orientation 16-MHz Single-Element Ultrasound Needle Transducers for Image-Guided Neurosurgical Intervention. *IEEE Trans. Ultrason. Ferroelectr. Freq. Control* **2016**, *63*, 233–244. [CrossRef]
16. McPhillips, R.; Qiu, Z.; Jiang, Y.; Mahboob, S.O.; Wang, H.; Meggs, C.; Schiavone, G.; Sanmartin, D.R.; Eljamel, S.; Desmulliez, M.P.; et al. Ex-vivo navigation of neurosurgical biopsy needles using microultrasound transducers with M-mode imaging. In Proceedings of the 2015 IEEE International Ultrasonics Symposium (IUS), Taipei, Taiwan, 21–24 October 2015; pp. 1–4. [CrossRef]
17. Jiang, Y.; Meggs, C.; Button, T.; Schiavone, G.; Desmulliez, M.P.; Qiu, Z.; Mahboob, S.; McPhillips, R.; Demore, C.E.; Casey, G.; et al. 15 MHz single element ultrasound needle transducers for neurosurgical applications. In Proceedings of the 2014 IEEE Symposium (IUS) Ultrasonics, Chicago, IL, USA, 3–6 September 2014; pp. 687–690. [CrossRef]
18. Pesteie, M.; Lessoway, V.; Abolmaesumi, P.; Rohling, R.N. Automatic Localization of the Needle Target for Ultrasound-Guided Epidural Injections. *IEEE Trans. Med. Imaging* **2018**, *37*, 81–92. [CrossRef]
19. Chiang, H.K.; Zhou, Q.; Mandell, M.S.; Tsou, M.-Y.; Lin, S.-P.; Shung, K.K.; Ting, C.-K. Eyes in the needle: Novel epidural needle with embedded high-frequency ultrasound transducer—epidural access in porcine model. *Anesthesiology* **2011**, *114*, 1320–1324. [CrossRef]
20. Liao, S.-W.; Su, F.-W.; Ting, C.-K.; Yang, C.-F.; Yang, C.-W.; Tsou, M.-Y.; Chiang, H.K. Intercostal Nerve Block Using an Innovative Intraneedle Ultrasound Transducer: A Proof-of-Concept study. *Ultrasound Med. Biol.* **2021**, *47*, 1881–1892. [CrossRef]
21. Sejdić, E.; Djurović, I.; Jiang, J. Time–frequency feature representation using energy concentration: An overview of recent advances. *Digit. Signal Process.* **2009**, *19*, 153–183. [CrossRef]
22. Liu, B.; Qian, S.; Hu, W. Identification of Denatured Biological Tissues Based on Time-Frequency Entropy and Refined Composite Multi-Scale Weighted Permutation Entropy during HIFU Treatment. *Entropy* **2019**, *21*, 666. [CrossRef]
23. Callejas, A.; Palma, R.; Hernández-Figueirido, D.; Rus, G. Damage Detection Using Ultrasonic Techniques in Concrete-Filled Steel Tubes (CFSTs) Columns. *Sensors* **2022**, *22*, 4400. [CrossRef] [PubMed]
24. Abdallah, F.W.; Chan, V.W.; Koshkin, A.; Abbas, S.; Brull, R. Ultrasound-guided sciatic nerve block in overweight and obese patients: A randomized comparison of performance time between the infragluteal and subgluteal space techniques. *Reg. Anesth. Pain Med.* **2013**, *38*, 547–552. [CrossRef] [PubMed]
25. Vlassakov, K.; Vafai, A.; Ende, D.; Patton, M.E.; Kapoor, S.; Chowdhury, A.; Macias, A.; Zeballos, J.; Janfaza, D.R.; Pentakota, S.; et al. A prospective, randomized comparison of ultrasonographic visualization of proximal intercostal block vs paravertebral block. *BMC Anesthesiol.* **2020**, *20*, 13. [CrossRef]
26. Sviggum, H.P.; Ahn, K.; Dilger, J.A.; Smith, H.M. Needle echogenicity in sonographically guided regional anesthesia: Blinded comparison of 4 enhanced needles and validation of visual criteria for evaluation. *J. Ultrasound Med.* **2013**, *32*, 143–148. [CrossRef] [PubMed]
27. Derasse, M.; Lefebvre, S.; Liistro, G.; Reychler, G. Chest Expansion and Lung Function for Healthy Subjects and Individuals With Pulmonary Disease. *Respir. Care* **2021**, *66*, 661–668. [CrossRef]
28. Coltheart, M. The persistences of vision. *Philos. Trans. R. Soc. B* **1980**, *290*, 57–69. [CrossRef]

biosensors

MDPI

Article

Detecting Hypoxia Through the Non-Invasive and Simultaneous Monitoring of Sweat Lactate and Tissue Oxygenation

Cindy Cheng, Sayan Ganguly, Pei Li and Xiaowu Tang *

Department of Chemistry, University of Waterloo, 200 University Ave West, Waterloo, ON N2L 3G1, Canada; c29cheng@uwaterloo.ca (C.C.); sganguly@uwaterloo.ca (S.G.); p2li@uwaterloo.ca (P.L.)
* Correspondence: tangxw@uwaterloo.ca; Tel.: +1-519-888-4567 (ext. 38037); Fax: +1-519-746-0435

Abstract: Hypoxia, characterized by inadequate tissue oxygenation, may result in tissue damage and organ failure if not addressed. Current detection approaches frequently prove insufficient, depending on symptoms and rudimentary metrics such as tissue oxygenation, which fail to comprehensively identify the onset of hypoxia. The European Pressure Ulcer Advisory Panel (EPUAP) has recognized sweat lactate as a possible marker for the early identification of decubitus ulcers, nevertheless, neither sweat lactate nor oxygenation independently provides an appropriate diagnosis of hypoxia. We have fabricated a wearable device that non-invasively and concurrently monitors sweat lactate and tissue oxygenation to fill this gap. The apparatus comprises three essential components: (i) a hydrogel-based colorimetric lactate biosensor, (ii) a near-infrared (NIR) sensor for assessing tissue oxygenation, and (iii) an integrated form factor for enhanced wearability. The lactate sensor alters its hue upon interaction with lactate in sweat, whereas the NIR sensor monitors tissue oxygenation levels in real-time. The device underwent testing on phantom exhibiting tissue-mimicking characteristics and on human sweat post aerobic and anaerobic activities. Moreover, the device was demonstrated to be capable of real-time "on-body" simultaneous monitoring of sweat lactate spikes and tissue oxygenation (StO_2) drops, which showed strong correlation during a hypoxia protocol. This innovative technology has a wide range of potential applications, such as post-operative care, sepsis detection, and athletic performance monitoring, and may provide economical healthcare solutions in resource-limited regions.

Keywords: hypoxia; wearable sensor; sweat lactate; tissue oxygenation

1. Introduction

Wearable sensors have become a significant growing sector for consumer, sports, and military applications, providing distinctive opportunities for in situ, real-time, and non-invasive health and performance monitoring [1]. The consumer electronics industry is presently saturated with instances of wearable technology in wrist-worn, textile, or strap-mounted configurations, with inflexible active materials and cumbersome power sources [2,3]. Available measures are typically restricted to elevated physiological signals, including temperature, heart rate, skin conductance, and respiratory rate [4]. Recent advancements in conformal and flexible electronics have yielded remarkable demonstrations of research-grade prototypes possessing comparable physiological measurement capabilities, yet with significantly different form factors [5]. Simultaneously, considerable focus is directed towards monitoring analyte presence in biofluids, hence broadening the scope of information obtainable from the wearer [6–8]. Sweat is an optimal subject for analysis owing to its suitability for non-invasive collection [9]. Numerous sweat-sensing devices have been showcased in the forms of patches, watches, and temporary tattoos, capable of measuring analytes including lactate, glucose, ammonia, ethanol, trace metals, hydration biomarkers, pH, and other metal ions found in sweat [10]. Complex systems that integrate measurements of many biomarkers and physiological signals on a single platform have also been documented, demonstrating the possibility for completely autonomous integrated systems [11–14].

Sweat presents a viable substitute for blood in routine analyses conducted on the sports field. Sweat is produced during physical exercise, its collection is non-invasive, and it contains numerous biomarkers indicative of physiological states, such as lactate [15]. Although recent research suggests that sweat lactate may effectively substitute blood analysis for assessing sports performance, this assertion remains contentious [16]. Numerous lactate sensors have been documented in the literature, aiming to deliver reliable on-body data for physiological evidence [17], including colorimetric, electrochemiluminescent, and electrochemical measurements [18]. Promphet et al. have created a colorimetric textile-based sensor for lactate monitoring on the body [19]. Chen et al. have introduced a versatile electrochemiluminescence platform capable of responding to concentrations up to 20 mM [20]. The sensor was effectively utilized in two on-body tests, yielding several discrete results. Nonetheless, the deterioration and/or leaching of the luminophore material over prolonged measurements are a concern, since it may result in erroneous outcomes over time.

Lactate is a by-product of glycolysis, produced during anaerobic glucose metabolism in response to the body's elevated energy requirements. It is a crucial indicator for assessing oxidative stress levels, muscular health, and tissue hypoxia [21]. The net lactate concentration in these biofluids is significantly influenced by the physiological condition and the location of anaerobic metabolism within the human body [22]. Lactate monitoring is crucial for some populations, particularly those susceptible to hypoxic situations, such as athletes and military personnel, to enhance training and endurance performance [23]. Elevated lactate levels in the body can rapidly disrupt fluid pH and lead to several adverse health consequences, including seizures, sepsis, renal failure, malignancies, cerebral stroke, and panic disorders [24]. The lactate concentration in eccrine sweat is influenced by various factors, including the kind of effort, duration, intensity, and sweat rate [25]. Medium-intensity activity for a brief period, or heat-induced perspiration, might produce a lactate concentration of up to 25 mM, whereas strenuous outdoor activities may provide concentrations as high as 115 mM [26–28]. Conversely, it has been demonstrated that sweat lactate concentration inversely correlates with sweat rate [29]. Evidence indicates that sweat lactate levels exhibit minor variation with age [30], sex [31], and sampling site (attributable to differences in sweat gland density) [32]. In addition to eccrine sweat glands, two further types of sweat glands have been identified in the body: apocrine and apoeccrine [33]. Although these glands produce perspiration, lactate is rarely found in it, thus we focus on the forearm eccrine glands, a common lactate testing site. Alam et al. have determined that blood lactate concentrations in healthy individuals at rest range from 0.5 to 2 mM [34]. During moderate to medium-intensity exercise, blood lactate concentration exhibits no substantial variation in magnitude and does not correlate well with sweat lactate concentration [35]. Goodwin et al. demonstrated that upon surpassing the lactate threshold (LT), blood lactate concentration increases exponentially, potentially reaching a peak of 11 mM [33]. Karpova et al. conducted a study demonstrating a favorable association between lactate levels in blood and perspiration after high-intensity exercise [36]. Laboratory and commercial sensing platforms use lactate oxidase (LOx) or dehydrogenase. Previous colorimetric skin-interfacing wearable sensors convey sweat to a lactate-specific enzyme assay area via complex serpentine microfluidic channels [37,38]. The lactate present in the sweat undergoes a reaction that produces a colored compound, the intensity of which is measured by mobile image analysis [39,40].

This study delivers an innovative method for the non-invasive detection of hypoxia by concurrently monitoring sweat lactate and tissue oxygenation, a combination that, to our knowledge, has not been previously reported. Although prior research has examined tissue oxygenation and lactate levels separately as indicators of hypoxia, their simultaneous real-time assessment provides a more thorough comprehension of the condition. Our engineered wearable device seamlessly combines a colorimetric lactate biosensor with a near-infrared (NIR) sensor, delivering real-time, non-invasive analysis of metabolic and tissue oxygenation conditions. The innovation resides in the integration of these two indicators into a singular device, facilitating improved precision in identifying hypoxic states and providing expanded therapeutic applications, including post-operative management and athletic surveillance. This dual-sensing technique enhances hypoxia diagnosis and facilitates economical, accessible healthcare solutions, especially in resource-constrained environments.

2. Materials and Methods

Materials. Sodium L-lactate (~98%), Alginate (sodium salt), Calcium chloride ($CaCl_2$), L-Lactate Oxidase (LOX), Horseradish peroxidase (HRP), and 3,3′,5,5′-Tetramethylbenzidine (TMB) were purchased from Sigma-Aldrich, Canada. Cellulose nanocrystal (CNC) was purchased from CelluForce Inc., Montreal, QC, Canada. Sodium hydroxide (NaOH, ≥97.0%) was purchased from Fisher Scientific (Canada). All aqueous solutions were prepared using highly purified water from a Millipore Milli-Q water system (≥18 MΩ cm). All chemicals were used without further purification.

UV-vis spectroscopy. The absorbance of lactate-sensing hydrogels was measured at 570 nm using a UV–vis spectrophotometer (Molecular Devices SpectraMax M2, Marshall Scientific, Hampton, NH, USA).

Preparation of lactate-sensing hydrogel. A homogenous alginate solution at 4 wt% was prepared by adding alginate powder to PBS (pH 7.4) and stirring for 2 h at 60 °C. The solution was then cooled to room temperature before use. CNC powder was added to DI water and sonicated briefly to make a homogeneous 2 wt% suspension. TMB was added drop-wise to the CNC suspension until the surface of the CNC nanoparticles was saturated. The TMB-loaded CNCs were collected by centrifugation and the precipitate was resuspended in DI water. The alginate and TMB/CNC solutions were then mixed at a 1:1 ratio. Right before performing the colorimetric assay, LOX and HRP solutions were added. The final concentrations in the pre-gel solution were 2 wt% alginate, 1 wt% TMB/CNC, 50 μg/mL LOX, and 2 μg/mL HRP. The pre-gel solution was then pipetted into a mold, covered with an ion-permeable membrane, and submerged in a 3 wt% $CaCl_2$ solution to crosslink.

NIR sensor electronics and readout software. The optical readout electronics comprising a small PCB board (3 cm × 2 cm) with the driving circuits of two photodetectors (PDs), two light-emitting diodes (LEDs), and a USB interface, along with its driver and application software, were donated by Mespere Lifesciences, Inc. (Waterloo, ON, Canada). The LEDs and PDs operate at wavelengths of 660 nm and 905 nm. The optical electronics are part of medical devices that have received regulatory approval for human use in Canada and the United States of America.

Enclosure for device integration. An enclosure was designed using SolidWorks and then 3D printed via Direct Light Processing (DLP) using Acrylonitrile Butadiene Styrene (ABS) resin (Photon Mono 4K, Anycubic, Shenzhen, China). The enclosure consists of two parts: one part houses the optical readout electronics, which is isolated from the second part containing hydrogel chambers, using a transparent waterproof double-sided adhesive.

Optical tissue phantom. Pro-Jet™ 900 NP dye (4 mg) was weighed out in a scintillation vial, then 20 mL ethanol was added, and the mixture was sonicated for 30 min to make a 0.2 mg/mL dye solution. TiO_2 powder (2 g) was then added to the Pro-Jet™ 900 NP solution (20 mL) and sonicated for 30 min. A high-speed mixer at 1200 RPM was used to mix Castin' Craft Clear Polyester Resin (2 L) with the TiO_2/Pro-Jet™ 900 NP solution (20 mL), and followed by the addition of Castin' Craft® Clear Catalyst (50 drops/L). The mixture was then degassed in a vacuum oven to remove air bubbles and carefully poured into a polyvinyl chloride (PVC) mold with a diameter of 15 cm and height of 10 cm. Further degassing was done to remove any remaining bubbles. The mixture was left 48 h to solidify before the solid-state tissue phantom was released from the mold and polished. The optical phantom possesses optical absorption and scattering properties similar to those of human tissue.

3. Results and Discussions

3.1. Lactate-Sensing Hydrogel

Figure 1a illustrates the construction of the alginate-based hydrogel and its lactate detection mechanism. The hydrogel was created by integrating essential components: TMB-loaded cellulose nanocrystals (CNC), lactate oxidase (LOX), horseradish peroxidase (HRP), and sodium alginate. Alginate was selected for its biocompatibility and capacity to create a durable hydrogel matrix through ionic crosslinking with calcium ions (Ca^2), so ensuring structural integrity and preserving the active components inside the gel. LOX and HRP, encapsulated inside the matrix, are crucial for the enzymatic cascade required for lactate detection. To prevent the leaching of TMB, which is a small molecule, TMB was first loaded onto negatively charged CNC. CNCs are nanoparticles about 20 nm in diameter and 200 nm in length [41]. LOX and HRP are macromolecules with molecular weights of approximately 80 kDa and 40 kDa, respectively, which considerably exceed the sizes of small analytes like lactate (90 Da) and oxygen (32 Da). The alginate-based hydrogel used in this investigation was engineered to entrap the enzymes while allowing the efficient transport of small analyte molecules. The mesh size of the alginate network was adjusted by optimizing the alginate content and crosslinking density during preparation [42,43]. Furthermore, the enzymes LOX and HRP were pre-mixed with alginate prior to gelation. Non-covalent interactions, such as hydrogen bonding and electrostatic interactions between the enzymes and alginate, are expected to be strong. Consequently, the enzymes are further retained through polymer entanglement. No noticeable leakage of LOX and HRP was observed.

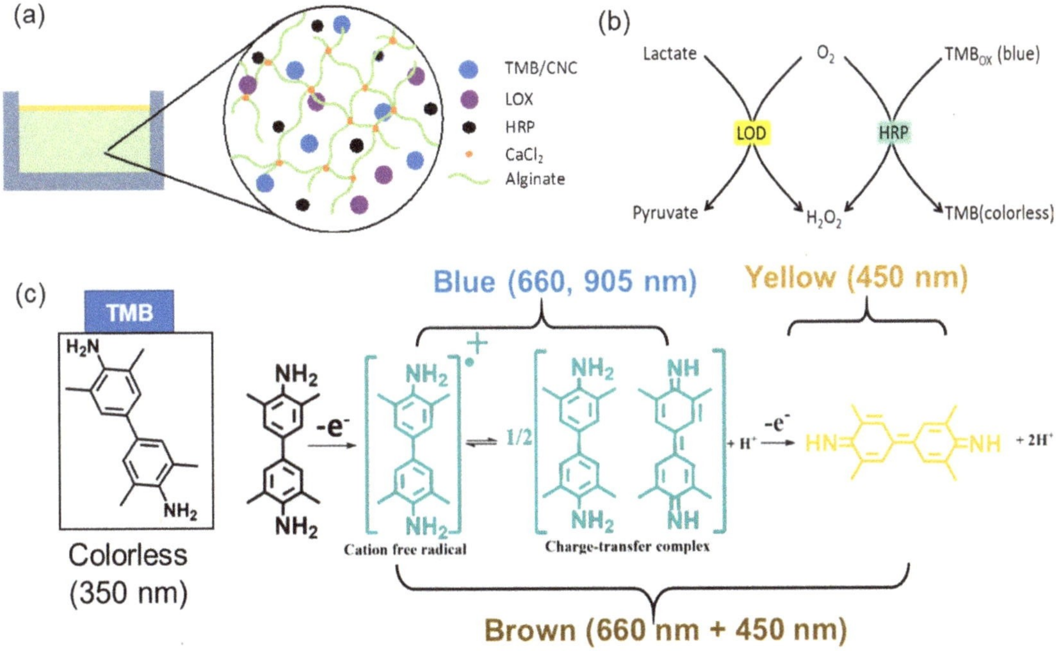

Figure 1. Schematic illustration of (**a**) the lactate-sensing hydrogel composition and (**b**) the colorimetric sensing mechanism. Oxidation of lactate, catalyzed by LOX, generates H_2O_2 as a byproduct; TMB reacts with H_2O_2, catalyzed by HRP, results in a color change. (**c**) Optical absorption peaks of TMB (350 nm) and TMB_{ox} (one-electron oxidation, 660 and 960 nm; two-electron oxidation, 450 nm).

Upon lactate diffusion into the hydrogel, LOX facilitates its oxidation to pyruvate and hydrogen peroxide (H_2O_2) (Figure 1b). The presence of H_2O_2 is essential, serving as a substrate for HRP, which oxidizes TMB from its colorless reduced form to its blue oxidized form (TMB_{ox}), with the intensity of the blue color directly correlating to lactate concentration, facilitating visual or spectrophotometric detection. The colorimetric transition depicted in Figure 1c entails TMB initially absorbing at 350 nm in its reduced state, subsequently undergoing oxidation to generate a blue cation/free radical, which exhibits absorption peaks at 660 nm and 905 nm. This intermediate can undergo further oxidation to form a brown charge-transfer complex, which absorbs at both 660 and 450 nm, and ultimately transitions into a fully oxidized yellow state, absorbing at 450 nm [44]. The specific color transitions, from colorless to blue, brown, and yellow, directly indicate lactate concentration, enabling real-time, non-invasive monitoring of lactate levels in sweat.

3.2. *UV-Vis Spectroscopic Validation of the TMB Assay in Hydrogel*

Unlike the commonly performed liquid-based TMB assay, the rate of reaction in a hydrogel, which is less studied, is dictated by mass transfer. The enzymes, LOX and HRP, and the TMB/CNCs are trapped in the solid phase of the gel. The diffusion rates of lactate and H_2O_2 in the hydrogel are controlled by alginate concentration and the crosslinking density. It is essential to fabricate hydrogels with enzymes (LOX and HRP) and chromogenic dyes (TMB) uniformly distributed in the gel matrix to prevent TMB from leaching out of while maintaining its reactivity, and to ensure a capacity and rate of colorimetric change suitable for monitoring sweat lactate. Before integration into the wearable device, lactate-sensing hydrogels were first formed in a 96-well plate to evaluate their performance using a UV-vis spectrometer.

As shown in Figure 2a, a row of wells was filled with the lactate-sensing hydrogels, dosed with various concentrations of lactate, and photographed every 3 min. For dosing, a drop (50 uL) of lactate solution in PBS was added to each well. The lactate concentrations were 0, 10, 20, 40, 60, 100 mM, respectively. Higher concentrations of lactate were observed to induce more significant and faster colour change. Figure 2b shows the representative absorption spectra and their corresponding colours. In the absence of lactate, the gel appeared transparent with a tint of blue, due to minimal absorption at 660 nm and 905 nm. The data presented in Figure 2b indicate that the rate of change in light absorption at 660 nm is directly proportional to the increase in lactate concentration, as this wavelength corresponds to the blue cation/free radical form of TMB, which is formed during the initial oxidation stage [45]. As lactate concentration increases further, the absorption at 905 nm increases as the hydrogel transitions to a brown charge-transfer complex [46]. This dual-wavelength analysis allows for the precise quantification of lactate concentration based on the spectral changes observed during the oxidation of TMB. The fully oxidized TMB species ultimately absorbs at 450 nm, indicating complete lactate saturation in the system, which should be avoided during real-time sensing. The hydrogel formulation was deemed optimal since the rate of colorimetric change is adequate, with a clear color transition from transparent (no lactate) to bright or dark blue in about 10 min at lactate concentrations up to 60 mM. For reference, sweat lactate levels range from 5 to 20 mM during resting or low activity and can reach up to 50 mM during intense physical activity. This detailed spectrometric analysis confirms that the hydrogel can reliably detect varying lactate concentrations by measuring the rate of change in light absorption at 660 nm and 905 nm, enabling the real-time, non-invasive monitoring of lactate levels.

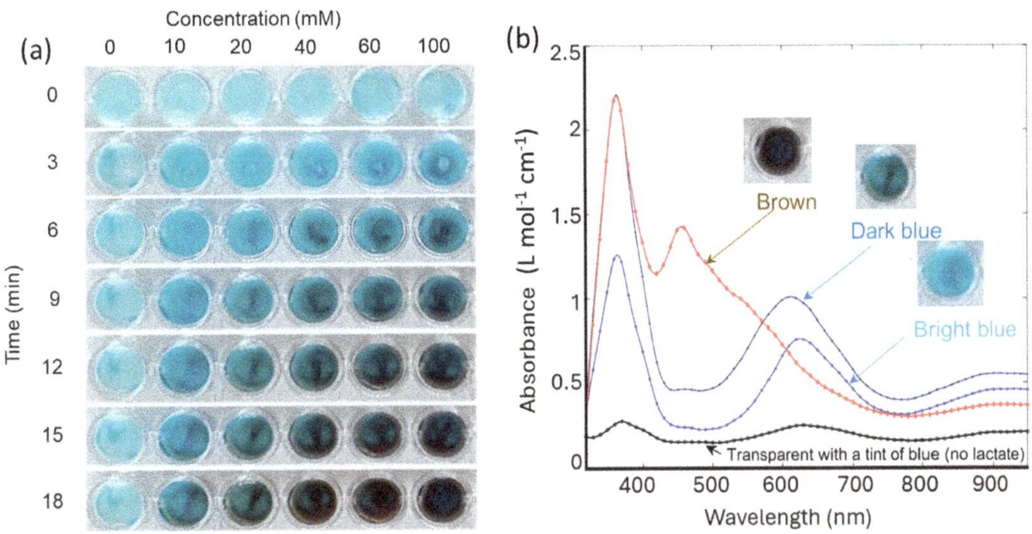

Figure 2. (**a**) Images of gels in a multi-well plate in response to various lactate concentrations at various time points. (**b**) Representative optical absorption spectra of gels in response to lactate.

3.3. Device Design and Fabrication

The device depicted in Figure 3a incorporates modern sensor technologies into a compact, wearable format for real-time, non-invasive, and concurrent monitoring of lactate and tissue oxygenation. It consists of two major functional components: the hydrogel-based lactate sensor and the near-infrared (NIR) sensor, both crucial for monitoring biochemical and physiological alterations in the body [47]. The device enclosure was precisely designed to facilitate secure integration of various components, guaranteeing durability, adaptability, and user-friendliness for ongoing wearable monitoring. This enclosure was 3D printed using a DLP printer, ensuring easy customization and high-resolution fabrication. The device enclosure comprised three essential parts: a top cover, a PCB cover, and a gel tray, all of which were assembled to provide structural integrity and facilitate the placement of the hydrogel sensor.

The NIR sensor comprises two light emitter (LED1 and LED2) and photodetector (PD1 and PD2) pairs, specifically engineered to engage with the hydrogel sensor and tissue for precise data gathering. The top cover and the PCB cover click together to form a waterproof shell around the electronics (Figure 3b). Four openings on the PCB cover allowed light to pass through and were sealed with a transparent waterproof adhesive, which also enabled the secure attachment of the gel tray (Figure 3c). The gel tray with four rectangular chambers was designed specifically to accommodate the lactate-sensing hydrogel, allowing for easy sample change and optimal interaction with sweat. The snug fit of the enclosure with the PCB board, and the rims separating the gel chambers, are critical to prevent light leakage and environmental light interference.

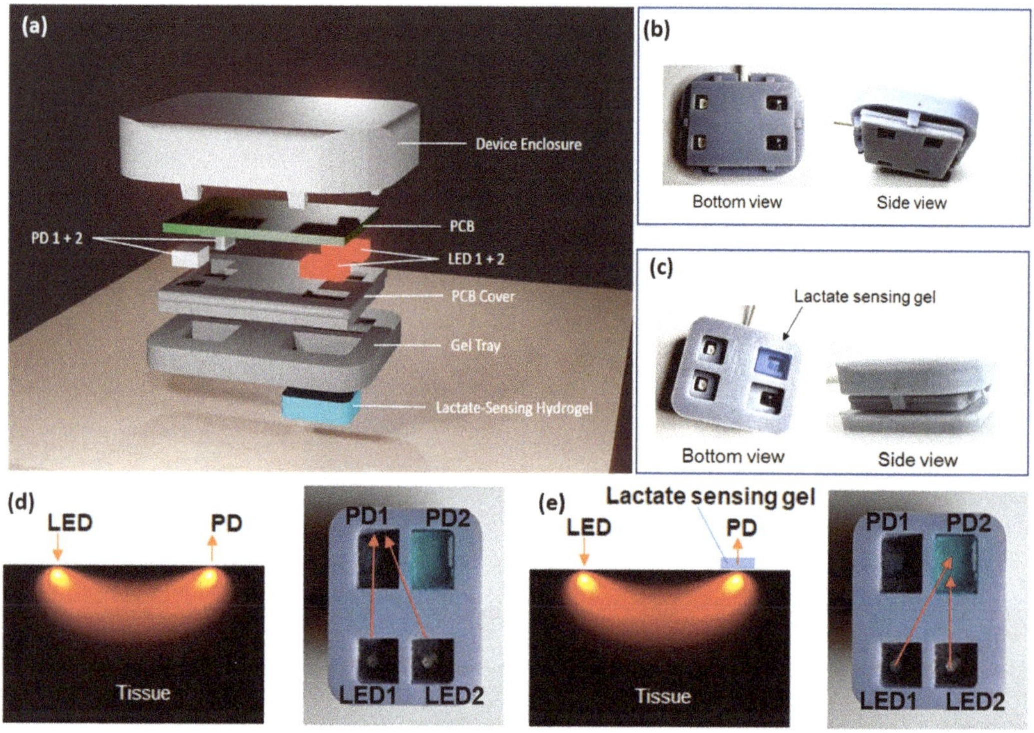

Figure 3. (**a**) An exploded view of the device prototype. Pictures of an assembled wearable device, (**b**) without, and (**c**) with the gel. Optical pathways from LEDs to PDs (**d**) without and (**e**) with the lactate sensing hydrogel on tissue. Left: simulation in tissue; right: bottom view of the wearable device.

3.4. Principle of Operation

The optical pathways between the LED/PD pairs are illustrated in Figure 3d,e. When the device is placed on skin, PD1 measures the attenuation of light emitted by LED1 through skin and body tissue underneath, while PD2 measures the attenuation of light emitted by LED2 through both tissue and the hydrogel. LED1 and LED2 were turned on alternately. The following equations describe the light intensities detected by the PD1 (I_1) and PD2 (I_2):

$$I_1 = I_0 \varnothing(\mu_a, \mu_s) \quad (1)$$

$$I_2 = I_0 \varnothing(\mu_a, \mu_s) \cdot e^{-\mu_{a,g} l_g} \quad (2)$$

where I_0 is the light intensity emitted by the LED, μ_a, μ_s, $\varnothing(\mu_a, \mu_s)$ are the absorption coefficient, scattering coefficient, and overall light attenuation of the tissue underneath the device, and $\mu_{a,g}$ and l_g are the absorption coefficient and thickness of the lactate-sensing hydrogel. The light intensity emitted by LED1 and LED2 is set to be identical, I_0, by the LED driver electronics.

By using the pair of LED1 and PD1 alone (Figure 3d, Equation (1)), without the lactate-sensing hydrogel in the optical pathway, the NIR sensor can accurately detect the tissue oxygenation level (StO_2) and blood volume index (BVI) [48], without being affected by lactate. In human blood, oxygen is primarily carried by the protein hemoglobin. Blood oxygenation (SpO_2) is, thus, defined as the ratio of oxygenated hemoglobin (HbO_2) to total hemoglobin (HbT), while tissue oxygenation (StO_2) is the average of arterial and venous blood SpO_2. HbO_2 and Hb have distinctively different absorption spectra, dominating over other chromophores found in tissue in the near-infrared (NIR) window. This allows for the use of NIR spectroscopy (NIRS) to determine their concentrations [49]. In our device, each LED and PD can emit and detect at both 660 nm and 905 nm. By measuring I_1 at the two wavelengths, the concentrations of both Hb and HbO_2 can be determined. Subsequently, HbT (the sum of Hb and HbO_2) and StO_2 can be calculated. HbT is proportional to the blood volume index (BVI).

Due to the close proximity of the two LED/PD pairs, the light attenuation by tissue, $\varnothing(\mu_a, \mu_s)$, between LED1/PD1 and LED2/PD2, can be assumed to be the same. Therefore, by taking the ratio of I_2 to I_1 (i.e., Equation (2) divided by Equation (1)), the absorption coefficient of the lactate-sensing hydrogel, $\mu_{a,g}$, can be calculated as shown in the equation below:

$$\mu_{a,g} = -\ln(I_2/I_1)/l_g, \tag{3}$$

where the hydrogel thickness (l_g) is molded to be 2 mm. Since $\mu_{a,g}$ is linearly proportional to the amount of TMB_{ox} generated in the sensing gel in response to lactate, we were able to quantitatively measure lactate concentration, without interference from the light attenuation by the skin and tissue. The capability of our device to monitor both a biochemical indicator (lactate) in sweat and physiological parameters (StO_2 and BVI) in tissue concurrently, without cross-talk, provides a comprehensive understanding of hypoxia and metabolic stress in real time.

3.5. Device Validation on a Solid-State Tissue Phantom

Once assembled, the integrated device was first tested on a phantom with tissue-mimicking optical properties to characterize its performance. This phantom mimicked the optical characteristics of human skin and tissue, ensuring that the device's measurements of light attenuation, tissue oxygenation, and lactate detection were reliable before proceeding to human studies. This initial characterization phase was critical in validating the device's ability to simultaneously monitor sweat lactate and tissue oxygenation under controlled conditions.

Figure 4a shows the device positioned on the top of a cylindrical optical phantom. The dimensions of the phantom are sufficiently large that its light absorption and scattering can be assumed to occur within a homogeneous tissue without boundaries. To mimic sweat on skin, a small volume (50 μL) of lactate solution in PBS (pH 6.0) was dropped onto the phantom before placing the integrated device on top. Figure 4b shows the real-time readout of the device in response to different lactate concentrations (0 mM, 10 mM, 40 mM, and 80 mM). With the lactate concentration increases, the normalized absorption markedly rises more rapidly over time and saturates at a higher value. This behavior is anticipated, as elevated lactate concentrations induce faster generation and accumulation of TMB_{ox} in the hydrogel, leading to a rapid increase in its optical absorption at both 660 nm and 905 nm. We took use of both wavelengths to eliminate background noise and baseline fluctuations. While responding significantly to lactate, the device showed highly stable concurrent readings of tissue oxygenation levels (StO_2) and BVI at all lactate concentrations (Figure 4c,d). This stability is crucial for guaranteeing that the integrated device can precisely and simultaneously monitor multiple parameters (sweat lactate, tissue oxygenation, and tissue blood volume) without cross-sensitivity between the results, making our device first of its kind and potentially addressing critical challenges in applications such as exercise physiology monitoring and sepsis management.

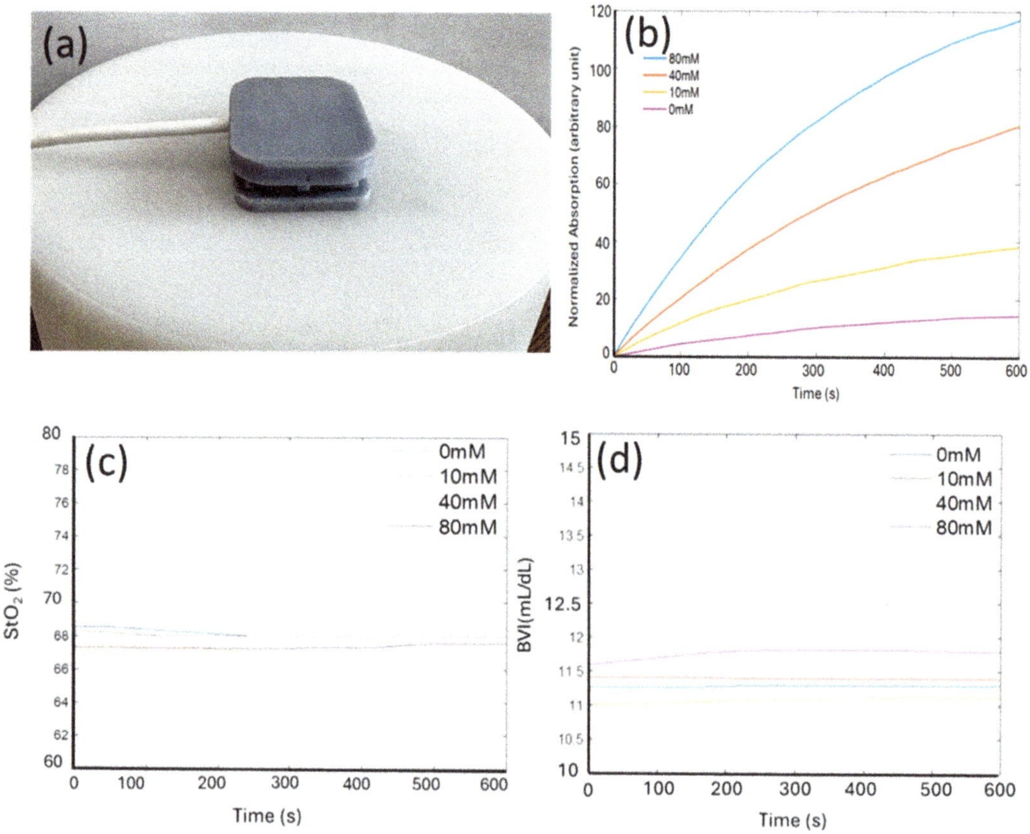

Figure 4. (**a**) Picture of the device placed on top of an optical phantom with optical properties mimicking those of human tissue (StO_2 = 68%). (**b**) Normalized absorption of the lactate sensing hydrogels, extracted from the optical readout, in response to various lactate solutions (0, 10, 40, 80 mM) in phosphate buffer (pH 6.0) respectively. (**c**,**d**) Stable StO_2 and BVI reading regardless of lactate concentration.

3.6. On-Body Measurements of Sweat Lactate

The integrated device's ability to monitor sweat lactate under aerobic and anaerobic conditions was evaluated. Figure 5a shows the experimental setup, in which the device was worn on the left thigh of a researcher using a medical-grade adhesive, while exercising on a stationary bike. The optical electronics are part of medical devices that have received regulatory approval for human wear and the alginate-based hydrogel is biocompatible, ensuring no harm to human skin or tissue. The device was connected to a laptop via a flexible cable, plugged into its USB port.

The researcher followed a two-phase exercise protocol that included both aerobic and anaerobic exercises. The aerobic phase involved light-intensity exercise to induce moderate perspiration. This was followed by the anaerobic exercise phase, which consisted of a 10-min graded incremental exercise test (GXT) on a stationary bike [50]. In the GXT protocol, the starting exercise intensity level (W_{start}) was set at one-fourth of the maximum exercise intensity (W_{max}), expressed as $W_{start} = W_{max}/4$. The exercise intensity increased incrementally each minute, with the incremental change (ΔW) calculated as $\Delta W = (W_{max} - W_{start})/8$. The exercise intensity at any given time t during the test was defined by the equation $W(t) = W_{start} + \Delta W \cdot t$. The researcher continued this incremental exercise until reaching the maximum intensity level (W_{max}).

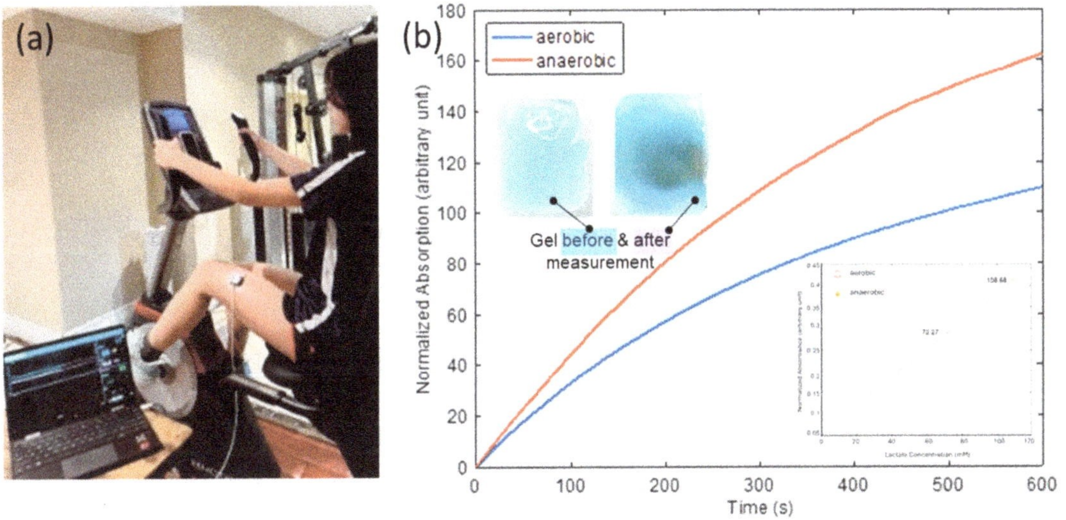

Figure 5. (**a**) Picture of the test setup showing the device worn on the researcher's left thigh while on a stational bike, with the laptop running data collection software. (**b**) Normalized absorption vs. time response curves for the aerobic and anaerobic conditions. Inset-left up: appearance of gel before and after the test; inset-right down: lactate detection calibration curve.

The recorded normalized optical absorption curves are shown in Figure 5b. The starting times (t = 0) on the curves were immediately after the completion of the aerobic and anaerobic phases, respectively. The device's response was markedly more pronounced under anaerobic conditions, as evidenced by the steeper increase in absorption. This was expected, as lactate generation is known to elevate in anaerobic conditions, presumably due to heightened glycolysis in the absence of oxygen. This result indicates that our device can differentiate the aerobic condition, with ample oxygen availability, and anaerobic conditions, characterized by restricted oxygen levels, by monitoring sweat lactate. The inset in the upper left of Figure 5b shows images of the lactate-sensing hydrogel before the device was placed on the researcher and after it was removed following the completion of the whole two-phase exercise protocol. Prior to lactate exposure, the hydrogel exhibited a lighter hue, however, following the test, it becomes more saturated, signifying its interaction with lactate over time. This visual alteration aligns with the absorption data, substantiating the notion that lactate concentration directly influences the optical absorption of the gel.

Furthermore, the quantitative determination of lactate concentration was attempted. A calibration curve was generated using the experimental data gathered on the optical phantom. The initial slopes (first 100 s) of the response curves in Figure 4b were found to be linearly proportional to lactate concentration, as shown in the lower right inset of Figure 5b. Calibration curves were established using a standardized volume of 50 μL, as this ensures uniform and complete coverage of the sensing gel surface. Note that the gels were fully hydrated, therefore, no swelling is expected upon contact with sweat. Equilibrium-driven lactate diffusion into the sensing gel is anticipated, rather than a net inflow of sweat. In practical situations, the initial rate of response was used to determine lactate concentration, regardless of sweat volume, assuming that sufficient sweat uniformly covers the surface of the sensing hydrogel, as is the case during the generation of calibration curves. The design of the system inherently addresses normalization against perspiration volume, as the colorimetric signal produced depends on lactate concentration rather than the absolute volume of sweat absorbed. The hydrogel matrix's capacity and geometry were specifically designed to maintain consistent enzymatic reaction dynamics and diffusion over time, thus mitigating the effects of fluctuating perspiration volumes within a practical range.

By extracting the slopes of the aerobic and anaerobic curves in the first 100 s, sweat lactate levels were determined to be 72 mM and 109 mM after aerobic and anaerobic exercises, respectively, validating the correlation between elevated sweat lactate level and cardio hypoxia. This experimental setup and our results demonstrated the device's practical applicability in physical activity, emulating real-world scenarios like fitness monitoring or clinical diagnostics.

3.7. On-Body Continuous Monitoring of Multiple Parameters

Figure 6a depicts an experimental setup where a researcher was seated within a sauna tent to stimulate perspiration. A pressure cuff affixed to the upper right arm was used to induce localized and temporary hypoxia in the forearm (Figure 6b) in a controlled manner, while the device concurrently measured local sweat lactate, tissue oxygenation and tissue blood volume (Figure 6c–e). The pressure-cuff protocol is a clinically safe and well-established method for creating blood flow restriction [51].

Figure 6c shows the normalized absorption by the lactate-sensing hydrogel over time. A distinct rise in lactate was observed after the onset of pressure at around 200 s, along with multiple lactate spikes during the pressure-on period, indicated by red arrows in Figure 6c, where steep rises in optical absorption were recorded. Concurrently, the device captured the drastic decrease and recovery of StO_2 in response to the pressure being applied and released, respectively (Figure 6d). Figure 6e shows that tissue blood volume increased initially in response to the onset of pressure, and then gradually decreased to normal. Our results are in excellent agreement with the physiological changes expected during blood flow restriction exercise using a pressure cuff. The initial increase in tissue blood volume is due to the sudden full restriction of venous outflow and partial restriction of arterial inflow. The blood inflow and outflow were later balanced. The applied pressure created local hypoxia in the skin and tissue underneath the device, as confirmed by the measured StO_2 levels. Sweat lactate is produced by local metabolism in sweat glands and surrounding tissues. Consequently, sweat lactate spikes were observed coinciding with the drop in StO_2 levels, indicating local hypoxia. While tissue oxygenation recovered immediately after the release of pressure, elevated sweat lactate production persisted for about 5 min after the release of pressure, at which point the optical absorption of the hydrogel plateaued. It indicates that sweat lactate is a good indicator of cellular hypoxia, which is related to tissue oxygenation but distinctly different physiologically.

Table 1 presents a comparative examination of the analytical performance of modern lactate detection systems across different sensor types, detection methodologies, and specimen kinds. Electrochemical sensors, specifically amperometric and voltametric techniques, prevail in the domain, with detection ranges extending from 0 to 1000 mM in blood serum and more focused ranges of 0 to 220 mM in sweat. These sensors are recognized for their swift response times and comparatively low expense. Nonetheless, they are often constrained by specificity, as certain sensors, particularly those employed for brain and peripheral tissue [52], demonstrate restricted linear ranges (0 to 3 mM), potentially affecting their adaptability in various physiological contexts [53]. Colorimetric methods, including those utilizing organic electrochemical transistors (OECT) and colorimetric combined with NIR techniques [54], demonstrate potential with reduced detection ranges, particularly in sweat (0–1 mM for OECT and 0–80 mM for the current study), underscoring their applicability for wearable non-invasive monitoring. Chemical detection methods, such as HILIC-ELSD and UV-Vis [55], provide commendable sensitivity and specificity; however, they are constrained by intricate sample preparation and elevated detection limits, generally in the microgram per milliliter (μg/mL) range, rendering them less appropriate for continuous real-time monitoring [56]. The present study's integrated colorimetric and NIR detection technique provides a harmonious method, merging the sensitivity of NIR with the simplicity of colorimetric alterations, facilitating efficient detection in sweat and presenting a feasible solution for non-invasive lactate monitoring. The comparative research reveals the trade-offs among sensitivity, detection range, and application in various physiological fluids, highlighting the possibility for future developments in wearable, non-invasive lactate detection.

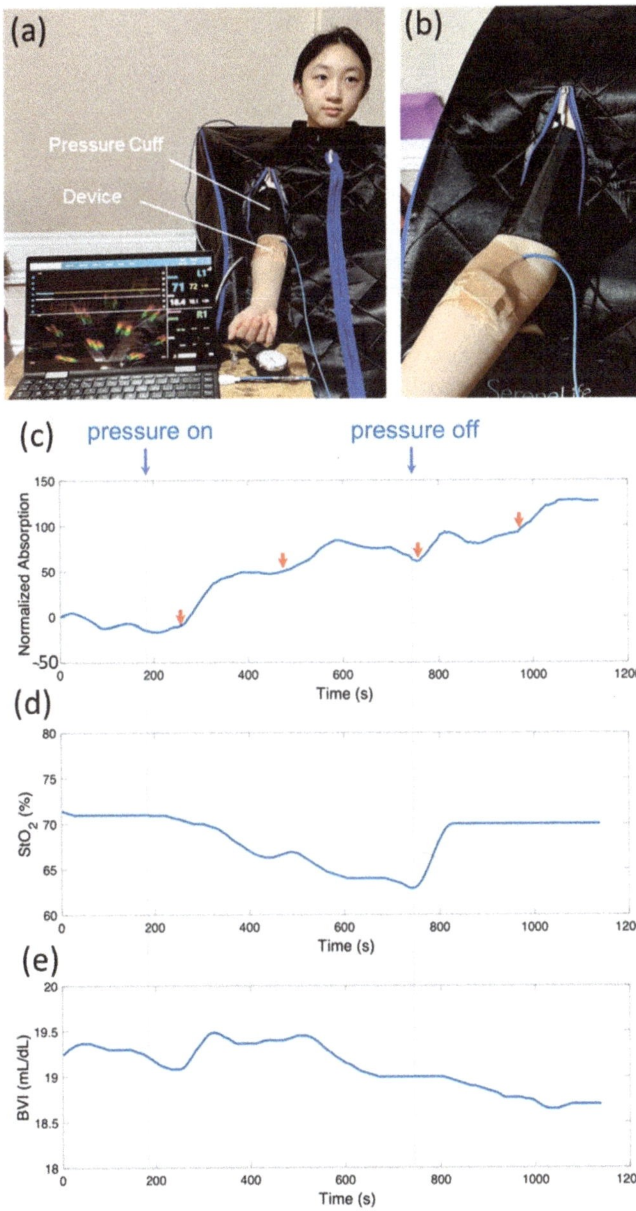

Figure 6. (**a**) Picture of a student researcher sitting in a sauna tent to stimulate sweating while a pressure cuff was used to create temporary hypoxia in the forearm. (**b**) A zoomed-in view of the pressure cuff worn on the upper arm and the wearable device on the forearm. (**c**–**e**) Simultaneous monitoring of sweat lactate production (top panel), tissue oxygenation level (middle panel), and BVI (bottom panel).

Table 1. Comparative analysis of the analytical performance of contemporary lactate detection devices.

Sensor Type	Detection Method	Linear Range (mM)	Specimen	Ref.
Electrochemical	Amperometric	0–1000	Blood serum	[53]
Electrochemical	Amperometric	0–3	Brain and peripheral tissue	[57]
Electrochemical	Voltametric	10–250	Serum	[58]
Electrochemical	Amperometric	1–220	Sweat	[52]
Electrochemical	mperometric	1–50		[59]
OECT	Colorimetric	0–1	Sweat	[54]
Electrochemical	Amperometric	0–24	Sweat	[60]
Chemical	HILIC-ELSD	20–400 μg/mL	Human milk	[55]
Chemical	UV-Vis	0.1–1.7	PBS	[56]
Chemical	Colorimetric + NIR	0–80	Sweat	This work

4. Conclusions

This study presented a comprehensive wearable system that can concurrently detect sweat lactate, tissue oxygenation, and tissue blood volume in a non-invasive and continuous fashion. This apparatus amalgamates two contemporary sensor technologies: a hydrogel-based colorimetric assay and a near-infrared sensor, within a compact design. The optical electronics within the gadget fulfill two functions: assessing tissue oxygenation and delivering a quantitative output of the colorimetric assay. The hydrogel-based colorimetric assay was initially assessed using UV-vis spectroscopy to adjust its composition for enzyme (LOX and HRP), chromogenic dye (TMB), and polymer (alginate) concentrations. The hydrogel exhibited sensitivity, biocompatibility, and stability. The variation in optical absorbance at 660 nm and 905 nm was directly related to sweat lactate content. The complete gadget was later evaluated on an optical tissue phantom and on a researcher's leg while exercise on a stationary bicycle. The device accurately detected sweat lactate while consistently monitoring tissue oxygenation and distinguishing between aerobic and anaerobic states. The system was ultimately verified for on-body continuous monitoring of sweat lactate, a biomarker of cellular hypoxia, alongside tissue oxygenation and tissue blood volume, which signify tissue blood and oxygen supply, by employing a pressure cuff for blood flow restriction.

Although increased sweat lactate has been recognized as a potential marker for the early identification of hypoxia, facilitating timely intervention to avert substantial tissue damage, it lacks specificity. Besides hypoxia, many variables, including heat stress and dehydration, can also lead to increased sweat lactate levels. The simultaneous presence of increased sweat lactate and inadequate tissue oxygenation can distinctly indicate the initiation of anaerobic metabolism induced by hypoxia. Currently, no technology is available that can concurrently monitor both sweat lactate and tissue oxygenation levels, as integrating these two measurements without interference in a compact, wearable format has proven to be exceedingly difficult. The device examined in this study effectively resolved this difficulty. It possesses numerous potential applications in therapeutic settings, particularly for monitoring patients with conditions marked by compromised oxygen transport or metabolic irregularities. Furthermore, the technology could benefit athletes, astronauts, or individuals exposed to high altitudes, where rapid detection of hypoxia is crucial to prevent adverse health effects.

Author Contributions: Conceptualization, C.C. and X.T.; methodology, C.C. and X.T.; software, C.C. and X.T.; validation, C.C. and X.T.; formal analysis, C.C. and X.T.; investigation, X.T.; resources, X.T.; data curation, C.C.; writing—original draft preparation, C.C., X.T. and S.G.; writing—review and editing, C.C., X.T., S.G. and P.L.; visualization, C.C.; supervision, X.T.; project administration, X.T.; funding acquisition, X.T. All authors have read and agreed to the published version of the manuscript.

Funding: This work is supported by a discovery grant from the Natural Sciences and Engineering Research Council (NSERC) of Canada to Dr. Tang.

Institutional Review Board Statement: Not applicable.

Informed Consent Statement: Not applicable.

Data Availability Statement: Data will be available on reasonable request to the corresponding author.

Conflicts of Interest: The authors declare no conflict of interest.

References

1. Lou, Z.; Wang, L.; Shen, G. Recent advances in smart wearable sensing systems. *Adv. Mater. Technol.* **2018**, *3*, 1800444. [CrossRef]
2. Liu, T.; Inoue, Y.; Shibata, K. Development of a wearable sensor system for quantitative gait analysis. *Measurement* **2009**, *42*, 978–988. [CrossRef]
3. Ates, H.C.; Nguyen, P.Q.; Gonzalez-Macia, L.; Morales-Narváez, E.; Güder, F.; Collins, J.J.; Dincer, C. End-to-end design of wearable sensors. *Nat. Rev. Mater.* **2022**, *7*, 887–907. [CrossRef] [PubMed]
4. Rana, M.; Mittal, V. Wearable sensors for real-time kinematics analysis in sports: A review. *IEEE Sens. J.* **2020**, *21*, 1187–1207. [CrossRef]
5. Yi, J.; Xianyu, Y. Gold nanomaterials-implemented wearable sensors for healthcare applications. *Adv. Funct. Mater.* **2022**, *32*, 2113012. [CrossRef]
6. Nasiri, S.; Khosravani, M.R. Progress and challenges in fabrication of wearable sensors for health monitoring. *Sens. Actuators A Phys.* **2020**, *312*, 112105. [CrossRef]
7. Lee, G.; Wei, Q.; Zhu, Y. Emerging wearable sensors for plant health monitoring. *Adv. Funct. Mater.* **2021**, *31*, 2106475. [CrossRef]
8. Chen, S.; Qi, J.; Fan, S.; Qiao, Z.; Yeo, J.C.; Lim, C.T. Flexible wearable sensors for cardiovascular health monitoring. *Adv. Healthc. Mater.* **2021**, *10*, 2100116. [CrossRef]
9. Xu, J.; Fang, Y.; Chen, J. Wearable biosensors for non-invasive sweat diagnostics. *Biosensors* **2021**, *11*, 245. [CrossRef]
10. Medagedara, M.; Peiris, T.; Wanasekara, N. Review of Recent Advances in Non-invasive, Flexible, Wearable Sweat Monitoring Sensors. *Instrumentation* **2020**, *7*, 36–50.
11. Saha, T.; Del Cano, R.; De la Paz, E.; Sandhu, S.S.; Wang, J. Access and management of sweat for non-invasive biomarker monitoring: A comprehensive review. *Small* **2023**, *19*, 2206064. [CrossRef] [PubMed]
12. Pour, S.R.S.; Calabria, D.; Emamiamin, A.; Lazzarini, E.; Pace, A.; Guardigli, M.; Zangheri, M.; Mirasoli, M. Microfluidic-Based Non-Invasive Wearable Biosensors for Real-Time Monitoring of Sweat Biomarkers. *Biosensors* **2024**, *14*, 29. [CrossRef] [PubMed]
13. Thaveesangsakulthai, I.; Chatdarong, K.; Somboonna, N.; Pombubpa, N.; Palaga, T.; Makmuang, S.; Wongravee, K.; Hoven, V.; Somboon, P.; Torvorapanit, P. A large scale study of portable sweat test sensor for accurate, non-invasive and rapid COVID-19 screening based on volatile compound marker detection. *Sci. Rep.* **2024**, *14*, 20148. [CrossRef] [PubMed]
14. Ghaffari, R.; Yang, D.S.; Kim, J.; Mansour, A.; Wright, J.A., Jr.; Model, J.B.; Wright, D.E.; Rogers, J.A.; Ray, T.R. State of sweat: Emerging wearable systems for real-time, noninvasive sweat sensing and analytics. *ACS Sens.* **2021**, *6*, 2787–2801. [CrossRef]
15. Morros, S.C.; Garcia-Rey, S.; Álvarez-Herms, J.; Basabe-Desmonts, L.; Lopez, F.B. A physiological perspective of the relevance of sweat biomarkers and their detection by wearable microfluidic technology: A Review. *Anal. Chim. Acta* **2024**, *1327*, 342988. [CrossRef]
16. Qiao, Y.; Qiao, L.; Chen, Z.; Liu, B.; Gao, L.; Zhang, L. Wearable sensor for continuous sweat biomarker monitoring. *Chemosensors* **2022**, *10*, 273. [CrossRef]
17. Ibrahim, N.F.A.; Sabani, N.; Johari, S.; Manaf, A.A.; Wahab, A.A.; Zakaria, Z.; Noor, A.M. A comprehensive review of the recent developments in wearable sweat-sensing devices. *Sensors* **2022**, *22*, 7670. [CrossRef]
18. Luo, T.-T.; Sun, Z.-H.; Li, C.-X.; Feng, J.-L.; Xiao, Z.-X.; Li, W.-D. Monitor for lactate in perspiration. *J. Physiol. Sci.* **2021**, *71*, 26. [CrossRef]
19. Promphet, N.; Rattanawaleedirojn, P.; Siralertmukul, K.; Soatthiyanon, N.; Potiyaraj, P.; Thanawattano, C.; Hinestroza, J.P.; Rodthongkum, N. Non-invasive textile based colorimetric sensor for the simultaneous detection of sweat pH and lactate. *Talanta* **2019**, *192*, 424–430. [CrossRef]
20. Chen, M.-M.; Cheng, S.-B.; Ji, K.; Gao, J.; Liu, Y.-L.; Wen, W.; Zhang, X.; Wang, S.; Huang, W.-H. Construction of a flexible electrochemiluminescence platform for sweat detection. *Chem. Sci.* **2019**, *10*, 6295–6303. [CrossRef]
21. Zhang, Z.; Wang, B.; Nie, Q.; Luo, Q.; Gong, H. Portable muscle oxygenation monitor based on near infrared spectroscopy. *Front. Optoelectron. China* **2009**, *2*, 248–252. [CrossRef]

22. Walenta, S.; Wetterling, M.; Lehrke, M.; Schwickert, G.; Sundfør, K.; Rofstad, E.K.; Mueller-Klieser, W. High lactate levels predict likelihood of metastases, tumor recurrence, and restricted patient survival in human cervical cancers. *Cancer Res.* **2000**, *60*, 916–921. [PubMed]
23. Cureton, E.L.; Kwan, R.O.; Dozier, K.C.; Sadjadi, J.; Pal, J.D.; Victorino, G.P. A different view of lactate in trauma patients: Protecting the injured brain. *J. Surg. Res.* **2010**, *159*, 468–473. [CrossRef] [PubMed]
24. Derbyshire, P.J.; Barr, H.; Davis, F.; Higson, S.P. Lactate in human sweat: A critical review of research to the present day. *J. Physiol. Sci.* **2012**, *62*, 429–440. [CrossRef] [PubMed]
25. Buono, M.J.; Lee, N.V.; Miller, P.W. The relationship between exercise intensity and the sweat lactate excretion rate. *J. Physiol. Sci.* **2010**, *60*, 103–107. [CrossRef]
26. Anastasova, S.; Crewther, B.; Bembnowicz, P.; Curto, V.; Ip, H.M.; Rosa, B.; Yang, G.-Z. A wearable multisensing patch for continuous sweat monitoring. *Biosens. Bioelectron.* **2017**, *93*, 139–145. [CrossRef] [PubMed]
27. Mitsubayashi, K.; Suzuki, M.; Tamiya, E.; Karube, I. Analysis of metabolites in sweat as a measure of physical condition. *Anal. Chim. Acta* **1994**, *289*, 27–34. [CrossRef]
28. Saha, T.; Fang, J.; Mukherjee, S.; Dickey, M.D.; Velev, O.D. Wearable osmotic-capillary patch for prolonged sweat harvesting and sensing. *ACS Appl. Mater. Interfaces* **2021**, *13*, 8071–8081. [CrossRef] [PubMed]
29. Ohkuwa, T.; Tsukamoto, K.; Yamai, K.; Itoh, H.; Yamazaki, Y.; Tsuda, T. The relationship between exercise intensity and lactate concentration on the skin surface. *Int. J. Biomed. Sci. IJBS* **2009**, *5*, 23. [CrossRef]
30. Meyer, F.; Laitano, O.; Bar-Or, O.; McDougall, D.; Heigenhauser, G. Effect of age and gender on sweat lactate and ammonia concentrations during exercise in the heat. *Braz. J. Med. Biol. Res.* **2007**, *40*, 135–143. [CrossRef]
31. Green, J.; Bishop, P.; Muir, I.; Lomax, R. Gender differences in sweat lactate. *Eur. J. Appl. Physiol.* **2000**, *82*, 230–235. [CrossRef] [PubMed]
32. Patterson, M.J.; Galloway, S.D.; Nimmo, M.A. Variations in regional sweat composition in normal human males. *Exp. Physiol.* **2000**, *85*, 869–875. [CrossRef] [PubMed]
33. Goodwin, M.L.; Harris, J.E.; Hernández, A.; Gladden, L.B. Blood lactate measurements and analysis during exercise: A guide for clinicians. *J. Diabetes Sci. Technol.* **2007**, *1*, 558–569. [CrossRef] [PubMed]
34. Alam, F.; RoyChoudhury, S.; Jalal, A.H.; Umasankar, Y.; Forouzanfar, S.; Akter, N.; Bhansali, S.; Pala, N. Lactate biosensing: The emerging point-of-care and personal health monitoring. *Biosens. Bioelectron.* **2018**, *117*, 818–829. [CrossRef] [PubMed]
35. Sakharov, D.; Shkurnikov, M.; Vagin, M.Y.; Yashina, E.; Karyakin, A.; Tonevitsky, A. Relationship between lactate concentrations in active muscle sweat and whole blood. *Bull. Exp. Biol. Med.* **2010**, *150*, 83. [CrossRef] [PubMed]
36. Karpova, E.V.; Laptev, A.I.; Andreev, E.A.; Karyakina, E.E.; Karyakin, A.A. Relationship between sweat and blood lactate levels during exhaustive physical exercise. *ChemElectroChem* **2020**, *7*, 191–194. [CrossRef]
37. Choi, J.; Bandodkar, A.J.; Reeder, J.T.; Ray, T.R.; Turnquist, A.; Kim, S.B.; Nyberg, N.; Hourlier-Fargette, A.; Model, J.B.; Aranyosi, A.J. Soft, skin-integrated multifunctional microfluidic systems for accurate colorimetric analysis of sweat biomarkers and temperature. *ACS Sens.* **2019**, *4*, 379–388. [CrossRef]
38. Choi, J.; Kang, D.; Han, S.; Kim, S.B.; Rogers, J.A. Thin, soft, skin-mounted microfluidic networks with capillary bursting valves for chrono-sampling of sweat. *Adv. Healthc. Mater.* **2017**, *6*, 1601355. [CrossRef]
39. Baysal, G.; Önder, S.; Göcek, İ.; Trabzon, L.; Kızıl, H.; Kök, F.N.; Kayaoğlu, B.K. Microfluidic device on a nonwoven fabric: A potential biosensor for lactate detection. *Text. Res. J.* **2014**, *84*, 1729–1741. [CrossRef]
40. Koh, A.; Kang, D.; Xue, Y.; Lee, S.; Pielak, R.M.; Kim, J.; Hwang, T.; Min, S.; Banks, A.; Bastien, P. A soft, wearable microfluidic device for the capture, storage, and colorimetric sensing of sweat. *Sci. Transl. Med.* **2016**, *8*, 366ra165. [CrossRef] [PubMed]
41. Wu, Y.; Lin, Z.Y.W.; Wenger, A.C.; Tam, K.C.; Tang, X.S. 3D bioprinting of liver-mimetic construct with alginate/cellulose nanocrystal hybrid bioink. *Bioprinting* **2018**, *9*, 1–6. [CrossRef]
42. Zhang, X.; Wang, K.; Hu, J.; Zhang, Y.; Dai, Y.; Xia, F. Role of a high calcium ion content in extending the properties of alginate dual-crosslinked hydrogels. *J. Mater. Chem. A* **2020**, *8*, 25390–25401. [CrossRef]
43. Jang, J.; Seol, Y.-J.; Kim, H.J.; Kundu, J.; Kim, S.W.; Cho, D.-W. Effects of alginate hydrogel cross-linking density on mechanical and biological behaviors for tissue engineering. *J. Mech. Behav. Biomed. Mater.* **2014**, *37*, 69–77. [CrossRef] [PubMed]
44. Fu, G.; Hou, R.; Mou, X.; Li, X. Integration and quantitative visualization of 3,3′,5,5′-tetramethylbenzidine-probed enzyme-linked immunosorbent assay-like signals in a photothermal bar-chart microfluidic chip for multiplexed immunosensing. *Anal. Chem.* **2021**, *93*, 15105–15114. [CrossRef] [PubMed]
45. Vaquer, A.; Barón, E.; de la Rica, R. Wearable analytical platform with enzyme-modulated dynamic range for the simultaneous colorimetric detection of sweat volume and sweat biomarkers. *ACS Sens.* **2020**, *6*, 130–136. [CrossRef]
46. Li, F.; Jiang, J.; Shen, N.; Peng, H.; Luo, Y.; Li, N.; Huang, L.; Lu, Y.; Liu, L.; Li, B. Flexible microfluidic colorimetric detection chip integrated with ABTS·+ and Co@ MnO_2 nanozyme catalyzed TMB reaction systems for bio-enzyme free detection of sweat uric acid. *Anal. Chim. Acta* **2024**, *1299*, 342453. [CrossRef]
47. Lee, H.K.; Yang, Y.J.; Koirala, G.R.; Oh, S.; Kim, T.-i. From Lab to Wearables: Innovations in Multifunctional Hydrogel Chemistry for Next-Generation Bioelectronic Devices. *Biomaterials* **2024**, *310*, 122632. [CrossRef]
48. Murkin, J.M.; Arango, M. Near-infrared spectroscopy as an index of brain and tissue oxygenation. *Br. J. Anaesth.* **2009**, *103* (Suppl. 1), i3–i13. [CrossRef]

49. Scheeren, T.; Schober, P.; Schwarte, L. Monitoring tissue oxygenation by near infrared spectroscopy (NIRS): Background and current applications. *J. Clin. Monit. Comput.* **2012**, *26*, 279–287. [CrossRef]
50. Jamnick, N.A.; Botella, J.; Pyne, D.B.; Bishop, D.J. Manipulating graded exercise test variables affects the validity of the lactate threshold and V˙ O 2 peak. *PLoS ONE* **2018**, *13*, e0199794. [CrossRef]
51. Wernbom, M.; Schoenfeld, B.J.; Paulsen, G.; Bjørnsen, T.; Cumming, K.T.; Aagaard, P.; Clark, B.C.; Raastad, T. Commentary: Can blood flow restricted exercise cause muscle damage? Commentary on blood flow restriction exercise: Considerations of methodology, application, and safety. *Front. Physiol.* **2020**, *11*, 243. [CrossRef]
52. Zhou, L. Molecularly Imprinted Sensor based on Ag-Au NPs/SPCE for Lactate Determination in Sweat for Healthcare and Sport Monitoring. *Int. J. Electrochem. Sci.* **2021**, *16*, 211043. [CrossRef]
53. Regiart, M.; Ledo, A.; Fernandes, E.; Messina, G.A.; Brett, C.M.; Bertotti, M.; Barbosa, R.M. Highly sensitive and selective nanostructured microbiosensors for glucose and lactate simultaneous measurements in blood serum and in vivo in brain tissue. *Biosens. Bioelectron.* **2022**, *199*, 113874. [CrossRef] [PubMed]
54. Currano, L.J.; Sage, F.C.; Hagedon, M.; Hamilton, L.; Patrone, J.; Gerasopoulos, K. Wearable Sensor System for Detection of Lactate in Sweat. *Sci. Rep.* **2018**, *8*, 15890. [CrossRef]
55. Ten-Doménech, I.; Martínez-Pérez-Cejuela, H.; Lerma-García, M.J.; Simó-Alfonso, E.F.; Herrero-Martínez, J.M. Molecularly imprinted polymers for selective solid-phase extraction of phospholipids from human milk samples. *Microchim. Acta* **2017**, *184*, 3389–3397. [CrossRef]
56. Mustafa, Y.L.; Leese, H.S. Fabrication of a Lactate-Specific Molecularly Imprinted Polymer toward Disease Detection. *ACS Omega* **2023**, *8*, 8732–8742. [CrossRef] [PubMed]
57. Booth, M.A.; Gowers, S.A.; Hersey, M.; Samper, I.C.; Park, S.; Anikeeva, P.; Hashemi, P.; Stevens, M.M.; Boutelle, M.G. Fiber-based electrochemical biosensors for monitoring pH and transient neurometabolic lactate. *Anal. Chem.* **2021**, *93*, 6646–6655. [CrossRef] [PubMed]
58. Moussa, F.B.; Achi, F.; Meskher, H.; Henni, A.; Belkhalfa, H. Green one-step reduction approach to prepare rGO@ AgNPs coupled with molecularly imprinted polymer for selective electrochemical detection of lactic acid as a cancer biomarker. *Mater. Chem. Phys.* **2022**, *289*, 126456. [CrossRef]
59. Xuan, X.; Perez-Rafols, C.; Chen, C.; Cuartero, M.; Crespo, G.A. Lactate biosensing for reliable on-body sweat analysis. *ACS Sens.* **2021**, *6*, 2763–2771. [CrossRef] [PubMed]
60. Joly, K.; Gurarie, E.; Sorum, M.S.; Kaczensky, P.; Cameron, M.D.; Jakes, A.F.; Borg, B.L.; Nandintsetseg, D.; Hopcraft, J.G.C.; Buuveibaatar, B. Longest terrestrial migrations and movements around the world. *Sci. Rep.* **2019**, *9*, 15333. [CrossRef]

biosensors

MDPI

Article

Optimization of Paper-Based Alveolar-Mimicking SERS Sensor for High-Sensitivity Detection of Antifungal Agent

Hyunjun Park [1,†], Kyunghwan Chai [2,†], Eugene Park [1,†], Woochang Kim [1], Gayoung Kim [2], Joohyung Park [1,*], Wonseok Lee [3,*] and Jinsung Park [1,4,*]

[1] Department of Biomechatronic Engineering, College of Biotechnology and Bioengineering, Sungkyunkwan University, Suwon 16419, Republic of Korea; guswns1105@g.skku.edu (H.P.); delorean12@g.skku.edu (E.P.); wkim91@skku.edu (W.K.)

[2] Department of Biopharmaceutical Convergence, Sungkyunkwan University, Suwon 16419, Republic of Korea; kyungh2012@g.skku.edu (K.C.); kgy9173@g.skku.edu (G.K.)

[3] Department of Electrical Engineering, Korea National University of Transportation, Chungju 27469, Republic of Korea

[4] Department of MetaBioHealth, Sungkyunkwan University, Suwon 16419, Republic of Korea

* Correspondence: parkjoodori@skku.edu (J.P.); wslee@ut.ac.kr (W.L.); nanojspark@skku.edu (J.P.)

† These authors contributed equally to this work.

Abstract: Crystal violet (CV) is a disinfectant and antifungal agent used in aquaculture that plays a vital role in treating aquatic diseases and sterilizing water. However, its potential for strong toxicity, including carcinogenicity and mutagenicity, upon accumulation in the body raises concerns regarding its safe use. Therefore, there is a growing need for the quantitative detection of CV in its early application stages to ensure human safety. Recently, Raman spectroscopy-based surface-enhanced Raman scattering (SERS) detection research has been actively conducted; consequently, an alveolar-mimicking SERS paper (AMSP) inspired by the structure of the human lungs was developed. The AMSP was optimized through various factors, including paper type, reducing agent, reducing agent concentration, and reaction time. This optimization enhanced the surface area of interaction with the target substances and promoted hotspot formation, resulting in enhanced SERS performance. The substrate exhibited exceptional uniformity, reproducibility, and reliability. CV was successfully detected at a concentration of 1 nM in laboratory settings. Furthermore, the AMSP detected CV at 1 nM in real-world environmental samples, including fish farm water and human serum, confirming its potential as a practical detection and monitoring platform for CV in real-world samples.

Keywords: crystal violet; bioinspired; paper-based; Raman spectroscopy; surface-enhanced Raman scattering (SERS); high sensitivity

1. Introduction

Crystal violet (CV), a dye belonging to the triphenylmethane group, is used in various of chemical, biological, and industrial applications. In the medical field, CV helps control fungal infections and intestinal parasites and serves as an antifungal agent for burn victims [1]. Industrially, CV is employed as an antimicrobial solution to prevent mold growth in poultry feed and as a disinfectant and antifungal agent in aquaculture, where it is used to treat diseases in aquatic animals and sterilize water [2,3]. However, despite its widespread use, CV is highly toxic and has been associated with carcinogenic and mutagenic effects in humans, raising concerns regarding its use in aquaculture [4,5]. Specifically, when exposed to large amounts of CV in the body, it has been shown to accumulate in muscles, kidneys, liver, and other tissues, producing leucocrystal violet (LCV) as a metabolic byproduct [6,7]. Therefore, the quantitative detection of CV in its early application stages is crucial to ensure human safety in both environmental and medical contexts.

Consequently, research is actively focused on developing monitoring platforms to detect trace levels of CV in environmental samples. Over the past few decades, numerous platforms have been developed to address these issues. High-performance liquid chromatography (HPLC) [8,9], liquid chromatography coupled with mass spectrometry (LC-MS) [10,11], UV–visible spectroscopy (UV-vis) [12,13], electrochemical sensors [14–17], and fluorescence analysis [18,19] have been utilized widely in CV detection studies. Among these techniques, LC-MS is the most commonly adopted. However, LC-MS is limited by its complex sample preparation procedures and high sample consumption. Recently, the Raman spectroscopy-based surface-enhanced Raman scattering (SERS) detection platform has attracted considerable attention in the fields of environmental pollutant detection and medical diagnosis due to its exceptional sensitivity and straightforward sampling techniques [20–23]. However, current SERS-based platforms face limitations due to the complex design and lengthy fabrication time required to achieve the uniform arrangements of metal nanostructures essential for the SERS effect [24–27]. To address these challenges, it is essential to develop substrates with numerous nanogaps to maximize the SERS effect while simplifying the fabrication design.

In this study, human lungs and alveoli were used as inspiration to achieve efficient SERS performance. The human lung has a branching structure that divides into bronchi, maximizing the pathways through which air can reach the alveoli, thereby enhancing the efficiency of gas exchange [28,29]. Additionally, the lungs comprise millions of small air sacs called alveoli, dramatically increasing the surface area and, consequently, the efficiency of gas exchange within the body [30,31]. Inspired by this structure, an alveolar-mimicking SERS paper (AMSP) with a simple design was fabricated using cellulose acetate (CA) paper, in which gold nanoparticles were grown and functionalized; this approach was optimized to maximize the interaction area with the target substances and improve hotspot generation, thereby enhancing the SERS effect. The fabricated AMSP was optimized by varying several parameters, including the substrate type, the type and concentration of the reducing agent used for gold nanoparticle growth, and the reaction time. These adjustments were performed to identify the conditions that yielded the most effective SERS performance. This study confirmed the high-sensitivity detection of CV using AMSP. Furthermore, the detection capability of CV was verified in aquaculture water and human serum, where CV contamination is a concern. These findings demonstrate the potential of the AMSP as a sensitive on-site detection platform for CV and other harmful substances in the environment.

2. Experimental Section

2.1. Materials and Reagents

CA paper (Φ = 55 mm) was obtained from Hyundai Micro (Seoul, Repulic of Korea). Cellulose and glass fiber papers (Whatman filter paper) were purchased from Sigma-Aldrich (Burlington, MA, USA). O-methoxyamine hydrochloride (OMH), trisodium citrate (TSC), ascorbic acid (AA), hydroquinone (HQ), gold (III) chloride hydrate ($HAuCl_4$), rhodamine 6G (R6G), 4-aminothiophenol (4-ATP), and human serum were purchased from Sigma-Aldrich. A pH 10 buffer and 99.5% anhydrous ethanol were acquired from Daejung Reagents (Siheung, Republic of Korea). All aqueous solutions were prepared with Millipore deionized water (DI water) with a resistance of 18.2 MΩ at 25 °C. Aquaculture water was obtained from a fish farm at Sungkyunkwan University (Suwon, Repulic of Korea).

2.2. Fabrication of Paper-Based SERS Substrate

The method for fabricating paper-based SERS substrates was refined and modified based on a previous study [32]. A hole puncher (Φ = 6 mm) was utilized to make small pieces of filter paper (cellulose filter paper, cellulose acetate paper, and glass fiber paper). The pieces of the paper were washed with ethanol and rinsed with DI water. The pieces were then laid on a Petri dish and dried in the oven at 60 °C. A reducing agent (20 mM; OMH, TSC, AA, or HQ) was prepared in a pH 10 buffer. The pieces were immersed in

10 mL of the reducing agent solution in a Petri dish. Next, 2 mL of $HAuCl_4$ solution was slowly added. After reacting with the reducing agent, the synthesized nanoparticles were deposited onto the paper substrates, and any remaining solution was carefully removed using a pipette. Subsequently, the final paper-based substrates were thoroughly dried at room temperature (25 °C) before use.

2.3. Optimization for Fabrication of AMSP

To select the optimal paper type, 10 μL drops of 4-ATP at a concentration of 100 μM were dropped to SERS substrates prepared with cellulose, CA, and glass fiber paper. SERS measurements were conducted following the substrates being allowed to dry completely at room temperature.

Next, Raman measurements were conducted under consistent conditions using paper-based SERS substrates fabricated with each reducing agent (OMH, AA, HQ, and TSC) to select and optimize the type of reducing agent. For this, 10 μL drops of 100 μM R6G indicator were applied to each substrate and allowed to dry completely. Then, Raman measurements were performed to compare the Raman intensities of the indicators on each substrate.

OMH solutions ranging from 5 to 100 mM were prepared in a pH 10 buffer to optimize the concentration of OMH, which was selected as the reducing agent, and the reaction time. Pieces of the CA paper were immersed in OMH solutions of various concentrations. A 2 mL portion of a 10 mM $HAuCl_4$ solution was added to each sample, and the reaction was allowed to proceed for 60 min. The reaction time was optimized by reducing $HAuCl_4$ for different times (30–120 min). Subsequently, 10 μL drops of 100 μM R6G were applied to the substrates and allowed to dry completely, then Raman measurements were conducted.

Two methods—drop-casting and dipping—were compared to optimize the reaction conditions for the analyte. For the drop-casting method, sample solutions were applied to the AMSP in volumes of 2, 5, and 10 μL. In the dipping method, the same substrate was immersed in 1 mL of analyte solution. The substrates were allowed to react with the analyte for 30 min, and all the substrates were dried entirely before conducting the Raman measurements.

2.4. SERS Measurements on AMSP

All Raman spectra were collected using a Renishaw inVia Raman microscope (inVia Reflex, Renishaw, Wotton-under-Edge, UK). A 785 nm wavelength laser was focused with a ×20 object lens (Leica DM2700 M, Germany, and Renishaw Centrus Detector, Wotton-under-Edge, UK). SERS spectra were measured using a 785 nm laser with a power of 1.01 mW after 1 s of exposure, five accumulations (optimization), and ten accumulations (performance confirmation). All spectra were acquired in the range of 497–1623 cm^{-1}. Cosmic rays were removed from each spectrum, and the polynomial baseline was subtracted using WiRE 5.1 software.

Different concentrations of R6G aqueous solutions (100 pM–100 μM) were prepared to examine the performance of the SERS substrate. Each AMSP was immersed in 1 mL of the R6G solution for 30 min. SERS measurements were conducted on the dried substrates under each condition. A 785 nm laser was selected to minimize fluorescence interference, as longer wavelengths reduce background fluorescence in SERS measurements [33].

CV was detected using SERS under three conditions: DI water, aquaculture water, and 1% human serum. In the aquaculture condition, the water was filtered twice using a 0.22 μm syringe filter and stored for 12 h to eliminate unnecessary residues.

CV solutions of different concentrations (1 nM–1 mM) were prepared with each solvent (DI water, 1% human serum, and aquaculture water). The substrates were then immersed in 1 mL of the solution for 30 min and thoroughly dried.

2.5. Characterization of AMSP

The morphology of the AMSP was observed using a field-emission scanning electron microscope (FE-SEM; JSM-IT800, JEOL, Tokyo, Japan). SEM images were obtained at 15 kV HV and different magnifications.

Then, the reproducibility and uniformity analyses of AMSP were conducted using 100 μM R6G aqueous solution. Reproducibility analysis involved SERS measurements of the R6G solution in five independently prepared samples. Uniformity analysis involved the measurement of the Raman intensity of R6G at 30 randomly selected locations on the prepared substrate.

3. Results and Discussion

3.1. Fabrication and Detection Strategy with AMSP

The optimization factors and detection strategy for the AMSP used to detect CV are illustrated in Figure 1. The AMSP was designed to mimic the structure of the human alveoli, thereby increasing the surface area available for interaction with air. Gold nanoparticles, known for their rapid and simple synthesis and excellent stability, were chosen as the primary components for this purpose [34]. First, α-cellulose, cellulose acetate, and glass fiber papers were evaluated to determine the appropriate substrate type. Each paper substrate was functionalized with gold nanoparticles and assessed using Raman spectroscopy. Consequently, the SERS substrate based on CA paper showed the highest Raman intensity, and was therefore selected as the optimal substrate. Optimizing the reducing agent in gold nanoparticle synthesis is essential, as each agent's unique reduction potential influences nanoparticle formation, thereby impacting the substrate's SERS performance [35]. Gold nanoparticles were functionalized on the selected CA paper substrate using various substances (OMH, AA, HQ, and TSC) as direct reducing agents in the synthesis of gold nanoparticles. The performance of the resulting substrates was evaluated using 100 μM R6G as the Raman reporter. Variations in gold nanoparticle synthesis were expected due to the differing reduction potentials of the reducing agents. The substrate produced using the OMH-reducing agent exhibited the highest Raman intensity, which led to the selection of OMH as the optimal reducing agent. The subsequent optimization focused on the concentration of the OMH-reducing agent and the reaction time. The reducing agent influenced the synthesis of gold nanoparticles, thereby affecting the SERS performance. Similarly, variations in the concentration and reaction time of the same reducing agent affected nanoparticle formation. Various OMH concentrations and reaction times were selected, and performance evaluations were conducted using Raman spectroscopy.

An optimal AMSP for CV detection was fabricated considering the optimized factors. SEM images demonstrated that the resulting substrate successfully mimicked the structure of human alveoli. The increased optical interaction area with the Raman laser, along with the gold nanoparticles arranged at regular intervals in an alveolar-like structure, maximized the SERS effect. This configuration enhanced the Raman signal of the target molecule, leading to improved detection sensitivity.

Using the AMSP detection mechanism, a strategy for the sensitive detection of CV, an antifungal agent in aquatic systems, was developed. Gold nanoparticles adsorbed onto the fiber bundles of the paper formed hotspots. When the target substance was located in these regions, the SERS effect amplified the Raman signal of the target. The performance of the fabricated AMSP was initially evaluated, followed by an assessment of its ability to detect various concentrations of CV, not only under laboratory conditions but also in aquaculture water and serum samples. The specific Raman spectroscopic peaks of the CV appeared at 1170, 1384, and 1583 cm^{-1}, with the analysis focusing on the peak at 1384 cm^{-1} [36,37].

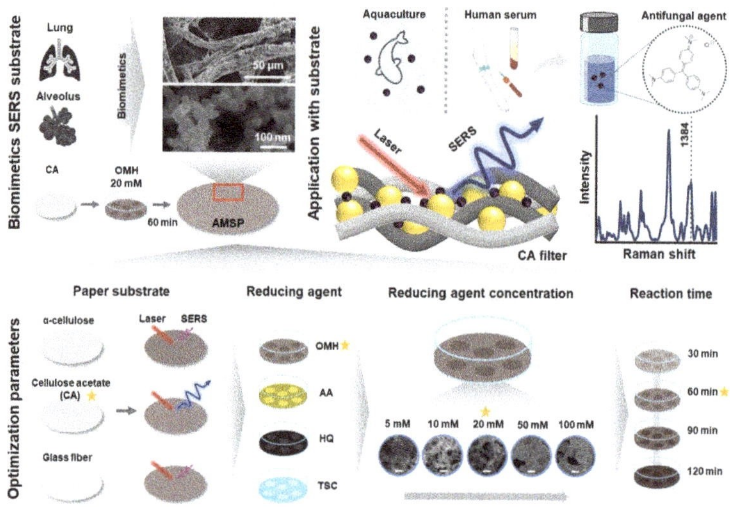

Figure 1. Schematic diagram illustrating the optimization of various parameters for the fabrication of AMSP and their performance evaluation. The optimization parameters included different types of paper, reducing agents, concentrations of reducing agents, and reaction times (Optimization factors are indicated by separate star symbols). Consequently, AMSP, mimicking the structure of lungs and alveoli, was fabricated. These substrates enabled the sensitive detection of CV in various samples. Scale bar in the figure: 100 nm.

3.2. Optimization of AMSP

The AMSP was optimized using two Raman reporters, 4-ATP and R6G (Figure 2). In this experiment, 4-ATP was used to optimize the SERS substrate owing to its high reactivity with amine groups and metal nanoparticles [38]. R6G, a fluorescent dye known to generate strong Raman signals, was used to evaluate its performance on paper-based substrates. Initially, several candidates, including α-cellulose, CA, and glass fiber paper, were assessed to identify the optimal paper substrate for SERS fabrication. Performance was evaluated as outlined in the Materials and Methods section. In the absence of a Raman reporter (control), both the α-cellulose substrate (Figure 2a) and CA substrate (Figure 2b) exhibited similar Raman spectra. However, after the introduction of 4-ATP, the CA substrate exhibited a significantly stronger 4-ATP Raman signal, highlighting its suitability for high-sensitivity applications. Conversely, the glass fiber substrate did not show any 4-ATP Raman signal (Figure 2c). This enhanced performance of the CA substrate can be attributed to its high chemical stability due to acetylation, which not only facilitates the stable physical adsorption of gold nanoparticles but also provides a consistent surface for uniform nanoparticle distribution [39,40]. Unlike CA paper, α-cellulose and glass fibers exhibit relatively high electrostatic properties and irregular surface morphologies, which result in lower and less uniform physical adsorption of nanoparticles. However, glass-fiber substrates, with their limited chemical interaction capabilities, are less effective in promoting the plasmon resonance required for Raman signal enhancement [41,42]. Thus, CA paper, with its acetylated structure, achieves optimal Raman intensity by supporting superior nanoparticle adsorption and plasmonic activity. Consequently, the CA-based substrate exhibited the highest Raman intensity because of its superior nanoparticle adsorption and plasmonic activity.

After confirming that the CA filter paper was the optimal substrate, optimization experiments were conducted to select the most suitable reducing agent for the synthesis of gold nanoparticles. The reducing agents were selected from those commonly used to reduce metal nanoparticles, and the experimental methods are detailed in the Materials and Methods section [32,43–45]. Figure 2d illustrates optical microscopy images of the paper-based substrates prepared with each reducing agent.

Performance evaluation was conducted on each substrate using the Raman reporter R6G at a concentration of 100 μM (Figure 2e). The Raman intensity at the characteristic Raman peak of R6G, 1504 cm^{-1}, was compared (Figure 2f). The substrate prepared using OMH as the reducing agent exhibited the highest Raman performance. This can be attributed to the relatively low reduction potential of TSC (−0.18 V) under the same conditions, which led to incomplete nanoparticle synthesis and the absence of an observable SERS effect [46]. In contrast, the reduction potentials of AA and HQ (0.5 and 0.8 V, respectively) were comparatively high, resulting in the aggregation of nanoparticles, thereby reducing the effectiveness of the SERS signal [47,48]. These results suggest that the varying reduction strengths of the reducing agents influence the surface potentials of the synthesized nanoparticles, which in turn affect their degree of physical adsorption onto paper substrates [49].

Subsequently, optimization experiments were conducted to determine the optimal concentration and reaction time for the selected OMH-reducing agent. The concentration and reaction time of the reducing agent can be considered critical factors in the synthesis of gold nanoparticles. A schematic of the concentration- and reaction-time-dependent optimization experiments is presented in Figure 2g. Performance evaluations for concentration conditions were consistently carried out using 100 μM R6G, and the results are displayed in Figure 2h. The SEM images of the substrates at varying concentrations are shown in Figure S1. These images include the bare CA paper substrate (control, Figure S1a) and the substrates prepared with OMH concentrations of 5, 10, 20, 50, and 100 mM (Figure S1b–f). No gold nanoparticles were observed in the control sample, whereas physical adsorption of gold nanoparticles onto the paper fibers was noticeable in the substrates treated with the reducing agent. The nanoparticles appeared unevenly distributed on the fibers at lower concentrations; this is likely because of the insufficient concentration of the reducing agent, which impeded proper nanoparticle synthesis [50,51]. Conversely, excessive nanoparticle formation led to aggregation at higher concentrations (50 and 100 mM), resulting in nonuniform adsorption. In contrast, the substrate prepared with 20 mM OMH contained uniformly distributed, well-formed gold nanoparticles. Significant differences in the size and uniformity were observed between the nanoparticles synthesized at 20 and 100 mM (Figure S2). Subsequently, the overall Raman spectra for each condition were examined, as demonstrated in Figure S3. Comparing the Raman intensities at the characteristic peak of R6G at 1504 cm^{-1}, the substrate prepared with 20 mM OMH exhibited the highest Raman performance (Figure 2h). This enhanced performance can be attributed to the variation in the gold nanoparticle size, which is influenced by the concentration of the reducing agent, as previously discussed. The observed differences are ascribed to these factors because the SERS performance is maximized through optimized particle size and interparticle gaps.

The reaction times evaluated were 30, 60, 90, and 120 min, representing the duration of the reaction between the reducing agent and $HAuCl_4$. The performance evaluation results for each condition are demonstrated in Figure S4a–d with whole Raman spectra. Figure 2i shows the comparison of Raman intensities at the characteristic peak of R6G at 1504 cm^{-1}, revealing that the substrate with a 60 min reaction time exhibited the highest Raman intensity. Thus, the optimal reaction time for synthesizing gold nanoparticles with the highest SERS efficiency was 60 min. Shorter reaction times resulted in insufficient nanoparticle formation, limiting the SERS effect. Conversely, extending the reaction time beyond 60 min led to increased particle size and aggregation, thus diminishing the enhancement efficiency [52,53]. Based on these optimization assessments, the final optimized substrate for this study was the substrate that reacted for 60 min using a 20 mM OMH-reducing agent on CA paper.

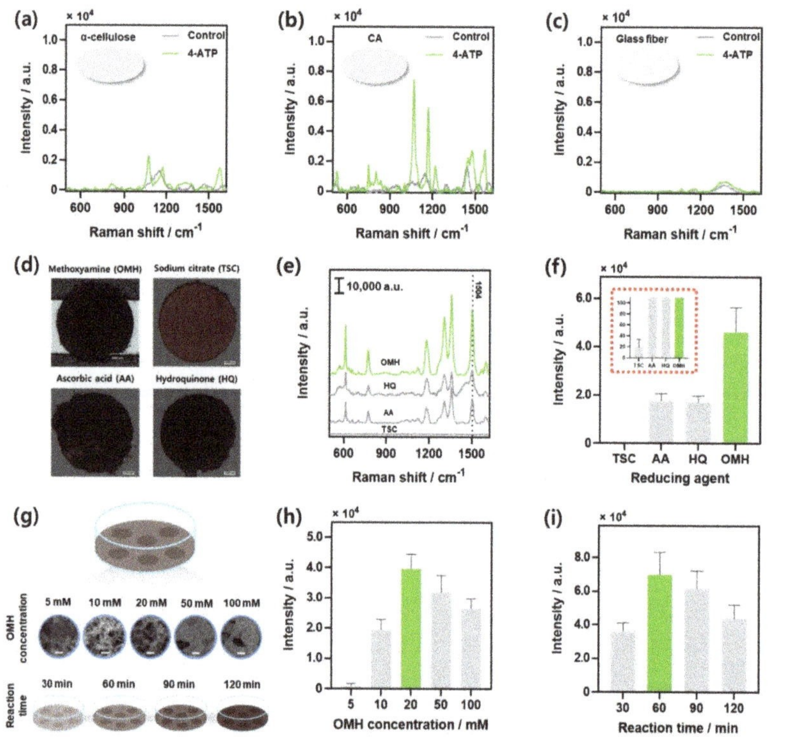

Figure 2. Optimization of paper-based SERS substrates. Raman spectra of 100 μM 4-ATP on SERS substrates made with OMH-reducing agent on different types of paper: (**a**) cellulose, (**b**) CA, and (**c**) glass fiber, respectively. (**d**) Images of SERS substrates prepared on CA paper with different reducing agents (OMH, TSC, AA, HQ). (**e**) Raman spectra of 100 μM R6G reacted on each substrate. (**f**) Comparison of Raman intensities at 1504 cm^{-1} peak of R6G. Red dotted inset image shows data with an adjusted y-axis scale. (**g**) Schematic of optimization process for OMH concentration and reaction time (scale bar: 100 nm). (**h**) Comparison of Raman intensities at 1504 cm^{-1} peak of R6G for AMSPs prepared with varying OMH concentrations. (**i**) Raman intensities at 1504 cm^{-9} peak of R6G for different reaction times using 20 mM OMH-reducing agent.

3.3. Performance Evaluation of Optimized AMSP

Performance evaluations were conducted to assess the suitability of the optimized paper-based substrate as a Raman spectroscopy platform. An actual photograph of the substrate fabricated under the optimized conditions is shown in Figure S5. This substrate utilizes the fiber structure of cellulose acetate paper, which is similar to the structure of the respiratory tract, as shown in Figure 3a. The structure of the trachea maximizes the airflow path, thereby increasing the efficiency of gas exchange within the lungs. Similarly, segmented fiber clusters increase the probability and efficiency of target molecule interactions. Figure 3b shows an SEM image of the gold nanoparticle structure designed to mimic alveoli. By forming and functionalizing nanoparticles on the fiber surface using OMH as a reducing agent, a substrate that mimics the structure of human alveoli is created. Alveoli have evolved to maximize the surface area in contact with capillaries and air, thereby increasing the efficiency of gas exchange. Similarly, gold nanoparticles maximize the surface area to facilitate efficient interactions with target molecules and improve the contact area with the Raman laser, thereby maximizing the SERS signal at the hotspot. Subsequently, the performance was evaluated using the alveoli-mimicking gold nanoparticle-coated cellulose acetate paper substrate (AMSP).

Prior to evaluation, the method for reacting the target substance with AMSP was optimized (Figure 3c). Two methods were compared: dropping the target substance in equal volumes onto the AMSP, allowing it to react before measurement (Figure 3c(i)), and dipping the AMSP into a prepared target substance solution, enabling sufficient reaction time (Figure 3c(ii)). These are the two most common sample preparation methods used for Raman measurements. The drop-casting method improves measurement efficiency by concentrating the target material as the sample dries, leading to localized accumulation [54]. Conversely, the dipping method enhances measurement efficiency by ensuring a uniform interaction between the target material and the entire substrate rather than being confined to specific areas [55]. For comparison, 100 μM R6G was used as the Raman indicator, and the results are shown in Figure 3d. When dropping the same volumes (2, 5, and 10 μL), the Raman intensities were 10,853.82 ± 1401.77, 16,318.33 ± 2022.82, and 2428.35 ± 1596.12, respectively (Figure 3d(i)). Conversely, the Raman intensity after sufficient dipping in the R6G solution and drying was 35,123.81 ± 5881.78 (Figure 3d(ii)). In general, when a sample is applied to a paper-based substrate by drop-casting, an increase in the reaction volume leads to a corresponding increase in the amount of the target, resulting in enhanced Raman intensity. However, this method is limited by an insufficient target loading relative to the surface area of the substrate and the potential for sample leakage beneath the substrate. In contrast, immersing the substrate in the solution allows for more efficient binding of the target material across the substrate, even at the same concentration. Consequently, the dipping method produced superior Raman intensity; thus, all subsequent experiments were performed using this optimized dipping procedure.

The Raman spectroscopic detection performance for R6G was evaluated at various concentrations using the optimized sampling conditions. Detection was conducted across a range from 100 μM to 100 pM, and the comparison of Raman intensities at the characteristic peak of R6G (1504 cm^{-1}) is presented in Figure 3e. The inset in Figure 3e shows the scaled data for the control, 100 pM, and 10 nM concentrations. The analysis showed that the control and 100 pM samples passed the *t*-test with a p-value < 0.0001, confirming R6G detection at 100 pM. This suggests that, under laboratory conditions using AMSP, the experimental detection limit for CV is 100 pM. Raman spectra for the 10 nM, 100 pM, and control samples are presented in Figure S6, showing the R6G peak at 1504 cm^{-1}. Based on these results, the analytical enhancement factor (AEF) of the AMSP was calculated as follows [56]:

$$\mathrm{AEF} = \frac{I_{SERS}/C_{SERS}}{I_{OR}/C_{OR}}$$

Here, I_{SERS} represents the SERS intensity, I_{OR} denotes the Raman intensity observed on the bare substrate, C_{SERS} is the concentration of the Raman indicator on the SERS substrate (i.e., the detection limit), and C_{OR} is the analyte concentration on the bare substrate. The calculated AEF for the AMSP was 5.93×10^6, and the Raman spectra results according to the data used for calculation and the concentration of the indicator R6G used are shown in Figure S7. The substrate demonstrated exceptional detection sensitivity at pM-level concentrations. Notably, its performance surpassed that of other paper-based SERS substrates presented in the literature (Table S1).

The uniformity and reproducibility of the fabricated substrates were evaluated to confirm their suitability for SERS detection. After reacting the R6G indicator with the AMSP, Raman mapping images (10 × 10) were acquired at 100 spots in the defined area (Figure 3f). The higher Raman intensities are represented in red, indicating uniformly high signals across the entire region. Subsequently, the Raman spectra measured at 30 random spots on a single substrate were displayed as a heatmap (Figure 3g). The distinct peaks at 1356 and 1504 cm^{-1} denote the strong SERS response of R6G on the AMSP substrate, with the spectrum color confirmed to be uniform across all measurements. Finally, five different substrates were fabricated and reacted with 100 μM R6G using the same method (Figure 3h). The Raman intensities at 1504 cm^{-1}, the specific peak of R6G, were compared across the substrates, showing a relative standard deviation (RSD) value of 3.29%; this demonstrated the excellent uniformity and reproducibility of the substrates.

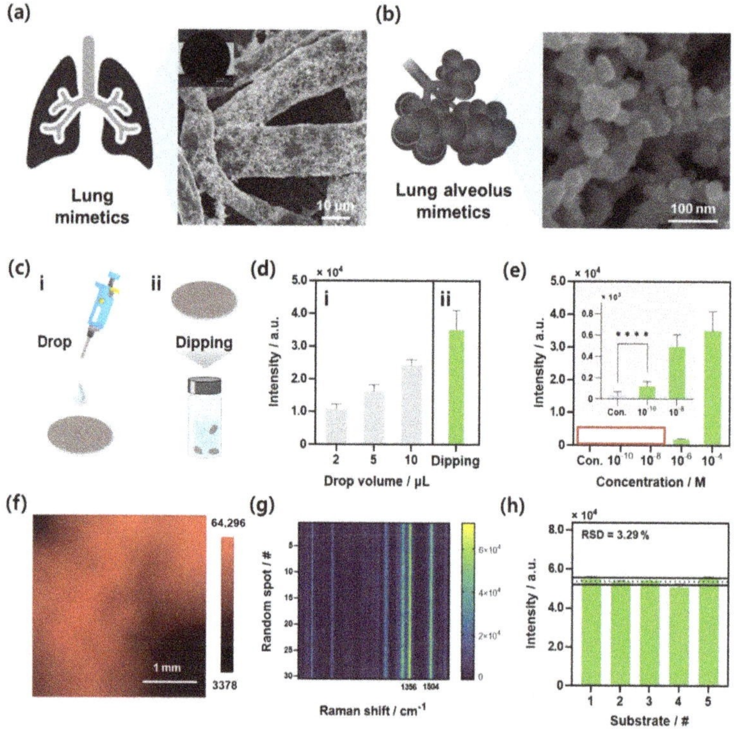

Figure 3. Performance evaluation of the optimized AMSP. (**a**) SEM image of AMSP developed in this study, and (**b**) SEM image of substrate mimicking alveolar structure. (**c**) Schematic of optimization of sample reaction conditions: (i) drop-casting and (ii) dipping. (**d**) Comparison of Raman intensities at 1504 cm^{-1} peak of 100 µM R6G for different reaction conditions: (i) results for drop volumes of 2, 5, and 10 µL, and (ii) results for dipping condition. (**e**) Detection results of R6G at different concen-trations (10^{-4} to 10^{-10} M) under optimized conditions; inset image shows results for low concentrations, with **** p-value < 0.0001. (**f**) Raman mapping image of 100 µM R6G indicator on AMSP. (**g**) Heatmap image of Raman spectra at 30 random spots (#: number unit) on same substrate, showing bright colors at specific peaks of 1356 and 1504 cm^{-1}. (**h**) Reproducibility assessment across five different AMSPs using measurements of 100 µM R6G indicator; RSD = 3.29%.

3.4. *CV Detection in Various Samples: DI Water, Aquaculture, and Human Serum*

Environments in which CV detection is critically required include fish farm waters in aquaculture and human biological systems. Therefore, to validate the CV detection capability of the AMSP, experiments were designed to detect CV in fish farm water, human serum samples, and DI water samples in a controlled laboratory setting (Figure 4). Initially, the experiments were conducted using purified DI water samples under laboratory conditions. CV sample solutions were prepared at concentrations ranging from 1 mM to 1 nM using DI water, and each sample was reacted with AMSP by immersion for 30 min (Figure 4a). The Raman intensity results for the specific CV Raman peak (1384 cm^{-1}) across different concentrations in DI water samples are shown in Figure 4d. The control and 1 nM concentrations satisfied the t-test criteria, with a p-value < 0.0001. This result serves as strong evidence that the low control signal highlights the advantages of CA paper and the appropriateness of the experimental approach. The comprehensive Raman spectra for each concentration are presented in Figure S8a-i, while the spectra for lower concentrations (10 μM, 1 nM, and control) are depicted separately in Figure S8a-ii.

Subsequently, water samples were collected from a campus fish farm, filtered, and used to prepare CV solutions at various concentrations. The AMSPs were reacted with each CV sample following the same procedure (Figure 4b). The Raman intensity results for the specific CV peak at 1384 cm^{-1} across various concentrations in the fish farm water samples are displayed in Figure 4e. In these samples, both the control and 1 nM concentration satisfied the *t*-test criteria, with a *p*-value < 0.0001, confirming that CV detection was experimentally successful at a concentration of 1 nM; this was confirmed to be a meaningful result compared with existing CV detection studies in a real environment [57,58]. The comprehensive Raman spectra for each concentration are provided in Figure S8b-i, with the inset showing the fish farm water used for the experiment. The spectra for lower concentrations (10 μM, 1 nM, and control) are displayed separately in Figure S8b-ii.

Finally, human serum samples were prepared by dilution to 10%. CV was then added to these samples to prepare concentrations ranging from 1 mM to 1 nM, and the substrate was allowed to react with the samples using the same method (Figure 4c). A comparison of the Raman intensities of the characteristic peaks for each CV concentration in the serum samples is shown in Figure 4f. The results indicated that CV detection at a concentration of 1 nM was successfully achieved even in serum samples with a *p*-value < 0.0001, satisfying the *t*-test criteria. The overall Raman spectra for each concentration are depicted in Figure S8c-i, and the spectra for low concentrations are shown separately in Figure S8c-ii. Compared to DI water or fish farm water, serum samples contain numerous impurities that can interfere with the reaction of the target substance, resulting in an overall lower Raman signal intensity, including that of the control [59]. Nevertheless, the AMSP demonstrated excellent CV detection performance across various environmental samples, achieving a sensitivity down to a concentration level of 1 nM. Table 1 provides a comparison with various other paper-based SERS substrates used for CV detection, demonstrating that the AMSP exhibits superior performance.

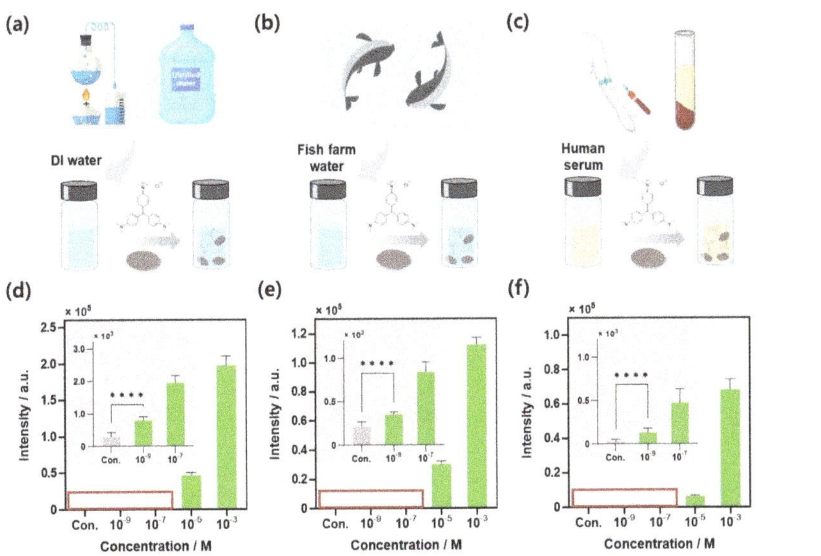

Figure 4. Detection of CV using AMSP. Schematics of sample preparation and reaction conditions for CV detection in (**a**) DI water, (**b**) fish farm water, and (**c**) human serum. Comparison of Raman intensities at 1384 cm^{-1}, a specific Raman peak of CV under each condition, (**d**) DI water, (**e**) fish farm water, (**f**) 10% human serum. The inset image of each data is the low-concentration result data of CV with the *y*-axis scale adjusted. (**** *p*-value < 0.0001.)

Table 1. CV detection reference using various paper-based SERS substrates.

Year	Substrate	Detection Limits	References
2019	Au nanocube	217 nM	[60]
2021	Ag nanostructure on paper-based	10 nM	[61]
2022	AuNPs on paper-based	100 nM	[62]
2022	Ag@Ag nanoparticles on paper-based	10 nM	[63]
2023	AgNPs on paper-based	100 nM	[64]
2024	**AMSP**	**1 nM**	**This work**

4. Conclusions

In this study, we developed a paper-based AMSP for CV detection, drawing inspiration from the ecological structure of bronchial and alveolar tissues in the lungs. The development process involved optimizing several parameters, such as the type of paper, the reducing agent and its concentration, and the reaction time, resulting in a substrate with superior SERS performance. Optimization experiments were conducted for the reaction with the target materials in addition to optimizing the substrate fabrication, demonstrating the high uniformity and reproducibility of AMSP for CV detection. Performance evaluation utilizing the R6G indicator with the proposed AMSP showed that experimental detection was successful at concentrations up to 100 pM. Under controlled laboratory conditions, the substrate could reliably detect CV at a concentration of 1 nM. Furthermore, its performance in real-world settings was evaluated by testing it with samples from fish farm water and human serum, achieving successful CV detection at the 1 nM level in all environments. Consequently, this study successfully developed a simple, optimized, and disposable paper-based SERS platform capable of effectively detecting CV in real-world environmental samples. This platform underscores the potential of the AMSP to provide high sensitivity across various field applications, highlighting its practical utility in varied sensing environments.

Supplementary Materials: The following supporting information can be downloaded at: https://www.mdpi.com/article/10.3390/bios14120566/s1, Figure S1: Low-magnification SEM images of paper-based substrates fabricated with different concentrations of OMH: (a) control, (b) 5 mM, (c) 10 mM, (d) 20 mM, (e) 50 mM, and (f) 100 mM.; Figure S2: High-magnification SEM images of paper-based substrates fabricated with different concentrations of OMH: (a) 20 mM and (b) 100 mM.; Figure S3: SERS spectra of 100 µM R6G in paper-based substrates fabricated with different concentrations of OMH. Each inset shows an optical image of a paper-based substrate. (a) 5 mM, (b) 10 mM, (c) 20 mM, (d) 50 mM, and (e) 100 mM.; Figure S4: SERS spectra of 100 µM R6G in paper-based substrates fabricated with different reaction times. Each inset shows an optical image of a paper-based substrate. (a) 30 min, (b) 60 min, (c) 90 min, and (d) 120 min.; Figure S5: Actual image of the AMSP.; Figure S6: Raman spectra data with y-axis scale adjusted for control, 100 pM, and 10 nM concentrations of the R6G indicator.; Figure S7: Raman spectra data with y-axis scale adjusted for control, 100 pM, and 10 nM concentrations of the R6G indicator.; Table S1: Comparison of enhancement factors (EF) of various fabricated paper-based SERS substrates.; Figure S8: Raman spectra of CV at various concentrations. (a-i) Raman spectra of CV in DI water at concentrations ranging from 10^{-3} to 10^{-9} M. (a-ii) Detailed Raman spectra for low concentrations of CV (10^{-7}, 10^{-9} M) and control in DI water. (b-i) Raman spectra of CV in fish farm water at concentrations ranging from 10^{-3} to 10^{-9} M, inset image shows water from aquaculture farm used in the experiment. (b-ii) Detailed Raman spectra for low concentrations of CV (10^{-7}, 10^{-9} M) and control in fish farm water. (c-i) Raman spectra of CV in 1% human serum at concentrations ranging from 10^{-3} to 10^{-9} M. (c-ii) Detailed Raman spectra for low concentrations of CV (10^{-7}, 10^{-9} M) and control in 10% human serum. Refs. [65–67] are cited in Supplementary Materials.

Author Contributions: H.P., Data curation, Writing—original draft and Visualization; K.C., Methodology, Software and Formal analysis; E.P., Validation, Investigation and Visualization; W.K., Resources and Visualization; G.K., Software and Formal analysis; J.P. (Joohyung Park), Resources, Writing—original draft and Funding acquisition; W.L., Conceptualization, Writing—review & editing and Project administration; J.P. (Jinsung Park), Writing—review & editing, Supervision and Project administration. All authors have read and agreed to the published version of the manuscript.

Funding: This research was funded by the National Research Foundation of Korea (NRF) under Grant No. NRF-2023R1A2C2004964, NRF-2022R1A2C4001990, RS-2023-00222737, RS-2024-00460957, RS-2024-00438542, RS-2024-00353529, the Korea Foundation for Women In Science, Engineering and Technology (WISET) grant funded by the Ministry of Science and ICT(MSIT) under the team research program for female engineering students (WISET-2024-136), and SMC-SKKU Future Convergence Research Program Grant.

Institutional Review Board Statement: Not applicable.

Informed Consent Statement: Not applicable.

Data Availability Statement: Data are contained within the article.

Conflicts of Interest: The authors declare no conflict of interest.

References

1. Šafařík, I.; Šafaříková, M. Detection of Low Concentrations of Malachite Green and Crystal Violet in Water. *Water Res.* **2002**, *36*, 196–200. [CrossRef] [PubMed]
2. Mani, S.; Bharagava, R.N. Exposure to Crystal Violet, Its Toxic, Genotoxic and Carcinogenic Effects on Environment and Its Degradation and Detoxification for Environmental Safety. *Rev. Environ. Contam. Toxicol.* **2016**, *237*, 71–104. [CrossRef] [PubMed]
3. Ghasemi, E.; Kaykhaii, M. Application of Micro-Cloud Point Extraction for Spectrophotometric Determination of Malachite Green, Crystal Violet and Rhodamine B in Aqueous Samples. *Spectrochim. Acta A Mol. Biomol. Spectrosc.* **2016**, *164*, 93–97. [CrossRef]
4. Littlefield, N.A.; Blackwell, B.N.; Hewitt, C.C.; Gaylor, D.W. Chronic Toxicity and Carcinogenicity Studies of Gentian Violet in Mice. *Toxicol. Sci.* **1985**, *5*, 902–912. [CrossRef]
5. An, L.; Deng, J.; Zhou, L.; Li, H.; Chen, F.; Wang, H.; Liu, Y. Simultaneous Spectrophotometric Determination of Trace Amount of Malachite Green and Crystal Violet in Water after Cloud Point Extraction Using Partial Least Squares Regression. *J. Hazard. Mater.* **2010**, *175*, 883–888. [CrossRef]
6. Saeed, A.; Sharif, M.; Iqbal, M. Application Potential of Grapefruit Peel as Dye Sorbent: Kinetics, Equilibrium and Mechanism of Crystal Violet Adsorption. *J. Hazard. Mater.* **2010**, *179*, 564–572. [CrossRef]
7. Xiao, X.; Chen, C.; Deng, J.; Wu, J.; He, K.; Xiang, Z.; Yang, Y. Analysis of Trace Malachite Green, Crystal Violet, and Their Metabolites in Zebrafish by Surface-Coated Probe Nanoelectrospray Ionization Mass Spectrometry. *Talanta* **2020**, *217*, 121064. [CrossRef]
8. Li, G.; Zhang, X.; Zhang, L.; Xu, S.; Li, C. Salt-Assisted Graphene Oxide Dispersive Solid Phase Microextraction for Sensitive Detection of Malachite Green and Crystal Violet by HPLC. *Chromatographia* **2015**, *78*, 979–985. [CrossRef]
9. Zhang, L.; Zhang, Y.; Tang, Y.; Li, X.; Zhang, X.; Li, C.; Xu, S. Magnetic Solid-Phase Extraction Based on Fe3O4/Graphene Oxide Nanoparticles for the Determination of Malachite Green and Crystal Violet in Environmental Water Samples by HPLC. *Int. J. Environ. Anal. Chem.* **2018**, *98*, 215–228. [CrossRef]
10. Favaro, G.; Confortin, D.; Pastore, P.; Brustolon, M. Application of LC-MS and LC-MS-MS to the Analysis of Photo-Decomposed Crystal Violet in the Investigation of Cultural Heritage Materials Aging. *J. Mass Spectrom.* **2012**, *47*, 1660–1670. [CrossRef]
11. Ascari, J.; Dracz, S.; Santos, F.A.; Lima, J.A.; Diniz, M.H.G.; Vargas, E.A. Validation of an LC-MS/MS Method for Malachite Green (MG), Leucomalachite Green (LMG), Crystal Violet (CV) and Leucocrystal Violet (LCV) Residues in Fish and Shrimp. *Food Addit. Contam. Part A* **2012**, *29*, 602–608. [CrossRef] [PubMed]
12. Ju, Y.; Fang, J.; Liu, X.; Xu, Z.; Ren, X.; Sun, C.; Yang, S.; Ren, Q.; Ding, Y.; Yu, K.; et al. Photodegradation of Crystal Violet in TiO_2 Suspensions Using UV–Vis Irradiation from Two Microwave-Powered Electrodeless Discharge Lamps (EDL-2): Products, Mechanism and Feasibility. *J. Hazard. Mater.* **2011**, *185*, 1489–1498. [CrossRef] [PubMed]
13. Orellana, S.; Soto, C.; Toral, M.I. UV–Vis, IR and 1H NMR Spectroscopic Studies and Characterization of Ionic-Pair Crystal Violet–Oxytetracycline. *Spectrochim. Acta A Mol. Biomol. Spectrosc.* **2010**, *75*, 437–443. [CrossRef] [PubMed]
14. Sajid, M.M.; Khan, S.B.; Shad, N.A.; Amin, N.; Zhang, Z. Visible Light Assisted Photocatalytic Degradation of Crystal Violet Dye and Electrochemical Detection of Ascorbic Acid Using a BiVO 4/FeVO 4 Heterojunction Composite. *RSC Adv.* **2018**, *8*, 23489–23498. [CrossRef]
15. Khan, M.Y.A.; Zahoor, M.; Shaheen, A.; Jamil, N.; Arshad, M.I.; Bajwa, S.Z.; Shad, N.A.; Butt, R.; Ali, I.; Iqbal, M.Z.; et al. Visible Light Photocatalytic Degradation of Crystal Violet Dye and Electrochemical Detection of Ascorbic Acid & Glucose Using BaWO4 Nanorods. *Mater. Res. Bull.* **2018**, *104*, 38–43. [CrossRef]

16. Elugoke, S.E.; Fayemi, O.E.; Adekunle, A.S.; Sherif, E.S.M.; Ebenso, E.E. Electrochemical Sensor for the Detection of Adrenaline at Poly (Crystal Violet) Modified Electrode: Optimization and Voltammetric Studies. *Heliyon* **2022**, *8*, e10835. [CrossRef]
17. Wang, W.; Tang, J.; Zheng, S.; Ma, X.; Zhu, J.; Li, F.; Wang, J. Electrochemical Determination of Bisphenol A at Multi-Walled Carbon Nanotubes/Poly (Crystal Violet) Modified Glassy Carbon Electrode. *Food Anal. Methods* **2017**, *10*, 3815–3824. [CrossRef]
18. Zhou, J.; Qing, M.; Ling, Y.; Wang, L.; Li, N.B.; Luo, H.Q. Double-Stranded DNA Nanobridge Enhanced Fluorescence of Crystal Violet/G-Quadruplex Complex for Detection of Lead Ions and Crystal Violet. *Sens. Actuators B Chem.* **2021**, *340*, 129968. [CrossRef]
19. Hu, Y.; Gao, Z. Yellow Emissive Se,N-Codoped Carbon Dots toward Sensitive Fluorescence Assay of Crystal Violet. *J. Hazard. Mater.* **2020**, *388*, 122073. [CrossRef]
20. Kim, W.; Kim, G.; Park, H.; Chai, K.; Park, J.; Park, J. Detecting and Tracking Thiram in Leakage Pathways Using Bioinspired Nanograss with Thuja Fruit-like Nanoparticles. *Sens. Actuators B Chem.* **2024**, *406*, 135405. [CrossRef]
21. Wei, H.; Hossein Abtahi, S.M.; Vikesland, P.J. Plasmonic Colorimetric and SERS Sensors for Environmental Analysis. *Environ. Sci. Nano* **2015**, *2*, 120–135. [CrossRef]
22. Huang, L.; Sun, H.; Sun, L.; Shi, K.; Chen, Y.; Ren, X.; Ge, Y.; Jiang, D.; Liu, X.; Knoll, W.; et al. Rapid, Label-Free Histopathological Diagnosis of Liver Cancer Based on Raman Spectroscopy and Deep Learning. *Nat. Commun.* **2023**, *14*, 48. [CrossRef] [PubMed]
23. Keshavarz, M.; Kassanos, P.; Tan, B.; Venkatakrishnan, K. Metal-Oxide Surface-Enhanced Raman Biosensor Template towards Point-of-Care EGFR Detection and Cancer Diagnostics. *Nanoscale Horiz.* **2020**, *5*, 294–307. [CrossRef]
24. Narasimhan, V.; Siddique, R.H.; Kim, U.J.; Lee, S.; Kim, H.; Roh, Y.G.; Wang, Y.M.; Choo, H. Glasswing-Butterfly-Inspired Multifunctional Scleral Lens and Smartphone Raman Spectrometer for Point-of-Care Tear Biomarker Analysis. *Adv. Sci.* **2023**, *10*, 2205113. [CrossRef]
25. Chang, Y.C.; Wu, S.H. Bi-Functional Al-Doped ZnO@SnO_2 Heteronanowires as Efficient Substrates for Improving Photocatalytic and SERS Performance. *J. Ind. Eng. Chem.* **2019**, *76*, 333–343. [CrossRef]
26. Kim, H.J.; Kim, B.; Lee, D.; Lee, B.H.; Cho, C. Improvement of Surface-Enhanced Raman Spectroscopy Response Characteristics of Nanoporous Ag Metal Thin Film with Surface Texture Structures. *J. Sens. Sci. Technol.* **2020**, *29*, 255–260. [CrossRef]
27. Kim, H.J.; Kim, B.; Lee, D.; Lee, B.H.; Cho, C. Fabrication of Surface-Enhanced Raman Scattering Substrate Using Black Silicon Layer Manufactured through Reactive Ion Etching. *J. Sens. Sci. Technol.* **2021**, *30*, 267–272. [CrossRef]
28. Knudsen, L.; Ochs, M. The Micromechanics of Lung Alveoli: Structure and Function of Surfactant and Tissue Components. *Histochem. Cell Biol.* **2018**, *150*, 661–676. [CrossRef]
29. Horan, M. Extracorporeal Membrane Oxygenation and Mechanical Support for the Circulation. *Anaesth. Intensive Care Med.* **2023**, *24*, 711–719. [CrossRef]
30. Aazmi, A.; Zhou, H.; Li, Y.; Yu, M.; Xu, X.; Wu, Y.; Ma, L.; Zhang, B.; Yang, H. Engineered Vasculature for Organ-on-a-Chip Systems. *Engineering* **2022**, *9*, 131–147. [CrossRef]
31. Soncini, R.; Klein, W. Surface Tension in Biological Systems–A Common Problem with a Variety of Solutions. *Comp. Biochem. Physiol. A Mol. Integr. Physiol.* **2023**, *284*, 111475. [CrossRef] [PubMed]
32. Kim, J.Y.; Koh, E.H.; Yang, J.Y.; Mun, C.; Lee, S.; Lee, H.; Kim, J.; Park, S.G.; Kang, M.; Kim, D.H.; et al. 3D Plasmonic Gold Nanopocket Structure for SERS Machine Learning-Based Microplastic Detection. *Adv. Funct. Mater.* **2024**, *34*, 2307584. [CrossRef]
33. Mandal, P.; Tewari, B.S. Progress in Surface Enhanced Raman Scattering Molecular Sensing: A Review. *Surf. Interfaces* **2022**, *28*, 101655. [CrossRef]
34. Budnyk, A.P.; Damin, A.; Agostini, G.; Zecchina, A. Gold Nanoparticle Aggregates Immobilized on High Surface Area Silica Substrate for Efficient and Clean SERS Applications. *J. Phys. Chem. C* **2010**, *114*, 3857–3862. [CrossRef]
35. Newman, J.D.S.; Blanchard, G.J. Formation of Gold Nanoparticles Using Amine Reducing Agents. *Langmuir* **2006**, *22*, 5882–5887. [CrossRef]
36. Liu, Y.; Yu, H.; Cheng, Y.; Guo, Y.; Yao, W.; Xie, Y. Non-Destructive Monitoring of Staphylococcus Aureus Biofilm by Surface-Enhanced Raman Scattering Spectroscopy. *Food Anal. Methods* **2020**, *13*, 1710–1716. [CrossRef]
37. Harraz, F.A.; Ismail, A.A.; Bouzid, H.; Al-Sayari, S.A.; Al-Hajry, A.; Al-Assiri, M.S. Surface-Enhanced Raman Scattering (SERS)-Active Substrates from Silver Plated-Porous Silicon for Detection of Crystal Violet. *Appl. Surf. Sci.* **2015**, *331*, 241–247. [CrossRef]
38. Quynh, L.M.; Nam, N.H.; Kong, K.; Nhung, N.T.; Notingher, I.; Henini, M.; Luong, N.H. Surface-Enhanced Raman Spectroscopy Study of 4-ATP on Gold Nanoparticles for Basal Cell Carcinoma Fingerprint Detection. *J. Electron. Mater.* **2016**, *45*, 2563–2568. [CrossRef]
39. Quintana, R.; Persenaire, O.; Lemmouchi, Y.; Bonnaud, L.; Dubois, P. Grafted d/l-Lactide to Cellulose Acetate by Reactive Melt Processing: Its Role as CA/PLA Blend Compatibilizer. *Eur. Polym. J.* **2014**, *57*, 30–36. [CrossRef]
40. Bhattacharya, A.; Misra, B.N. Grafting: A Versatile Means to Modify Polymers: Techniques, Factors and Applications. *Prog. Polym. Sci.* **2004**, *29*, 767–814. [CrossRef]
41. Salama, A.; Abouzeid, R.E.; Owda, M.E.; Cruz-Maya, I.; Guarino, V. Cellulose–Silver Composites Materials: Preparation and Applications. *Biomolecules* **2021**, *11*, 1684. [CrossRef] [PubMed]
42. Gopiraman, M.; Deng, D.; Saravanamoorthy, S.; Chung, I.M.; Kim, I.S. Gold, Silver and Nickel Nanoparticle Anchored Cellulose Nanofiber Composites as Highly Active Catalysts for the Rapid and Selective Reduction of Nitrophenols in Water. *RSC Adv.* **2018**, *8*, 3014–3023. [CrossRef] [PubMed]

43. Park, H.; Kim, W.; Kim, M.; Lee, G.; Lee, W.; Park, J. Eco-Friendly and Enhanced Colorimetric Detection of Aluminum Ions Using Pectin-Rich Apple Extract-Based Gold Nanoparticles. *Spectrochim. Acta A Mol. Biomol. Spectrosc.* **2021**, *245*, 118880. [CrossRef] [PubMed]
44. Borsley, S.; Edwards, W.; Mati, I.K.; Poss, G.; Diez-Castellnou, M.; Marro, N.; Kay, E.R. A General One-Step Synthesis of Alkanethiyl-Stabilized Gold Nanoparticles with Control over Core Size and Monolayer Functionality. *Chem. Mater.* **2023**, *35*, 6168–6177. [CrossRef] [PubMed]
45. Mechler, A.; Torriero, A.A.J.; Nafady, A.; Lee, C.Y.; Bond, A.M.; O'Mullane, A.P.; Bhargava, S.K. The Formation of Gold Nanoparticles Using Hydroquinone as a Reducing Agent through a Localized PH Change upon Addition of NaOH to a Solution of $HAuCl_4$. *Colloids Surf. A Physicochem. Eng. Asp.* **2010**, *370*, 35–41. [CrossRef]
46. Pacioni, N.L.; Borsarelli, C.D.; Rey, V.; Veglia, A.V. Synthetic Routes for the Preparation of Silver Nanoparticles: A Mechanistic Perspective. *Eng. Mater.* **2015**, 13–46. [CrossRef]
47. Matsui, T.; Kitagawa, Y.; Okumura, M.; Shigeta, Y. Accurate Standard Hydrogen Electrode Potential and Applications to the Redox Potentials of Vitamin C and NAD/NADH. *J. Phys. Chem. A* **2015**, *119*, 369–376. [CrossRef]
48. Yu, X.; Manthiram, A. Electrochemical Energy Storage with an Aqueous Quinone-Air Chemistry. *ACS Appl. Energy Mater.* **2018**, *1*, 2424–2428. [CrossRef]
49. Yoo, S.; Nam, D.H.; Singh, T.I.; Leem, G.; Lee, S. Effect of Reducing Agents on the Synthesis of Anisotropic Gold Nanoparticles. *Nano Converg.* **2022**, *9*, 5. [CrossRef]
50. Sau, T.K.; Pal, A.; Jana, N.R.; Wang, Z.L.; Pal, T. Size Controlled Synthesis of Gold Nanoparticles Using Photochemically Prepared Seed Particles. *J. Nanopart. Res.* **2001**, *3*, 257–261. [CrossRef]
51. Kim, H.S.; Seo, Y.S.; Kim, K.; Han, J.W.; Park, Y.; Cho, S. Concentration Effect of Reducing Agents on Green Synthesis of Gold Nanoparticles: Size, Morphology, and Growth Mechanism. *Nanoscale Res. Lett.* **2016**, *11*, 230. [CrossRef] [PubMed]
52. Yazdani, S.; Daneshkhah, A.; Diwate, A.; Patel, H.; Smith, J.; Reul, O.; Cheng, R.; Izadian, A.; Hajrasouliha, A.R. Model for Gold Nanoparticle Synthesis: Effect of pH and Reaction Time. *ACS Omega* **2021**, *6*, 16847. [CrossRef] [PubMed]
53. Mahmoud, M.A.; El-Sayed, M.A. Aggregation of Gold Nanoframes Reduces, Rather than Enhances, Sers Efficiency Due to the Trade-off of the Inter- and Intraparticle Plasmonic Fields. *Nano Lett.* **2009**, *9*, 3025–3031. [CrossRef] [PubMed]
54. Park, J.; Chai, K.; Kim, W.; Yoon, T.; Park, H.; Kim, W.; You, J.; Na, S.; Park, J. Highly Enhanced Hg2+ Detection Using Optimized DNA and a Double Coffee Ring Effect-Based SERS Map. *Biosens. Bioelectron.* **2024**, *264*, 116646. [CrossRef] [PubMed]
55. Mai, Q.D.; Nguyen, H.A.; Dinh, N.X.; Thu Thuy, N.T.; Tran, Q.H.; Thanh, P.C.; Pham, A.T.; Le, A.T. Versatile and High Performance In-Paper Flexible SERS Chips for Simple and in-Situ Detection of Methylene Blue in River Water and Thiram on Apple Skin. *Talanta* **2023**, *253*, 124114. [CrossRef]
56. Le Ru, E.C.; Blackie, E.; Meyer, M.; Etchegoint, P.G. Surface Enhanced Raman Scattering Enhancement Factors: A Comprehensive Study. *J. Phys. Chem. C* **2007**, *111*, 13794–13803. [CrossRef]
57. Yu, S.; Yuan, X.; Yang, J.; Yuan, J.; Shi, J.; Wang, Y.; Chen, Y.; Gao, S. A Chemometric-Assisted Method for the Simultaneous Determination of Malachite Green and Crystal Violet in Water Based on Absorbance–PH Data Generated by a Homemade PH Gradient Apparatus. *Spectrochim. Acta A Mol. Biomol. Spectrosc.* **2015**, *150*, 403–408. [CrossRef]
58. Sadeghi, S.; Nasehi, Z. Simultaneous Determination of Brilliant Green and Crystal Violet Dyes in Fish and Water Samples with Dispersive Liquid-Liquid Micro-Extraction Using Ionic Liquid Followed by Zero Crossing First Derivative Spectrophotometric Analysis Method. *Spectrochim. Acta A Mol. Biomol. Spectrosc.* **2018**, *201*, 134–142. [CrossRef]
59. Lu, D.; Lin, X.; Chen, C.; Lu, Y.; Feng, S.; Huang, Z.; You, R.; Chen, J.; Wu, Y. Interference-Free SERS Tags for Ultrasensitive Quantitative Detection of Tyrosinase in Human Serum Based on Magnetic Bead Separation. *Anal. Chim. Acta* **2020**, *1138*, 150–157. [CrossRef]
60. Lin, S.; Lin, X.; Song, X.; Han, S.; Wang, L.; Hasi, W. Fabrication of Flexible Paper-Based Surface-Enhanced Raman Scattering Substrate from Au Nanocubes Monolayer for Trace Detection of Crystal Violet on Shell. *J. Raman Spectrosc.* **2019**, *50*, 1074–1084. [CrossRef]
61. Khan, G.A.; Demirtaş, Ö.; Demir, A.K.; Aytekin, Ö.; Bek, A.; Bhatti, A.S.; Ahmed, W. Fabrication of Flexible, Cost-Effective, and Scalable Silver Substrates for Efficient Surface Enhanced Raman Spectroscopy Based Trace Detection. *Colloids Surf. A Physicochem. Eng. Asp.* **2021**, *619*, 126542. [CrossRef]
62. Beeram, R.; Banerjee, D.; Narlagiri, L.M.; Soma, V.R. Machine Learning for Rapid Quantification of Trace Analyte Molecules Using SERS and Flexible Plasmonic Paper Substrates. *Anal. Methods* **2022**, *14*, 1788–1796. [CrossRef] [PubMed]
63. Batool, A.; Khan, G.A.; Ahmed, W. Seed-Mediated Growth of Highly Concentrated Silver Nanoparticles on a Flexible Substrate for Applications in SERS-Based Trace Detection. *Vib. Spectrosc.* **2022**, *123*, 103438. [CrossRef]
64. Martins, N.C.T.; Fateixa, S.; Trindade, T. Chitosan Coated Papers as Sustainable Platforms for the Development of Surface-Enhanced Raman Scattering Hydrophobic Substrates. *J. Mol. Liq.* **2023**, *375*, 121388. [CrossRef]
65. Sun, H.; Li, X.; Hu, Z.; Gu, C.; Chen, D.; Wang, J.; Li, B.; Jiang, T.; Zhou, X. Hydrophilic-Hydrophobic Silver Nanowire-Paper Based SERS Substrate for in-Situ Detection of Furazolidone under Various Environments. *Appl. Surf. Sci.* **2021**, *556*, 149748. [CrossRef]

66. Qi, X.; Cheng, Y.; Xu, R.; Li, X.; Zhang, Z.; Chen, L.; Shao, Y.; Gao, Z.; Zhu, M. Designing of a Functional Paper-Tip Substrate for Sensitive Surface-Enhanced Raman Spectroscopy (SERS) Detection. *Anal. Chim. Acta* **2023**, *1280*, 341872. [CrossRef]
67. Eskandari, V.; Sahbafar, H.; Karooby, E.; Heris, M.H.; Mehmandoust, S.; Razmjoue, D.; Hadi, A. Surface-Enhanced Raman Scattering (SERS) Filter Paper Substrates Decorated with Silver Nanoparticles for the Detection of Molecular Vibrations of Acyclovir Drug. *Spectrochim. Acta A Mol. Biomol. Spectrosc.* **2023**, *298*, 122762. [CrossRef]

biosensors

MDPI

Article

Novel Multiparametric Bioelectronic Measurement System for Monitoring Virus-Induced Alterations in Functional Neuronal Networks

Heinz-Georg Jahnke [1,*,†], Verena te Kamp [2,†], Christoph Prönnecke [1], Sabine Schmidt [1], Ronny Azendorf [1], Barbara Klupp [2], Andrea A. Robitzki [3] and Stefan Finke [2]

1 Centre for Biotechnology and Biomedicine, Biochemical Cell Technology, Leipzig University, Deutscher Platz 5, 04103 Leipzig, Germany; sabine.schmidt@bbz.uni-leipzig.de (S.S.)
2 Institute of Molecular Virology and Cell Biology, Friedrich-Loeffler-Institut, Südufer 10, 17493 Greifswald, Germany; barbara.klupp@fli.de (B.K.)
3 Division Management for Biology, Chemistry and Process Engineering, Karlsruhe Institute of Technology (KIT), Hermann-von-Helmholtz-Platz 1, 76344 Eggenstein-Leopoldshafen, Germany
* Correspondence: heinz-georg.jahnke@bbz.uni-leipzig.de; Tel.: +49-341-973-1246; Fax: +49-341-973-1249
† These authors contributed equally to this work.

Abstract: Development and optimisation of bioelectronic monitoring techniques like microelectrode array-based field potential measurement and impedance spectroscopy for the functional, label-free and non-invasive monitoring of in vitro neuronal networks is widely investigated in the field of biosensors. Thus, these techniques were individually used to demonstrate the capabilities of, e.g., detecting compound-induced toxicity in neuronal culture models. In contrast, extended application for investigating the effects of central nervous system infecting viruses are rarely described. In this context, we wanted to analyse the effect of herpesviruses on functional neuronal networks. Therefore, we developed a unique hybrid bioelectronic monitoring platform that allows for performing field potential monitoring and impedance spectroscopy on the same microelectrode. In the first step, a neuronal culture model based on primary hippocampal cells from neonatal rats was established with reproducible and stable synchronised electrophysiological network activity after 21 days of cultivation on microelectrode arrays. For a proof of concept, the pseudorabies model virus PrV Kaplan-ΔgG-GFP was applied and the effect on the neuronal networks was monitored by impedance spectroscopy and field potential measurement for 72 h in a multiparametric mode. Analysis of several bioelectronic parameters revealed a virus concentration-dependent degeneration of the neuronal network within 24–48 h, with a significant early change in electrophysiological activity, subsequently leading to a loss of activity and network synchronicity. In conclusion, we successfully developed a microelectrode array-based hybrid bioelectronic measurement platform for quantitative monitoring of pathologic effects of a herpesvirus on electrophysiological active neuronal networks.

Keywords: impedance spectroscopy; field potential monitoring; microelectrode arrays; primary hippocampal neurons; pseudorabies virus model

1. Introduction

Investigation of viruses in their hosts with regard to pathological effects has always been at the core focus of research since viruses were discovered. Even though most viruses simply use host cells to replicate themselves and then destroy the host cells, there are many known virus-induced effects that subtly alter cellular properties and processes like induction of mitophagy [1], alteration of host miRNA expression [2] or epigenotype, as well as gene expression pattern [3]. Moreover, all behaviour of the host could be transformed, like a decrease in cognitive functioning [4]. For example, lentiviral expression of the rabies virus glycoprotein altered the neuronal network structure and even enhanced spatial memory [5], suggesting that the neurotropic rabies virus is able to infect the central and peripheral neurons [6] and is thus able to modify neuronal functionality through specific interactions. In humans, rabies encephalitis is accompanied by symptoms like hallucinations, aggressive behaviour, strong sweating and salivation, hydrophobia or even an addiction to bite [7], further suggesting that neuronal functions and communication are affected. Despite these phenomenological effects in the virus, hosts are widely known and investigated, while the exact mechanisms on a (sub)cellular level in neurons and the central nervous system are poorly understood. Therefore, an actual focus of research in this field is the development of in vitro models for investigating molecular virology mechanisms. In this context, the swine herpesvirus (PrV) is a very interesting model for virology and neurology because it is able to infect synaptically connected neurons and, therefore, provides a great opportunity for investigating neuroinvasion from attachment and entry to replication, intracellular trafficking and egress, transneuronal spread and immune responses of the host [8,9]. Now that in vitro model systems are getting better and better, and the biological toolbox for mechanism analysis continuously expands, there is a high demand for functional monitoring technologies, especially for in vitro models of neuronal networks. In this context, microelectrode-based field potential monitoring is perfectly suited for the spatial resolved analysis of electrophysiological network activity [10–12]. Another appropriate method is impedance spectroscopy, which is highly suited for monitoring morphological changes and cellular degeneration [13–15]. The combined use of these label-free, non-invasive monitoring techniques would allow for a comprehensive multimodal analysis of complex virus-induced effects in neuronal networks. While hybrid measurement systems are already available in the field of cardiomyocyte models [16], these systems are only based on individual large-area electrodes for the impedimetric measurement and separated microelectrodes for the field potential monitoring. Due to the technical challenges, systems that offer high-resolution impedance spectroscopy, as well as field potential monitoring on the same microelectrode while also comprising a sufficient number of electrodes per culture unit, are still not yet described or even commercially available. In this context, our aim was to develop a hybrid, i.e., multimodal and multiparametric measurement system for combined high-resolution impedance spectroscopy and field potential monitoring on microelectrode arrays for spatial resolved high-content analysis of neuronal networks. For a proof of concept, we used the pseudorabies model virus PrV Kaplan-ΔgG-GFP on rat primary hippocampal networks for analysing the effect of this virus in detail.

2. Materials and Methods

2.1. Fabrication of Microelectrode Arrays

Microelectrode arrays were produced in our clean room facility using the lift-off technique as previously described in [10]. Briefly, glass substrates were cleaned and structured with photoresist (AR-P 3510, Allresist GmbH, Strausberg, Germany) using a Mask Aligner (MA6, SÜSS MicroTec GmbH, Garching, Germany) to form conducting paths and electrodes. A 350 nm thick gold coating was applied by a sputtering process in a CREAMET 500 high vacuum device (CREAVAC GmbH, Dresden, Germany). The passivation layer was fabricated using the photoresist SU8-2 (KAYAKU Advanced Materials, Newton, MA, USA). For final cleaning and hydrophilisation, microelectrode arrays were plasma treated in an argon atmosphere with 400 mA for 450 s (CREAMET 500, CREAVAC GmbH, Dresden, Germany).

2.2. Bioelectronic Measurement System

For the development of the hybrid measurement setup, a specific adapted data acquisition (DAQ) system was obtained from Sciospec Scientific GmbH (Bennewitz, Germany). Briefly, the system comprised two 96-channel data acquisition cards with an integrated 200× amplifier and a low pass filter. Furthermore, a high-precision impedance analyser module was integrated. For automated measurement, LabView 2018 (National Instruments)-based software was developed for impedance measurement (IMATadvanced v1) and field potential monitoring (FiPRAT v1). Impedance spectra were recorded from 500 Hz to 5 MHz (51 frequency points), with a signal amplitude of ±10 mV. Field potential streams were recorded with 20 kHz for at least 6 min for each time point.

2.3. Primary Neuronal Cell Culture and Virus Used for Infection

All performed work described in this article has been carried out in accordance with the German legal and ethical requirements of appropriate procedures. All animal work followed national and international guidelines in accordance with Directive 86/609/EEC for animal experiments in and under the responsibility of the Friedrich-Loeffler-Institut. One day old Sprague Dawley rats from our own animal house (hosted by the Friedrich-Loeffler-Institut) were decapitated; the skull was opened and the brain was transferred to an HBSS (Life Technologies, Carlsbad, CA, USA)-containing petri dish. The cerebellum and hypothalamus were removed and the hemispheres were divided. The midbrain of each hemisphere was removed; the hippocampus was separated and mechanically minced with a scalpel. The hippocampi were transferred to a 15 mL tube with 10 mL ice cold HBSS (two hippocampi per tube) and centrifuged (2 min, 200× *g*, 4 °C). The supernatant was removed and 1 mL 0.25% trypsin/EDTA (Life Technologies), as well as 0.5 mL DNase I (Applichem, Darmstadt, Germany), was added, followed by incubation at 37 °C for 30 min under rotation (150 rpm). The cells were carefully resuspended in 2 mL NB-A medium (Life Technologies) containing 10% foetal calf serum (Life Technologies) using a fire-rounded glass pipette and centrifuged for 5 min at 500× *g* and 4 °C. The cells were resuspended in NB-A medium without supplements, combined and counted using a LUNA automated cell counter (Logos Biosystems, Villeneuve, France). Cells were seeded on Matrigel™ (Corning Inc., Corning, NY, USA) pre-coated (1:20 dilution, 3 h at room temperature) microelectrode arrays (250,000 cells/MEA). Medium was changed the next day and half of the medium was changed every second day thereafter.

For infection of neuron cultures, a previously described recombinant GFP expressing PrV Kaplan-ΔgG-GFP with the non-essential gG gene deleted was used [17].

2.4. Data Analysis and Statistics

Impedance spectra data were analysed using a self-developed software (IDAT v4, LabView), which calculates the cellular contribution from the impedance magnitude spectra ((|Z|cell − |Z|cellfree)/|Z|cell-free × 100%), and extracts the maximum value for each spectrum and traces it over time. The electrical equivalent circuit model fitting is based on a non-linear optimisation function with constraints provided by LabView. Determination of the cellular parameters was achieved using a two-step fitting procedure. In the first step, blank impedance spectra (without cells) were used to determine the system parameters. In the second step, these determined system parameters were set as the constant for fitting the cell parameters.

Recorded field potential streams were analysed using a self-developed software (FiPRAT, LabView). First, action potentials were detected by an automated threshold-based algorithm. In detail, the basic noise level is dynamically determined based on a median analysis as previously described [18]. Instead of a multiplication factor of four for getting the absolute threshold level, we used six for a safe exclusion of false positive spikes. Based on the detected action potentials, descriptive parameters were analysed within intervals of 5 min for each time point. Network synchronicity was analysed by spike population coherence [19] with a time constant (tau) of 50 ms. GraphPad Prism 9 was used for all statistical analyses. To test for significant effects between multiple groups, an ANOVA test with a Tukey post hoc test (recommended by GraphPad Prism) was performed.

3. Results and Discussion

3.1. Development of the Hybrid Bioelectronic Measurement System

For the comprehensive, spatially resolved, label-free bioelectronic analysis of neuronal networks, microelectrode arrays are a prerequisite. Furthermore, the combination of different bioelectronic monitoring techniques like field potential measurement and impedance spectroscopy are excellently suited for spatially resolving electrophysiological activity, as well as cellular morphology and the viability of neuronal in vitro cultures. While both techniques are widely used individually, the concept of using both techniques on the same microelectrode (Figure 1A) has not been satisfactorily demonstrated for real-time measurement applications thus far. One major reason for this is the opposite nature of these techniques with, on the one hand, an amplification and detection of weak extracellular potentials in the µV range, and on the other hand, an active measurement where signals in the range of mV are applied for the frequency-dependent analysis of the response.

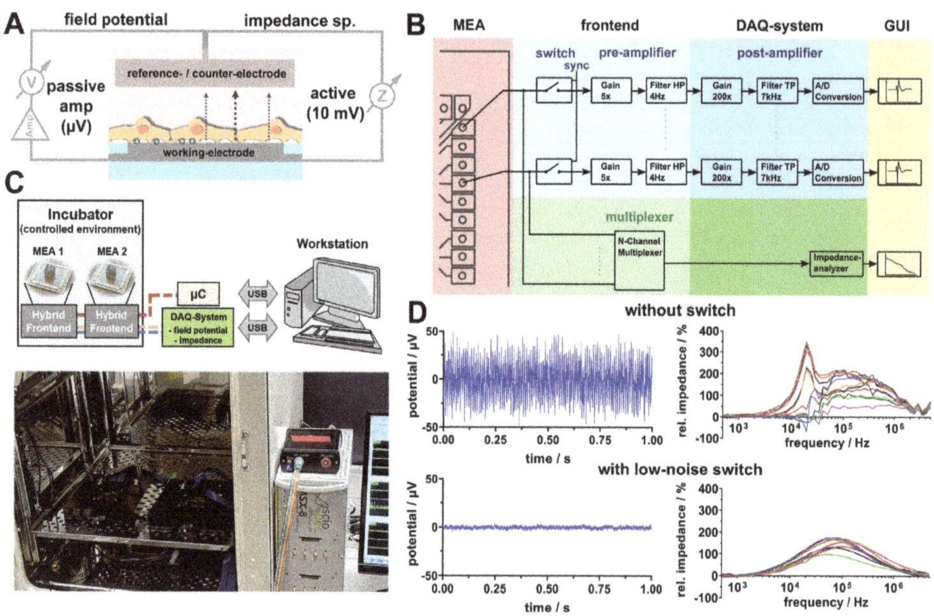

Figure 1. Description of hybrid measurement system. (**A**) Scheme of combined field potential and impedance spectroscopy measurement on the same microelectrode. (**B**) Scheme of field potential circuit pathway and high-precision impedance spectroscopy pathway based on channel multiplexing. Connection of both pathways on the measurement electrode is realised by a low noise switch. (**C**) Scheme and photo of whole measurement setup. (**D**) Measurement signals (field potential baseline and cell signal represented by relative impedance) of directly connected (without switch) measurement pathways and pathways connected over low noise switches.

Moreover, using microelectrodes, which have a high intrinsic impedance, poses major challenges on the measurement equipment for each technique, like the high-precision impedance analyser, as well as all connecting and multiplexing circuits. In this context, we developed a measurement path for each technique that could be integrated in a single frontend for the connection of microelectrode arrays (Figure 1B). First experiments with both measurement paths connected directly to the same electrode revealed clear interferences for both measurement techniques, even when they were performed in sequence (Figure 1D). Thus, the field potential baseline of microelectrodes (100 µm diameter) showed maximum signals of up to 40 µV and a noise level (standard deviation) of >15 µV, which is definitely too high for a sensitive detection of neuronal action potentials [10]. In accordance therewith, high noise levels were observed within a wide frequency range of impedance spectra and, thereof, derived frequency-dependent cell signals (Figure 1D). In the search for a solution, we investigated several switch circuits for isolating both measurement pathways from each other, which, in fact, was quite difficult because those circuits also easily induce noise, especially in the field potential amplifier pathway. Finally, we were able to identify the low noise switch circuit ADG1212 from analog devices that offer a low on (1 pF)/off (2.6 pF) capacitance, low charge injection (1 pC) and low channel-to-channel crosstalk of 90 dB, resulting in minimum noise and measurement signal disturbances. These switches were introduced for each channel in front of both measurement pathways (measurement electrodes) and synchronised by a microcontroller (Figure 1B). These switches were also introduced for switching between reference electrode (field potential) and counter-electrode (impedance) paths. Based on this circuit design, a hybrid frontend that provides a field potential baseline of microelectrodes (100 μm diameter) with maximum signals < 3 μV and a noise level (standard deviation) of 1.8 μV (Figure 1D) was realised. For the realisation of a complete measurement setup, the hybrid frontend was designed with the aim for use in a controlled environment for cell culture (incubator). With respect to a stable temperature of 37 °C during continuous operation, the heat dissipation by introduced energy was in focus. Therefore, the 60-channel field potential amplifier was divided into two stages, where only the first stage was integrated into the frontend (amplification factor of 5) and the second stage (amplification factor of 200) was integrated into the custom-made DAQ system obtained from Sciospec Scientific Instruments. Based on this approach, the whole energy consumption of the frontend could be limited to 2.1 W. Using a standard commercially available cell incubator, two frontends could be placed in parallel for an increased measurement throughput. The frontends are connected by flat ribbon cable to the DAQ system and to a microcontroller box outside of the incubator. The DAQ system and microcontroller box are connected by USB cable to the workstation (Figure 1C). For performing automated measurements with both measurement techniques on both frontends in parallel, controlling software was developed for field potential monitoring (FiPRAT) and impedance spectroscopy (IMATadvanced) (Supplementary Figure S1)

3.2. Establishment of Primary Hippocampal Culture Model on Optimised Microelectrode Arrays

In addition to the measurement setup, the microelectrode arrays are the most crucial factor for a robust and sensitive monitoring. Therefore, we started with a microelectrode layout that was successfully used in a previous study for the characterisation of human-induced pluripotent stem cell (hiPSC)-derived neuronal network maturation [10]. This array comprises 54 measurement microelectrodes with a diameter of 50 μm on a cultivation area of 0.67 cm^2 and a reference electrode with a total area of 1.77 mm^2 (Figure 2A). The size of the microelectrode was originally adapted for detecting action potentials with a signal amplitude of at least 15 μV, which was successfully used for the performed analysis and comparison of hiPSC-derived neuronal networks.

Since our new hybrid system offers an excellent low basic DAQ noise characteristic and consists of two different measurement techniques, we investigated in detail the influence of the electrode size. First, the amount of detectable action-potential-derived field potential spikes was analysed in depth for electrophysiological active primary hippocampal cultures. Therefore, microelectrode arrays with electrode diameters of 20 μm, 30 μm, 50 μm and 100 μm were fabricated. Regarding the reference/counter to the working electrode area ratio, even the largest measurement microelectrode (100 μm) leads to a ratio of 225, which ensures that the working electrode is the dominant electrode in the impedance measurement and that the reference electrode provides a stable reference potential for the field potential monitoring when connected to the technical ground. The determined basic noise levels were as follows (Figure 2B): the lowest was at 1.8 μV for 100 μm and twice as high at 3.6 μV for 50 μm, 6.2 μV for 30 μm and 8.5 μV for 20 μm. Based on this, we recorded field potential streams from electrophysiological active primary hippocampal cultures and analysed the number of detected action potentials (spikes) depending on the basic noise level, on which the detection threshold is calculated (Supplementary Figure S2). Astonishingly, doubling the basic noise level from 1.8 μV (100 μm) to 3.6 μV (50 μm) led to a reduction of recognised spikes by 90%. Additionally, we analysed the impedance for the selected electrode sizes in comparison to a reference frontend that was developed and optimized for analysis based on only impedance spectroscopy (Supplementary Figure S3 and Figure 2E). The analysis revealed no deviations in the impedance magnitude spectra as well as for the maximum achievable cell signals. Furthermore, a smaller electrode size correlates with decreased cell signals for both systems. This is in line with a previous study, where signals of HEK-293A cells were analysed for electrodes with a diameter between 50 µm and 200 µm [20], with a maximum signal amplitude for 100 µm large electrodes. Although smaller electrode sizes are in general favoured for the analysis of neuronal cultures to achieve higher specificity and separation of individual neuronal signals, we have chosen the 100 µm large electrodes with respect to the significant higher number of detectable action potentials, as well as higher impedimetric cell signal amplitudes. Thus, we fabricated new microelectrode arrays (MEA) with an electrode size of 100 µm and characterised the field potential signals from hippocampal cultures demonstrating detection of activity patterns (single spikes and bursts), including high-resolution of a single spike with a duration of 1–3 ms (Figure 2C) as well as achievable overall activity in detail (Figure 2D). First, we analysed several preparations regarding the time point for starting the experiments. Although the first electrophysiological activity was in general observable within the first two weeks, at least three weeks were needed for the establishment of robust spontaneous activity with high numbers of spike and burst signals (Figure 2C). Since biological variations within different preparations and between different MEAs naturally occur for the electrophysiological activity, we defined quality criteria for the use in experiments in the next step. Based on 28 MEAs with cells from four different preparations and the ability to observe virus-induced alterations in both directions, we set a lower and upper limit of 20% and 80% for the number of active electrodes and 100 spikes per minute (Figure 2D).

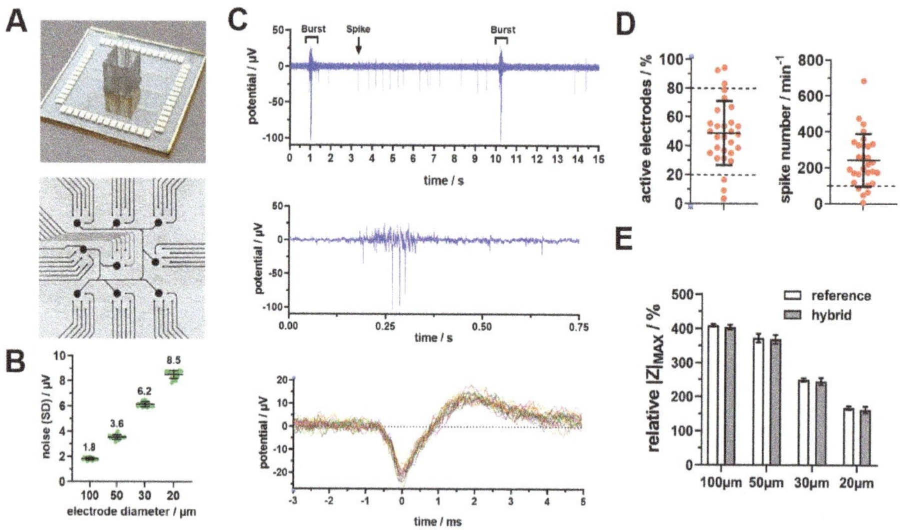

Figure 2. Microelectrode size optimisation and bioelectronic characterisation of primary hippocampal cultures. (**A**) Images of used microelectrode arrays with 54 measurement electrodes and a shared reference/counter-electrode with a bonded 8.2 mm × 8.2 mm culture chamber. (**B**) Noise (standard deviation) analysis for field potential base signal of microelectrodes with different diameter (n = 18 electrodes, mean ± sd). (**C**) Representative field potential signal of electrophysiological active rat hippocampal culture (day 21) at different time scales, demonstrating different signal patterns (single spikes and spike clusters called bursts). High-resolution single spikes are shown as an overlay of 30 spikes from a single electrode. (**D**) Basic characterisation of electrophysiological activity for defining quality criteria of used cultures in experiments. Dashed lines represent minimum and maximum criteria (n = 28, mean ± sd). (**E**) Maximum relative impedance (cell signal) for only an impedance reference frontend in comparison to the novel hybrid frontend (n = 8, mean ± sd).

3.3. *Impedance Spectroscopy-Based Monitoring of Virus-Induced Degeneration of the Neuronal Networks*

For a proof of concept, the pseudorabies model virus PrV Kaplan-ΔgG-GFP was used for investigating the effect on the established hippocampal neurons cultured on MEAs. Due to the GFP label introduced into this virus, the infection of the cultures could be easily proved and monitored over the whole time of the experiment (Supplementary Figure S4). For the analysis of the virus-amount-dependent effect, two concentrations (multiplicity of infection—MOI) were applied on day 21 after cell-seeding on MEAs, followed by 72 h of bioelectronic and microscopic monitoring of the virus-induced effects. The successful infection could be observed already after 12 h. While for an MOI of 0.1 this was generally observed in single cells, a widespread infection could be observed for an MOI of 1 (Supplementary Figures S4 and S5 and Figure 3A). During the following 24 h, the level of infected cells resembled each other. In combination with the transmitted light images, the morphological degeneration of the neuronal networks could be clearly observed. For an MOI of 1, first dissolution of the cell layer could be observed already after 24 h (Figure 3A), whereas for an MOI of 0.1, this was clearly visible only after 48 h (Supplementary Figure S5). After 72 h, a comprehensive dissolution of the whole cell layer was observable for both concentrations.

For bioelectronic monitoring of this degeneration process, we used impedance spectroscopy, which is widely used for in vitro analyses of cell degeneration and cytotoxic effects [14,21], in contrast to field potential monitoring.

For a comprehensive statistical analysis, we monitored six MEAs per group with cells from three different preparations (n = 6). The extracted cellular contribution to the impedance magnitude spectra (relative impedance spectra) revealed the high stability of the established neuronal cell cultures over the entire experimental time period of three days, as well as the time- and virus-amount-dependent cell layer degenerating effect (Figure 3B). From the relative impedance spectra of each electrode, the maximum cell signal was determined and traced over time (Figure 3C, left). The statistical analysis of the averaged time traces confirmed the high stability of the control cultures due to the small deviation of mainly $\pm$ 5%, except for a slight increase up to 113% between 48 h and 60 h, which is still a small change. In contrast, the virus application led to a concentration-dependent continuous decrease in the relative impedance after 24 h. For a statistical analysis of significant effects, values of selected time points were normalised to the appropriate control traces (Figure 3C, right). Thus, a significant decrease could be observed for an MOI of 1, beginning after 24 h (65.9 $\pm$ 7.2 %) and resulting in a decrease down to 7.5 $\pm$ 2.1 % after 72 h. For an MOI of 0.1, a first significant decrease was observed after 36 h (75.1 $\pm$ 7.7 %), followed by a continuous decrease down to 28.5 $\pm$ 5.3 % after 72 h. This observed time-dependent decrease in the impedimetric cell signal perfectly correlates with the microscopically observed degeneration of the cell layer. Additionally, we performed an electrical equivalent circuit-model-based fitting of the complex impedance dataset to receive more insights in the observed effects (Figure S6). The fitting of spectra from cell cultures could be successfully performed with a capacitance (C_{Cell}) and resistance (R_{Para}), parallel to an additional resistance in the series (R_{Ser}) (Figure S6A,B). The statistical analysis of these biological parameters revealed a virus-amount- and time-dependent increase in C_{Para}, as well as a decrease in R_{Ser} (Figure S6C,D). This proves the suitability of impedance spectroscopy as a monitoring technique for degeneration of cell layers and cytotoxic effects and correlates with previous studies of impedance spectroscopy-based neurodegeneration monitoring on neuronal cell lines [13,15,22]. In this context, the decrease of up to 92% means the extreme degeneration and damage of the whole cell layer, at least for the higher virus concentration. This is in line with previous studies using PrV Kaplan-ΔG-GFP [23].

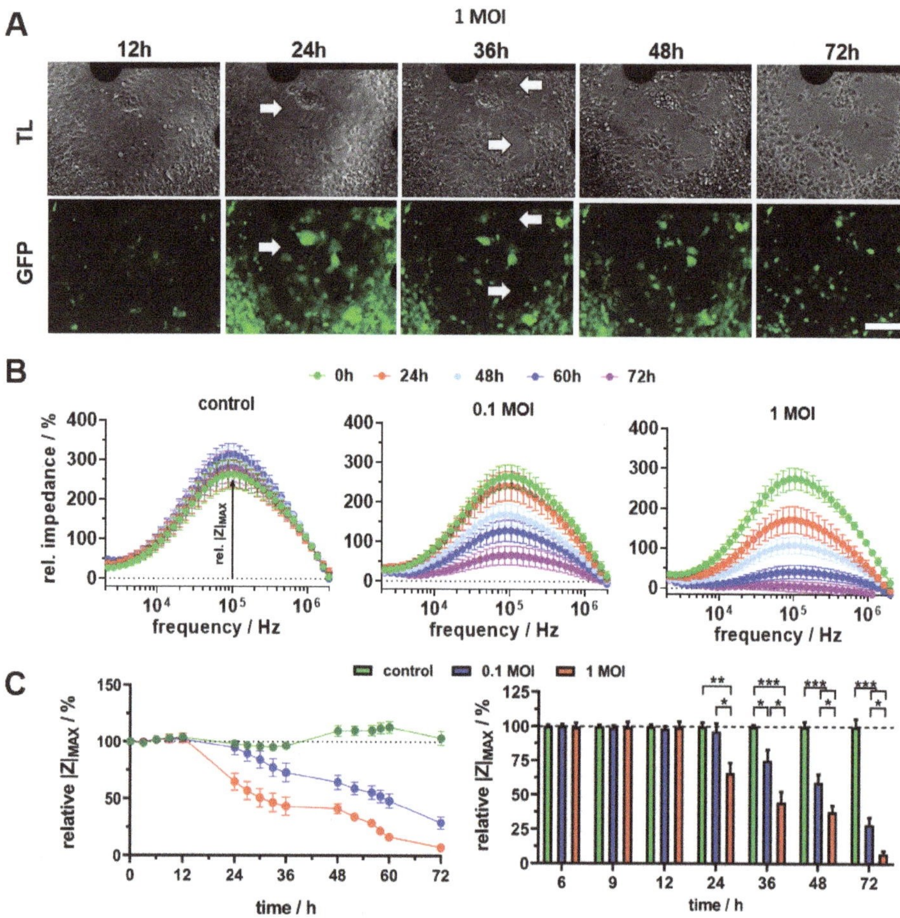

Figure 3. Virus-induced morphological degeneration of the neuronal network. (**A**) Transmitted light (TL) and fluorescence (GFP) microscopic images of neuronal cultures on an exemplarily MEA incubated with 1 MOI virus at discrete time points. White arrows mark initial areas of morphological dissolution/degeneration of the cell layer (scale bar = 200 µm). (**B**) Determined cellular contribution to impedance magnitude spectra (relative impedance) at selected time points (n = 6, mean ± sem). (**C**) Relative impedance spectra-derived maximum cell signal (relative impedance maximum) traces of control, 0.1 MOI and 1 MOI virus normalised to time point 0 h (**left**) as well as statistical analysis of values normalised to control for selected time points (**right**). (n = 6, mean ± sem, * $p < 0.05$, ** $p < 0.01$, *** $p < 0.001$).

3.4. Field-Potential-Based Analysis of Virus Effects on the Electrophysiological Activity of Neural Networks

In the next step, the concentration-dependent effect of virus infection on the electrophysiological activity was investigated by field potential monitoring. Therefore, the action potentials were automatically detected for all recorded data and plotted by the self-developed software FiPRAT for an initial review of the overall activity pattern (Supplementary Figures S7–S9). Here, the clear decrease in activity after 48 h (MOI = 0.1) and 24 h (MOI = 1), respectively, was striking. This correlates with the impedimetric monitoring of significant cell layer degeneration, which obviously should also result in a loss of electrophysiological activity. For an in-depth analysis, the recorded raw data of six MEAs per group with cells from three different preparations (n = 6) were analysed regarding several basic descriptive parameters describing the electrophysiological network activity (Figure 4A). First, the number of electrodes with electrophysiological activity were analysed. The relative values per MEA were normalised to time point 0 h (experiment start). The averaged time traces revealed a high stability of electrode numbers with neuronal activity over the whole 72 h, with maximum deviations of 3%. The virus application led to a decrease in electrode numbers with neuronal activity for an MOI of 1 and a more moderate but continuously decreasing number of electrodes for an MOI of 0.1. The statistical analysis for the significant effect of values normalised to the appropriate controls revealed a significant decrease for an MOI of 1 after 24 h and for an MOI of 0.1 after 36 h. For an MOI of 1, the electrode numbers with neuronal activity decreased to almost zero (1.4 ± 0.9 %) after 36 h and did not show any activity in any experiment after 72 h. For an MOI of 0.1, the electrode numbers with neuronal activity decreased to 39.1 ± 9.5 % after 48 h and 5.7 ± 4.5 % after 72 h.

In comparison to the impedimetric-based degeneration of the cell layer, the faster loss of the detectable network activity is comprehensible since this effect of loss of electrophysiologic functionality should clearly happen before a distinct cellular degeneration. For validating the presence of a neural network and the virus-induced breakdown, we fixated the cultures on the MEAs after the experiment (72 h) and stained them for the axonal marker neurofilament 200 (neuronal marker) and astrocyte marker GFAP (Figure 4B). The confocal laser scanning microscope-derived images revealed a pronounced and dense neuronal network with a high amount of long axonal structures as well as a homogeneously distributed support of astrocytes. In contrast, both virus concentrations led to a nearly complete disappearance of the neuronal network, as well as noticeable degeneration of the astrocyte support layer. This strengthens the impedimetric-derived observations for a virus-induced extensive cell layer degeneration as well as the field-potential-derived observation of a complete loss of electrode numbers with neuronal activity.

In the next step, the spike number (number of spikes per minute) was analysed. While after 24 h (MOI of 1) and 36 h (MOI of 0.1), respectively, there was another decrease in the spike number, dependent on virus concentration and time, more strikingly, an initial increase in spike numbers was noticeable. After 6 h, a significant increase could already be detected for an MOI of 0.1 (114.3 ± 3.9%) that further increased up to 134.6 ± 3.3% after 9 h and 132.4 ± 5.4% after 12 h. This effect was comparable for an MOI of 1 (9 h: 133.3 ± 8.9%, 12 h: 137.6 ± 6.2%). The analysis of the burst number (bursts per minute) revealed a comparable effect with a first significant increase after 9 h for an MOI of 1 (139.9 ± 9.2%), and both concentrations after 12 h (MOI of 0.1: 120.8 ± 4.6%, MOI of 1: 135.4 ± 9.7%). This is a very interesting observation and means that there is a distinct virus-induced increase in neuronal activity within the first 12 h after infection, before the virus-caused degenerative process leads to a breakdown of neuronal network activity.

Since there is a lack of comparable in vitro studies on neuronal network activity so far, we found only minor hints for such a virus-induced effect. In a previous study, the effect of the pseudorabies virus on a single rat's sympathetic neurons was analysed by patch clamp and revealed increased action potential firing rates within 8 h after infection [24].

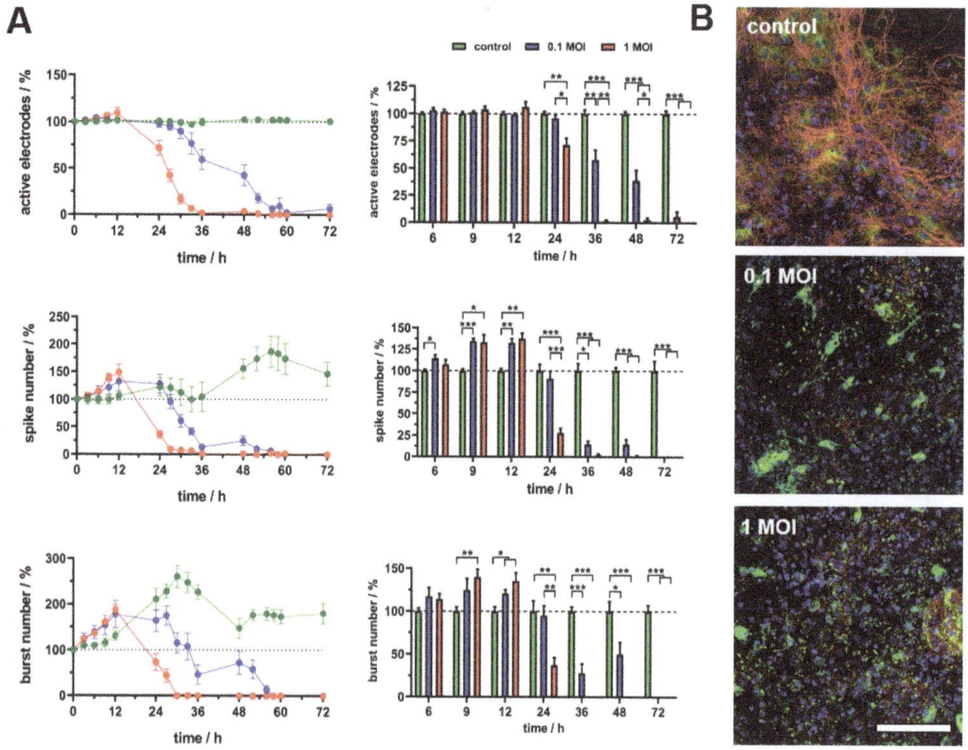

Figure 4. Virus effects on the electrophysiological activity of the neural network. (**A**) Time traces of discrete electrophysiological parameters for control, 0.1 MOI and 1 MOI that are normalised to time point 0 h (**left**) and statistical analysis of values normalised to control (**right**) for selected timepoints (n = 6, mean ± sem, * $p < 0.05$, ** $p < 0.01$, *** $p < 0.001$). (**B**) Immunocytochemical characterisation of neuronal cultures on MEAs after end of experiment (time point 72 h) with neuronal marker neurofilament 200 (red), astrocyte marker GFAP (green) and DAPI nuclei stain (blue) (scale bar 100 µm).

3.5. PrV Kaplan-ΔgG-GFP Virus Leads to Specific Time-Dependent Alteration of Neuronal Network Synchronicity

Beyond the analysis of individual electrode time traces and, thereof, derived activity parameters, the big advantage of microelectrode arrays is the clearly defined spatial resolution that allows the identification and analysis of network interactions. For demonstration, the field potential traces of six selected neighbouring electrodes were used (Figure 5A) to show partial and complete coordinated network communication, which can be described as synchronous signalling events.

Based on the knowledge of the electrode position, the activity can be spatially matched (Figure 5B zoom view, Supplementary Figures S10 and S11). The exemplarily shown maps for an MOI of 1 at 0 h and 12 h reveal a higher number of synchronously occurring spikes for 12 h, which means a higher network synchronicity. For a quantitative analysis of network synchronicity, the recorded raw data of six MEAs from three experiments were analysed. The time traces normalised to 0 h (Figure 5C) revealed a comparable effect to those observed for the spike and burst numbers. The virus application led to an initial significant increase in the synchronicity after 12 h (MOI of 0.1: 132.6 ± 13.5%, MOI of 1: 143.9 ± 13.9%), followed by a significant decrease after 24 h (MOI of 1) and 48 h (MOI of 0.1), respectively. While the later occurring decrease in network synchronicity comprehensibly correlates with the virus-induced degeneration of the network, the initially observed increase is, again, a very interesting observation. This means that the virus not only causes a temporal increase in neuronal activity, but also an increased synchronous network signalling.

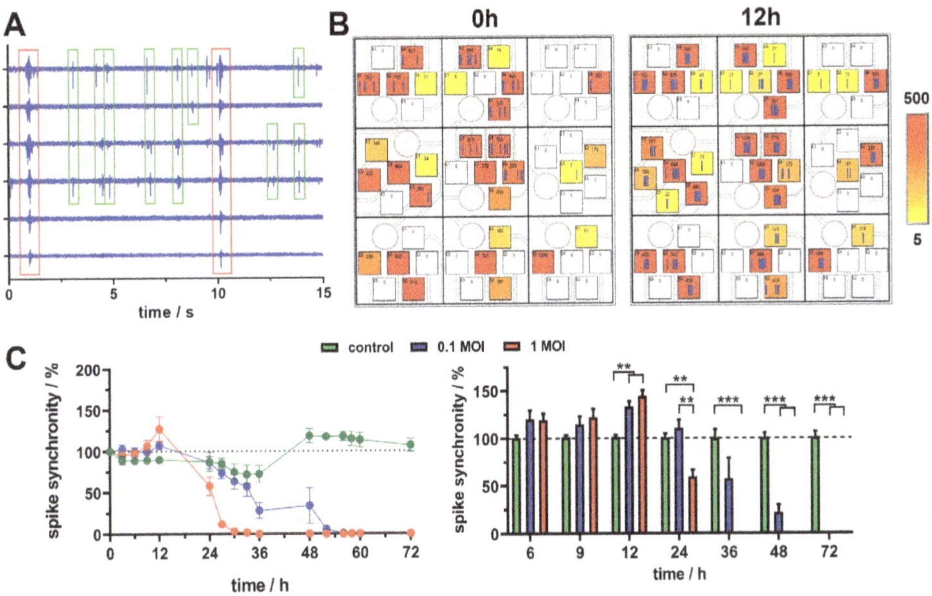

Figure 5. Analysis of neuronal network synchronicity. (**A**) Exemplary field potential traces of neighbouring electrodes with partially (green boxes) and completely (red boxes) synchronised spikes. (**B**) Spatial allocation of spike traces (blue vertical lines in square boxes, box wide corresponds to time window of two seconds) for active electrodes marked as coloured squares (colour code represents spike count per minute with a minimum threshold for activity of 5). Exemplarily selected time points for 1 MOI virus. (**C**) Time traces of spike synchronicity for control, 0.1 MOI and 1 MOI that are normalised to time point 0 h (**left**) and statistical analysis of values normalised to control (**right**) for selected time points (n = 6, mean ± sem, ** $p < 0.01$, *** $p < 0.001$).

4. Conclusions

In the presented study, we were able to demonstrate the successful development of a hybrid measurement system that is capable of high-resolved impedance spectroscopy and field potential monitoring on the same microelectrode. By using specific low noise switches for both measurement pathways, even high intrinsic impedance of small microelectrode measurement artefacts and high noise levels could be suppressed. Although there already is a developed field potential monitoring system described with, e.g., a much higher number of electrodes/electrode densities (e.g., CMOS MEAs), there is no other system described that offers the demonstrated capabilities and flexibilities for hybrid monitoring on microelectrode arrays (Supplementary Table S1). Moreover, the experiments were carried out using two 60-channel hybrid frontends in parallel to demonstrate the capabilities for parallelisation of the system. The current system can be easily scaled up to six parallel frontends for a more comfortable analysis of different experiment conditions. Moreover, the DAQ system is designed for further upscaling. Thus, thousands of measurement channels could be easily realised on demand.

For a demonstration of the system's capabilities, the pseudorabies model virus PrV Kaplan-ΔgG-GFP was applied on hippocampal cultures. The analysis revealed a structural and functional degeneration of the neuronal network within 24 to 48 h. More strikingly, we could detect a temporal increase in neuronal activity as well as synchronicity within the first 12 h after virus application. Therefore, this demonstrates an advantage for the complementary read-out of information on the same electrode and the same cell sample, in the sense of a multimodal monitoring in order to draw synergistic conclusions about the behaviour or reaction of neuronal networks to neuron-infecting viruses. Based on this successful proof of principle, the developed system can be used for the detailed long-term analysis of, e.g., the rabies virus (RABV) on different target hosts' in vitro models. Moreover, the multimodal hybrid monitoring platform could even be used for the analysis of complex virus effects in human neuronal networks that could be generated, e.g., from hiPSC.

Such in vitro systems could help to answer urgent medical questions of potential neurological risks from novel occurring viruses like coronavirus SARS-CoV-2.

Supplementary Materials: The following supporting information can be downloaded at: https://www.mdpi.com/article/10.3390/bios14060295/s1, References [25–28] are cited in the Supplementary Materials.

Author Contributions: Conceptualization, H.-G.J., A.A.R. and S.F.; Methodology, H.-G.J., V.t.K., S.S., R.A., B.K. and S.F.; Software, H.-G.J.; Formal analysis, H.-G.J.; Investigation, H.-G.J., V.t.K. and C.P.; Resources, S.S. and R.A.; Writing—original draft, H.-G.J.; Writing—review & editing, V.t.K., C.P., S.S., R.A., B.K., A.A.R. and S.F.; Visualization, H.-G.J.; Supervision, H.-G.J., A.A.R. and S.F.; Project administration, S.F.; Funding acquisition, A.A.R. All authors have read and agreed to the published version of the manuscript.

Funding: This work was funded by the Saxon Ministry of Science and the Fine Arts (SMWK, project All-OnME, Grant No. 100333200) and is tax-supported on the basis of the budget approved by the members of the parliament of the Free State of Saxony.

Institutional Review Board Statement: All performed work described in this article has been carried out in accordance with the German legal and ethical requirements of appropriate procedures. All animal work followed national and international guidelines in accordance with Directive 86/609/EEC for animal experiments in and under the responsibility of the Friedrich-Loeffler-Institute.

Data Availability Statement: Data is contained within the article or Supplementary Material, raw dataset is available on request from the authors.

Conflicts of Interest: The authors declare no conflicts of interest.

References

1. Zhang, L.; Qin, Y.; Chen, M. Viral strategies for triggering and manipulating mitophagy. *Autophagy* **2018**, *14*, 1665–1673. [CrossRef] [PubMed]
2. Ashraf, U.; Benoit-Pilven, C.; Lacroix, V.; Navratil, V.; Naffakh, N. Advances in Analyzing Virus-Induced Alterations of Host Cell Splicing. *Trends Microbiol.* **2019**, *27*, 268–281. [CrossRef]
3. Abdel-Hameed, E.A.; Ji, H.; Shata, M.T. HIV-Induced Epigenetic Alterations in Host Cells. *Adv. Exp. Med. Biol.* **2016**, *879*, 27–38. [CrossRef] [PubMed]
4. Yolken, R.H.; Jones-Brando, L.; Dunigan, D.D.; Kannan, G.; Dickerson, F.; Severance, E.; Sabunciyan, S.; Talbot, C.C., Jr.; Prandovszky, E.; Gurnon, J.R.; et al. Chlorovirus ATCV-1 is part of the human oropharyngeal virome and is associated with changes in cognitive functions in humans and mice. *Proc. Natl. Acad. Sci. USA* **2014**, *111*, 16106–16111. [CrossRef] [PubMed]
5. Ghassemi, S.; Asgari, T.; Mirzapour-Delavar, H.; Aliakbari, S.; Pourbadie, H.G.; Prehaud, C.; Lafon, M.; Gholami, A.; Azadmanesh, K.; Naderi, N.; et al. Lentiviral Expression of Rabies Virus Glycoprotein in the Rat Hippocampus Strengthens Synaptic Plasticity. *Cell. Mol. Neurobiol.* **2021**, *42*, 1429–1440. [CrossRef]
6. Potratz, M.; Zaeck, L.M.; Weigel, C.; Klein, A.; Freuling, C.M.; Muller, T.; Finke, S. Neuroglia infection by rabies virus after anterograde virus spread in peripheral neurons. *Acta Neuropathol. Commun.* **2020**, *8*, 199. [CrossRef]
7. Hueffer, K.; Khatri, S.; Rideout, S.; Harris, M.B.; Papke, R.L.; Stokes, C.; Schulte, M.K. Rabies virus modifies host behaviour through a snake-toxin like region of its glycoprotein that inhibits neurotransmitter receptors in the CNS. *Sci. Rep.* **2017**, *7*, 12818. [CrossRef]
8. Pomeranz, L.E.; Reynolds, A.E.; Hengartner, C.J. Molecular biology of pseudorabies virus: Impact on neurovirology and veterinary medicine. *Microbiol. Mol. Biol. Rev.* **2005**, *69*, 462–500. [CrossRef]
9. Mettenleiter, T.C. Pseudorabies (Aujeszky's disease) virus: State of the art. August 1993. *Acta Vet. Hung.* **1994**, *42*, 153–177.
10. Seidel, D.; Jahnke, H.G.; Englich, B.; Girard, M.; Robitzki, A.A. In vitro field potential monitoring on a multi-microelectrode array for the electrophysiological long-term screening of neural stem cell maturation. *Analyst* **2017**, *142*, 1929–1937. [CrossRef]
11. Obien, M.E.; Deligkaris, K.; Bullmann, T.; Bakkum, D.J.; Frey, U. Revealing neuronal function through microelectrode array recordings. *Front. Neurosci.* **2014**, *8*, 423. [CrossRef] [PubMed]

12. Novellino, A.; Scelfo, B.; Palosaari, T.; Price, A.; Sobanski, T.; Shafer, T.J.; Johnstone, A.F.; Gross, G.W.; Gramowski, A.; Schroeder, O.; et al. Development of micro-electrode array based tests for neurotoxicity: Assessment of interlaboratory reproducibility with neuroactive chemicals. *Front. Neuroeng.* **2011**, *4*, 4. [CrossRef] [PubMed]
13. Jahnke, H.G.; Rothermel, A.; Sternberger, I.; Mack, T.G.; Kurz, R.G.; Panke, O.; Striggow, F.; Robitzki, A.A. An impedimetric microelectrode-based array sensor for label-free detection of tau hyperphosphorylation in human cells. *Lab Chip* **2009**, *9*, 1422–1428. [CrossRef]
14. Eichler, M.; Jahnke, H.G.; Krinke, D.; Muller, A.; Schmidt, S.; Azendorf, R.; Robitzki, A.A. A novel 96-well multielectrode array based impedimetric monitoring platform for comparative drug efficacy analysis on 2D and 3D brain tumor cultures. *Biosens. Bioelectron.* **2015**, *67*, 582–589. [CrossRef] [PubMed]
15. Diemert, S.; Dolga, A.M.; Tobaben, S.; Grohm, J.; Pfeifer, S.; Oexler, E.; Culmsee, C. Impedance measurement for real time detection of neuronal cell death. *J. Neurosci. Methods* **2012**, *203*, 69–77. [CrossRef] [PubMed]
16. Zhang, X.; Guo, L.; Zeng, H.; White, S.L.; Furniss, M.; Balasubramanian, B.; Lis, E.; Lagrutta, A.; Sannajust, F.; Zhao, L.L.; et al. Multi-parametric assessment of cardiomyocyte excitation-contraction coupling using impedance and field potential recording: A tool for cardiac safety assessment. *J. Pharmacol. Toxicol. Methods* **2016**, *81*, 201–216. [CrossRef]
17. Fuchs, W.; Backovic, M.; Klupp, B.G.; Rey, F.A.; Mettenleiter, T.C. Structure-based mutational analysis of the highly conserved domain IV of glycoprotein H of pseudorabies virus. *J. Virol.* **2012**, *86*, 8002–8013. [CrossRef] [PubMed]
18. Quiroga, R.Q.; Nadasdy, Z.; Ben-Shaul, Y. Unsupervised spike detection and sorting with wavelets and superparamagnetic clustering. *Neural Comput.* **2004**, *16*, 1661–1687. [CrossRef] [PubMed]
19. Wang, X.J.; Buzsaki, G. Gamma oscillation by synaptic inhibition in a hippocampal interneuronal network model. *J. Neurosci.* **1996**, *16*, 6402–6413. [CrossRef]
20. Jahnke, H.G.; Schmidt, S.; Frank, R.; Weigel, W.; Pronnecke, C.; Robitzki, A.A. FEM-based design of optical transparent indium tin oxide multielectrode arrays for multiparametric, high sensitive cell based assays. *Biosens. Bioelectron.* **2019**, *129*, 208–215. [CrossRef]
21. Ceriotti, L.; Ponti, J.; Colpo, P.; Sabbioni, E.; Rossi, F. Assessment of cytotoxicity by impedance spectroscopy. *Biosens. Bioelectron.* **2007**, *22*, 3057–3063. [CrossRef] [PubMed]
22. Adil, O.; Shamsi, M.H. Electrochemical Impedance Immunoassay for ALS-Associated Neurofilament Protein: Matrix Effect on the Immunoplatform. *Biosensors* **2023**, *13*, 247. [CrossRef] [PubMed]
23. Jons, A.; Mettenleiter, T.C. Green fluorescent protein expressed by recombinant pseudorabies virus as an in vivo marker for viral replication. *J. Virol. Methods* **1997**, *66*, 283–292. [CrossRef] [PubMed]
24. McCarthy, K.M.; Tank, D.W.; Enquist, L.W. Pseudorabies virus infection alters neuronal activity and connectivity in vitro. *PLoS Pathog.* **2009**, *5*, e1000640. [CrossRef] [PubMed]
25. Emery, B.A.; Hu, X.; Khanzada, S.; Kempermann, G.; Amin, H. High-resolution CMOS-based biosensor for assessing hippocampal circuit dynamics in experience-dependent plasticity. *Biosens. Bioelectron.* **2023**, *237*, 115471. [CrossRef] [PubMed]
26. Miccoli, B.; Lopez, C.M.; Goikoetxea, E.; Putzeys, J.; Sekeri, M.; Krylychkina, O.; Chang, S.W.; Firrincieli, A.; Andrei, A.; Reumers, V.; et al. High-Density Electrical Recording and Impedance Imaging With a Multi-Modal CMOS Multi-Electrode Array Chip. *Front. Neurosci.* **2019**, *13*, 641. [CrossRef] [PubMed]
27. Middya, S.; Curto, V.F.; Fernandez-Villegas, A.; Robbins, M.; Gurke, J.; Moonen, E.J.M.; Kaminski Schierle, G.S.; Malliaras, G.G. Microelectrode Arrays for Simultaneous Electrophysiology and Advanced Optical Microscopy. *Adv. Sci.* **2021**, *8*, 2004434. [CrossRef]
28. Zhao, E.T.; Hull, J.M.; Mintz Hemed, N.; Ulusan, H.; Bartram, J.; Zhang, A.; Wang, P.; Pham, A.; Ronchi, S.; Huguenard, J.R.; et al. A CMOS-based highly scalable flexible neural electrode interface. *Sci. Adv.* **2023**, *9*, eadf9524. [CrossRef]

biosensors

MDPI

Review

Democratization of Point-of-Care Viral Biosensors: Bridging the Gap from Academia to the Clinic

Westley Van Zant and Partha Ray *

Department of Medicine, Division of Infectious Diseases and Global Public Health, University of California, San Diego, CA 92093, USA
* Correspondence: pray@health.ucsd.edu

Abstract

The COVID-19 pandemic and recent viral outbreaks have highlighted the need for viral diagnostics that balance accuracy with accessibility. While traditional laboratory methods remain essential, point-of-care solutions are critical for decentralized testing at the population level. However, a gap persists between academic proof-of-concept studies and clinically viable tools, with novel technologies remaining inaccessible to clinics due to cost, complexity, training, and logistical constraints. Recent advances in surface functionalization, assay simplification, multiplexing, and performance in complex media have improved the feasibility of both optical and non-optical sensing techniques. These innovations, coupled with scalable manufacturing methods such as 3D printing and streamlined hardware production, pave the way for practical deployment in real-world settings. Additionally, software-assisted data interpretation, through simplified readouts, smartphone integration, and machine learning, enables the broader use of diagnostics once limited to experts. This review explores improvements in viral diagnostic approaches, including colorimetric, optical, and electrochemical assays, showcasing their potential for democratization efforts targeting the clinic. We also examine trends such as open-source hardware, modular assay design, and standardized reporting, which collectively reduce barriers to clinical adoption and the public dissemination of information. By analyzing these interdisciplinary advances, we demonstrate how emerging technologies can mature into accessible, low-cost diagnostic tools for widespread testing.

Keywords: biosensors; accessible diagnostics; point-of-care (POC) testing; democratization

1. Introduction

In recent years, viral diagnostic technologies have evolved significantly, rapidly improving to meet the demand for quick, accurate, and accessible testing capabilities. Given the emergence of large-scale viral infections such as avian influenza (H5N1) [1], Ebola [2], and the Zika virus [3], the need for advancements in diagnostic technologies remains pressing. Effectively diagnosing emerging viruses, especially in the early stages of infection, is critical to prevent localized outbreaks from reaching a global scale [4]. The ongoing COVID-19 pandemic, caused by SARS-CoV-2, illustrates the severe consequences of failing to control localized endemic events before the pathogen spreads through global markets and travel systems. The pandemic also highlighted the demand for effective diagnostics to extend beyond the laboratory, with population-level testing pushing diagnostics toward at-home and point-of-care (POC) settings [5]. This has pushed development efforts to expand in several directions, with enhancements in sensitivity and accuracy balanced with cost reduction, miniaturization, scalability for mass production, and simplified sampling and data processing [6]. The World Health Organization (WHO) broadly characterizes ideal POC diagnostics according to their real-time connectivity, ease of specimen collection, affordability, sensitivity, specificity, user-friendliness, rapidness, and robustness [7]. Advancements in these characteristics address the challenges posed by emerging and ongoing infectious diseases and democratize access to diagnostics, both in their administration and in interpreting their results.

At its core, a biosensor typically includes a biorecognition element designed to capture diagnostic targets selectively. These targets include nucleic acids, viral proteins, antibodies, and entire virions [8,9]. The capture elements for these biomarkers can take various forms, such as antibodies (both whole and functional fragments) and proteins (Figure 1A), or nucleic acids (Figure 1B), like DNA, RNA, and aptamers [10]. Alternatively, artificial receptors like molecularly imprinted polymers (MIPs) can be templated with viral proteins as a synthetic alternative to protein or nucleic acid capture [11] (Figure 1C). Combining multiple biorecognition elements also enables the simultaneous detection of multiple analytes, including multiple viruses or antigens from one pathogen [12]. Accompanying the biorecognition element is a form of signal transduction that translates the capture of the biomarker into a measurable readout [13]. Optical signaling methods rely on changes in light properties, including refractive index, fluorescence, color changes, absorbance, or light interference [14]. Electrochemical signal transduction can appear in several forms, including amperometric readouts that measure current, potentiometric readouts of voltage, and changes in impedance [15]. Biosensors can also integrate mechanical signal transduction, exemplified by quartz crystal microbalance (QCM) and microcantilever sensors [16,17].

Furthermore, signal transduction may involve thermal or magnetic methods [18,19]. Signal amplification techniques can be employed to lower detection limits and enhance the readout of weak signals. Nanomaterials such as colloidal gold [20], quantum dots [21], and graphene nanotubes [22] can be combined with biorecognition elements to improve the quality and strength of faint signals. While some diagnostics like lateral flow assays (LFAs) provide a simple binary readout that users can interpret directly [23], other diagnostic methods require post-processing to convert the signal into a comprehensible output [24]. Various post-processing methods are employed, from software interfaces and smartphone apps that interpret results to more complex machine learning (ML) models designed to handle increasingly intricate datasets [25,26]. Moreover, viral diagnostics often utilize microfluidic sample delivery systems, which assist users by reducing sample volumes, directing the specimen to the sensing region, and minimizing the risk of human error associated with manual sample processing [27,28].

Multiple approaches to viral diagnostics can be taken, depending on the nature of the viral recognition target. The detection of viral genomes has become particularly ubiquitous due to the COVID-19 pandemic, and methods such as real-time reverse transcription polymerase chain reaction (RT-PCR), next-generation sequencing (NGS), and reverse-transcription loop-mediated isothermal amplification (RT-LAMP) have all undergone significant development in recent years [29,30]. While RT-PCR serves as the gold standard for viral detection due to its rapid time-to-result, high specificity, and low copy number requirements, it requires specialized instrumentation, trained personnel, and the production and storage of reagents such as polymerases and primers [31]. RT-LAMP does improve accessibility by eliminating the need for a thermal cycler, but it has also seen limited adoption in the clinic [32]. Other diagnostic methods target viral proteins or antibodies raised against them rather than genomic information. Methods like ELISA are the standard for protein-based diagnostics, but they also suffer from limitations in personnel, sample handling, and costly reagents like antibodies [33]. In this sense, there is a tradeoff in the diagnostic methods already demonstrated in point-of-care settings, where higher sensitivity often accompanies complex assay administration, expensive reagents, and a higher barrier for understanding assay readouts.

Across the scientific community, there is no shortage of promising publications effectively demonstrating advancements in viral detection from the standpoint of instrumentation and methodology. However, barriers still exist in bringing these new advancements to fruition in the clinic itself. In addition to logistical constraints like manufacturing costs and scalability, an underlying cause is the current prioritization dominating the research world. Development efforts centered around high-complexity instrumentation based on well-established local infrastructure can exclude disenfranchised communities from accessing novel diagnostic technologies [34]. Further contributing to this may be the cyclical nature of research funding, where pursuing novel assay developments rather than focusing on maturing established technologies can present a far more intellectually and financially viable approach when compared to addressing needs for cost improvements, multiplexing, and time-to-results. On the commercial side of research and development, diagnostic products are also frequently neglected because the contagious diseases prevalent in low-income countries do not yield substantial economic returns for biotech companies—this lack of profitability results in insufficient investment in their development [35]. Until this mindset largely shifts, this cycle of seeking new research areas rather than honing existing methodologies will likely continue.

Factors like the need to control local outbreaks, expand accessibility to disenfranchised communities, and conduct population-scale testing mean diagnostics must expand beyond central laboratories, regardless of how technically effective lab-oriented methods may be. One potential solution to this issue is establishing partnerships between academic institutions, industry, regulatory agencies, and government. In this model, the most promising diagnostic products developed in academic laboratories would receive targeted financial incentives from the government and be manufactured by the industrial partners with approval from the regulators. This collaboration could significantly advance the development of vital diagnostic products for infectious diseases. The feasibility of this approach has been demonstrated in the recent NIH Rapid Acceleration of Diagnostics (RADx) initiative, which provides companies with small business innovation research (SBIR) and small business technology transfer (STTR) grants to aid in the rapid development of COVID-19 diagnostics alongside universities and regulatory bodies [36,37]. Expanding such programs beyond COVID-19 could help address the diagnostic gaps for other infectious diseases, leveraging this new but now-proven collaborative model.

We must recognize the urgent need for solutions to infectious diseases. Access to effective diagnostics and treatments is crucial for the health and well-being of millions. By prioritizing this partnership, we can drive meaningful progress and tackle the critical health challenges that underserved populations face. In today's interconnected world, no geographical barrier can prevent infectious diseases from crossing borders. The rapid globalization of trade and travel has facilitated the movement of people and goods around the globe, increasing the likelihood of pathogens being transmitted from one location to another [38]. In this environment, outbreaks can escalate quickly, as seen in the case of diseases like COVID-19, which spread swiftly from one country to another. This highlights the vulnerabilities of public health systems worldwide.

In this review, we discuss a variety of advancements in viral diagnostics that help bring assays developed in academic labs closer to the clinic for real-world implementation. We highlight that across the several approaches to viral biosensing, from optical to non-optical methods, improvements in cost, sensor design, and assay administration have greatly improved the accessibility and breadth of novel diagnostics. While many of the advancements discussed have not yet been implemented in POC settings, they demonstrate the ever-moving push to bridge the gap between POC accessibility and laboratory diagnostic technology. Further, we discuss the production processes that improve the feasibility of pushing state-of-the-art diagnostics into clinical settings by lowering costs, enabling local component manufacturing, and simplifying both diagnostic administration and reporting through pre-functionalized sensors and out-of-the-box software tools. Finally, we discuss accessibility improvements in scientific reporting, which support the transition of novel diagnostics into the clinic through administrative shifts towards accessible and understandable literature and reporting.

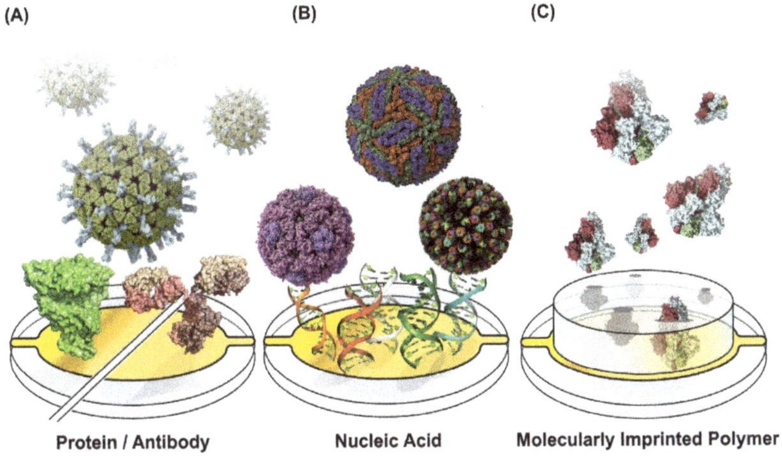

Figure 1. Capture Motif-Functionalized Biosensor Surfaces. Sensing surfaces are functionalized across the various viral diagnostic approaches with motifs for capturing proteins and whole virions. (**A**) Proteins can be surface-tethered capture moieties, with virus-binding proteins and antibodies providing specific binding (PDB IDs 4V7Q, 1R42, 1IGT). (**B**) Surface-linked nucleic acids can serve a similar viral capture function while providing additional stability and cost reduction. Their programmability enables multiplexed detection for sensing multiple analytes on one sensor surface (PDB IDs 5IRE, 1CD3, 3KZ4). (**C**) Molecularly imprinted polymers (MIPS) can be pre-produced to capture viruses with cavities imprinted to capture specific viral proteins (PDB ID 6VXX). Structures were rendered using UCSF ChimeraX 1.9 [39], and figures were generated using Adobe Illustrator version 24.

2. Optical Methods

2.1. Surface Plasmon Resonance

Surface Plasmon Resonance (SPR) is a label-free optical technique that utilizes photon–electron interactions with plasmonic materials. The method operates through the phenomenon of light under ideal matching conditions, undergoing total internal reflection at the interface of a thin metal film and a dielectric. Tracking the angle of light at which the total internal reflection occurs enables the real-time analysis of binding events since changes in the refractive index will shift the angle of incidence [40] (Figure 2A). SPR biosensing is commonplace in pharmaceutical research due to its ability to study binding affinities between biomolecules. It is well-suited to screening drugs at on and off-rates and affinities. Still, it can also detect the presence of antibodies and viral antigens. The absence of labels means that the interactions being studied are closer to their native interactive states, and the method is exceptionally sensitive [41]. The technique is not without flaws, however. Typical commercial instrumentation is often expensive, and costs per run are high due to the use of plasmonic metals (like gold and silver) in chip production. Still, instruments frequently costing hundreds of thousands of dollars cannot be feasibly implemented in clinics. Further, while SPR is straightforward to train, it requires the additional considerations of buffer preparation and system design to minimize bulk refractive index shifts and non-specific surface binding [42], which makes it adoption in clinics difficult. Further, multiplexed sensing is limited by the number of fluidic channels, which is a constraint that proves difficult to overcome with expensive and closed-source instrumentation.

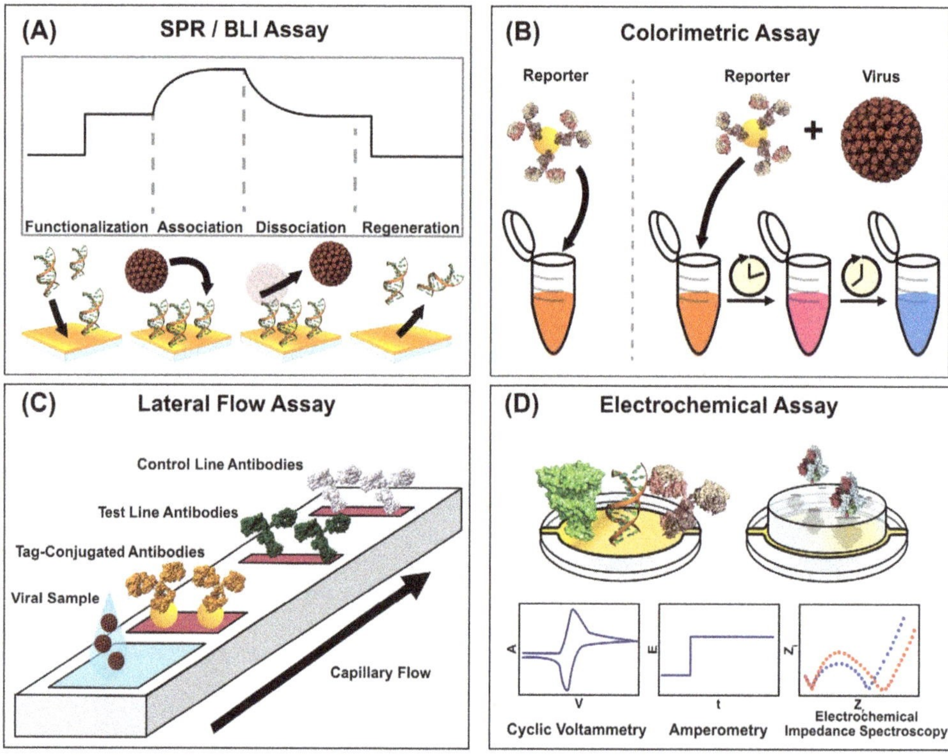

Figure 2. Schematics of Viral Biosensing Approaches. Viral biosensing can be achieved in a variety of ways. (**A**) Label-free biosensing methods can track the association of surface-immobilized capture motifs and analytes in real time, giving rise to kinetic and binding affinity parameters. (**B**) Colorimetric assays detect viral particles through the association of viral particles with reporter-conjugated capture moieties, producing a distinct visible readout with simple assay administration. Viral detection is coupled with a progressive, visible color change in the sample, while a virus-free sample remains as the same starting color. (**C**) Lateral flow assays are a clinical standard due to their relative simplicity and binary readout resulting from reporter-conjugated binding activity coupled with viral binding. The different antibodies, consisting of control, test, and tag-conjugated, are colored in white, green, and orange respectively to illustrate the multiple regions through which the sample flows. (**D**) Electrochemical assays can utilize the electrical signals associated with chemical binding events, and they can take several approaches, which include cyclic voltammetry, amperometry, and electrochemical impedance spectroscopy. (PDB IDs: 1D28, 4V7Q, 1IGT, 1R42, 6VXX). Structures were rendered in UCSF ChimeraX 1.9 [39], and figures were generated in Affinity Designer 2.

Viral SPR biosensing is by no means a novel technique, however. For instance, Riedel et al. demonstrated an SPR biosensor for detecting the Epstein–Barr virus in patient sera in 2014 [43]. Similarly, Florschütz et al. used surface-immobilized nucleotides to capture the barley stripe mosaic virus RNA (BSMV) in 2013 [44]. More recently, Yano et al. demonstrated a SARS-CoV-2 SPR biosensor that detected the nucleocapsid, claiming a

LOD as low as 85 fM, in which surface-immobilized anti-nucleocapsid antibodies were used for capture [45]. In contrast, secondary antibody-conjugated AuNPs were used for amplification. Surface-immobilized aptamers have also been used for viral detection, with sensors for the HIV-1 Tat protein [46] and the SARS-CoV-2 N-protein in complex media [47], which have been shown using commercial instrumentation.

While these studies demonstrate the method's ability to detect viral analytes, recent work has aimed to push SPR closer to clinical relevance. Recently, a portable SPR system for analyzing nasopharyngeal COVID-19 tests has been demonstrated, using covalently linked, RBD-binding nanobodies on the chip surface. The study utilized a small-footprint SPR spectrometer [48], which is a fraction of the size of other commercial SPR instruments, and machine learning was incorporated to help identify positive signal shifts. In another SARS-CoV-2 assay, a four-channel SPR chip was immobilized with the spike protein and the human ACE2 receptor to detect antibodies and viral particles in serum samples [49].

The size reduction of SPR instruments is gradually progressing, with instruments from the likes of OpenSPR (Nicoya, Kitchener, ON, Canada) [50], NanoSPR (NanoSPR, Chicago, IL, USA) [51], and Plasmetrix (Plasmetrix, Montreal, QC, Canada) [52] demonstrating how SPR instruments can easily find a place in the clinic where larger commercial instruments could not. In addition to their physical footprint, sensor surface advancements have also been made to improve the clinical adoptability of SPR. Wallace et al. showed the development of disposable, multiplexed plasmonic chip surfaces with multiple discrete sensing regions that could detect SARS-CoV-2, Norovirus, and Zika [12]. The pre-functionalized surface allowed them to be produced and shipped to clinics, ready to use out of the box; this is alongside improvements in the size of recent instrumentation. While this system was centered around hyperspectral polarimetry, the plasmonic nature of the chip design means it could be implemented in SPR systems, and more importantly, pre-functionalized surfaces reduce the complexity of running assays in a clinic. While SPR systems are already sufficiently sensitive to detect viral biomarkers in patient samples, there is still work regarding miniaturization, adaptable chip surfaces, and multiplexed sensing methodologies to push them closer to clinical adoption. There is an unavoidable cost regarding sensor chips due to their reliance on rare metals, but reusable surfaces and multiplexed sensing can help achieve more diagnoses per chip investment.

2.2. Biolayer Interferometry

Biolayer interferometry is gaining traction as another versatile tool for detecting viral infections. BLI systems characterize binding interactions through a recognition ligand coupled to a sensor tip. Interactions with analytes in solution occur at the tip, creating a thin layer that causes interference in light passing through the length of the sensor [53]. Light bouncing off a ligand-binding and a stationary reference surface causes an interference pattern, which induces a measurable wavelength shift. Since the wavelength shift is proportional to the thickness of the layer, active measurement enables real-time binding analysis [54], which can elucidate kinetic parameters as biomarkers bind. Measuring the biomolecular interactions of analytes directly in solution and minimizing sample workup procedures can make it well-suited to point-of-care applications. Like SPR, biolayer interferometry carries the benefits of sensitive, real-time, and label-free biosensing (Figure 2A). Still, careful surface design and considerations like non-specific binding and surface regeneration are required. BLI avoids the time-consuming steps of washing and visualization necessary for ELISA diagnostics, making it more consistent in the hands of clinicians. BLI has been used to develop viral inhibitors, such as therapeutic antibody epitope mapping [55] and the characterization of inhibitors for viral infections [56,57]. Still, its applicability extends into diagnostics as well. In addition to simplified sensor preparation, BLI's applicability in the clinic can be further bolstered with more robust capture probes on the sensor surface. Poolsup et al. demonstrated a BLI aptasensor to detect SARS-CoV-2 anti-nucleocapsid antibodies in human saliva [58]. By using aptamers rather than protein-based capture ligands, the sensor can be made far more robust, which is especially conducive for sensors that will be mass-manufactured and distributed to clinics [9].

Numerous studies have utilized BLI to characterize virus–ligand interactions, and several have tried to democratize novel viral diagnostics for clinical use. For instance, Dzimianski et al. demonstrated a single-use BLI assay for detecting antibodies against the SARS-CoV-2 spike protein while achieving isotype detection [59]. By producing sensors pre-functionalized with capture ligands, their approach demonstrates how simplification can reduce costs and quicken the time-to-result for BLI diagnostics. Additionally, this assay was used on a BLI instrument compatible with 96-well plates, offering markedly greater throughput.

Further, single-particle interferometric resonance imaging sensors (IRISs) utilize sensor chips arrayed with viral capture probes to count and size viral particles by capturing light transmitted from an LED imaging system [60]. This approach has proven versatile in viral detection, with studies utilizing SP-IRIS to detect VSV [61], H1N1 [62], and filamentous ebolaviruses [60]. In terms of accessibility, this method simplifies the diagnostic process by eliminating the purification steps and reducing sample volumes. Recent improvements in this methodology have further improved its operational simplicity by utilizing multi-wavelength sensing, eliminating the need to capture and process Z-stacked images [63]. Minimizing the need for trained personnel, this instrument could outcompete ELISA methods in detecting the monkeypox virus (MPXV) in 20 min. Doing so while fitting on a benchtop, the improvements in simplicity and time-to-result make this instrument well-suited for clinical use.

The recent developments in BLI have shown how well-fitting this method can be for clinical viral diagnostic applications. The continued iterative improvements in BLI sensing show how the development of novel diagnostic technologies does not end at demonstrating their efficacy but rather how continued refinement helps push proof-of-concept techniques into practical implementation in the clinic. While the quirks of novel biosensors can be worked around more easily in academic labs, where personnel are highly specialized and intimately involved in the development process, honing the technology for simplified use opens the door for untrained personnel to make use of the technology in the field.

2.3. Colorimetric and Fluorescent Sensors

Colorimetric sensors represent a class of highly accessible diagnostic tools already commonly used in clinical viral diagnostics. Unlike methods such as PCR, which rely on complex instrumentation and specialized training, colorimetric sensors translate the presence of viral biomarkers, such as antibodies, proteins, or nucleic acids, into visible color changes. This readout mechanism pairs target-specific capture motifs like antibodies or oligonucleotides with chemical or nanomaterial-based signal amplifiers, such as gold nanoparticle aggregation or enzymatic chromogenic reactions (Figure 2B). The resulting signals are detectable by the naked eye or with basic spectrophotometric devices. They represent a rapid, affordable, simple-to-administer testing strategy that retains strong accuracy. These tests are relatively simple and conducive to widespread field testing, evidenced by the widespread adoption of lateral flow assays (LFAs) [64] (Figure 2C). Unlike the more complex optical or electrochemical sensing methods, colorimetric sensors have already established a strong presence in clinical and at-home testing.

In contrast, fluorescent systems can offer improved sensitivity and quantitative abilities by coupling biomarker capture with a light-emissive signal. However, this can come with a reduction in simplicity compared to a simple colorimetric lateral flow assay, as instrumentation would be needed for precise quantification. Current clinical advancements in colorimetric and fluorescent platforms focus on refining sensitivity and usability, with CRISPR-Cas systems emerging as an approach for extremely accurate yet simple diagnostic systems. CRISPR-Cas systems' strong specificity and programmability enable significantly more sensitive sensing and the ability to tune existing diagnostics for new viruses and mutated variants.

The specificity of CRISPR-Cas-based colorimetric reporter systems has been utilized to perform highly multiplexed diagnostics for multiple viruses and subtypes, simplifying the administration of multi-virus testing by requiring only one substrate [65]. Further, this strong specificity can help reduce the number of sample workup and purification steps by enabling consistent viral sensing directly in serum [66]. With the time sensitivity accompanying viral infections, improvements in the administration and time-to-result of tests make them much more conducive for rapid field testing. Beyond CRISPR systems, other novel colorimetric reporting systems have also been demonstrated. Weerathunge et al. demonstrated a Murine Norovirus (MNV) colorimetric sensor, which utilized a gold 'nanozyme' that produces a blue color when the virus displaces color-suppressing aptamers [67]. Offering speed, sensitivity, and simplicity in administration, this example shows how alternatives to established capture and readout mechanisms in colorimetric tests can improve upon an already accessible diagnostic approach.

Despite their simplicity, colorimetric sensors have still exemplified how further refinements can help simplify the interpretation of assay results. Visual readouts are well-suited to the simple, binary detection of an infection, but granularity can be added to colorimetric readouts by coupling computational analysis with the assay. For instance, in a plasmonic nanoparticle-based LFA for SARS-CoV-2 detection, a bimodal detection approach was used to provide an initial visual readout followed by a quantifiable fluorometric reporter [68]. Two-step readouts like this could benefit clinical testing since the first reporter can enable quick responses like isolation and contact tracing. In contrast, the more accurate secondary readout can help clinicians plan treatment regimens. Quantifiable colorimetric readouts would previously require dedicated detection hardware, but recent work has aimed to democratize this process by using hardware that clinics will already have. Since the requirements for processing a colorimetric readout are essentially a light source, a camera, and a means of processing a signal, smartphones and computers can replace a dedicated detector, using their screens as a light source and their built-in cameras for detection [69]. For example, Kim et al. demonstrated a SARS-CoV-2 CRISPR-Cas12 trans-cleavage assay using a smartphone app to quantify the colorimetric readouts of N-protein detection [70]. Using a common dye for the reporting system, the assay was both cheap to produce and able to be run by anyone with a smartphone. Incorporating easy-to-use interfaces in such readout programs makes an additional dimension of viral diagnosis accessible to clinicians and patients without any modification to assay administration. Such assays could also be implemented using commodity hardware in clinics with a low financial barrier.

Colorimetric diagnostics' simplicity, mainly when administered in a lateral flow assay, has already achieved the accessibility that other diagnostics seek to attain. However, there is always room for refinement. The binary nature of colorimetric readouts leaves much to be desired regarding quantification, and shifts towards more granular readouts offer improvements in this area. Further, the development of multiplexed tests can improve the number of viruses being screened within a single assay, improving the efficiency of each assay.

2.4. Non-Optical Methods

While both optical and colorimetric biosensing technologies have seen significant improvements, which improve their clinical viability, electrochemical sensing methods have presented themselves as another viral diagnostic approach (Figure 2D that has also seen recent democratization efforts. These methods, including impedance spectroscopy, voltammetry, and amperometry, offer sensitive and quantitative diagnostic readouts packaged into small, cheap, and simple systems. Impedance spectroscopy monitors the changes in electrical resistance resulting from the binding of an analyte, such as a viral antigen, to a capturing surface. Voltammetry measures binding by analyzing current as a function of applied potential, which can enable viral detection directly through analyte oxidation or reduction and redox-active labels. Voltammetric sensors boast strong sensitivity, rapid measurement, and relative portability, and their sensitivity can be maximized by incorporating specific capture elements like antibodies [71]. Amperometry biosensors achieve detection by measuring current at a fixed potential, where analyte binding results in a proportional, measurable redox reaction at the electrode. The simplicity of amperometric biosensors makes them well-suited to compact and portable biosensing devices, such as those that could be used in clinical and field testing [72]. Nonetheless, there are still practical barriers to seeing electrochemical viral biosensing developments through to the stage of full clinical implementation. Often, electrochemical sensors require a means of additional amplification to improve their sensitivity. Further, surface biofouling becomes particularly prevalent when sensing complex media, as the absorption of proximal proteins, nucleic acids, and lipids to the sensor surface can reduce the long-term stability of the sensing system [73]. Several recent advancements in electrochemical viral biosensors have been made in their simplicity, ease of use, and robustness, bringing them closer to feasible implementation in the clinic.

Alongside the accessibility and cost improvements in optical biosensing, electrochemical viral sensors have also trended towards cheaper and more robust sensing surfaces. These improvements can be tailored to clinical adoption through reusable or disposable single-use sensing surfaces. It was recently demonstrated that gold leaf electrodes—made of extremely thin gold sheets—can be engineered into electrochemical viral biosensors by immobilizing oligonucleotides conjugated with redox-mediating compounds like methylene blue [74]. Since the electrodes could be assembled with gold leaf sheets, razors, stencils, and simple adhesives, they lack the costly cleanroom equipment and labor often required to manufacture gold electrodes. Further, incorporating isothermal amplification into the assay protocol enabled the detection of HIV and HPV without even a thermal cycler. This approach to electrode manufacturing is strongly conducive to mass manufacturing and self-assembly in grassroots testing environments. Advances in electrode materials have also pushed the democratization of electrochemical-based viral diagnostics. Electrodes utilizing graphene have also proven effective for viral biosensing due to their high surface area, conductivity, and compatibility with functionalization. For example, Huang et al. demonstrated a graphene-based Hepatitis B surface antigen sensor, which also touted simple, cleanroom-free manufacturing, that could detect the antigen in human serum [75]. Although the electrode was primarily described as an alternative for lab-based testing, this sensor's simple manufacturing and disposability could also make it well suited to the clinic as a cheap, pre-manufactured product. The benefits of graphene-based electrodes have been demonstrated more widely in the field, with graphene being incorporated into nanocomposite inks [76] and even bacterially derived cellulose [77]. Incorporating recognition elements into matrices like cellulose and inks enables deposition through screen printing for cheaper and simpler electrode fabrication as an alternative to simple metal electrodes. Screen printing enables the mass production of graphene electrodes, as their surfaces can be engineered to immobilize virus-recognizing moieties like antibodies [78]. While antibody-immobilized electrodes are less affordable and robust than capture moieties like nucleic acids, their specificity is strongly conducive to directly testing patient samples in media such as saliva or serum. Further, the automated nature of printing heads could be combined with multiple inks containing different capture elements to give way to a scalable means of producing multiplexed-capable electrochemical sensors against multiple viral targets.

Electrode technology is one area where electrochemical sensing has advanced, with significant progress in improving sensing hardware for clinical adoption. While academic researchers may find traditional electrochemical setups straightforward to assemble, clinical labs require hardware tailored to their needs—prioritizing speed, throughput, and ease of use. Compact, user-friendly designs are far more practical in clinical settings, and recent studies have demonstrated such solutions for viral sensing. For example, during the COVID-19 pandemic, multiple compact and affordable sensors were developed using accessible electronics and simple readouts. Salahandish et al. created a multiplexed electrochemical impedance spectroscopy (EIS) sensor with commercially available hardware [76], enabling the binary detection of SARS-CoV-2 and variant-specific identification. Similarly, another study demonstrated an amplification-free immunosensor targeting viral RNA using a portable, smartphone-operated electrochemical workstation [79]. This plug-and-play system required just two copies of SARS-CoV-2 per assay, proving that compact hardware can maintain high sensitivity. Further streamlining clinical viral sensing, wireless data transmission allows at-home testing with real-time results sent directly to clinics. One example used laser-engraved graphene electrodes in a SARS-CoV-2 diagnostic device that wirelessly transmitted multiplexed results [80]. Each sensor detected viral proteins or IgG/IgM antibodies, enabling high-sensitivity, multi-target testing. Importantly, these sensing surfaces are also disposable, making them promising candidates for pre-packaged and sterile single-use assays. Such devices could enhance telemedicine by bridging the gap between home and clinic. Moreover, integrating wireless sensors into IoT ecosystems and population monitoring platforms could significantly improve large-scale infection surveillance.

Electrochemical viral biosensing has emerged as a versatile and democratized diagnostic tool, combining sensitivity, portability, and cost-effectiveness to bridge the gaps between research and clinical implementation. Advances in impedance spectroscopy, voltammetry, and amperometry have enabled compact, quantitative systems. At the same time, innovations in electrode materials, such as affordable gold leaf and graphene surfaces, have simplified manufacturing and improved accessibility. Coupled with user-friendly hardware designs, including smartphone-compatible workstations and wireless telemedicine integration, these systems offer multiplexed, high-sensitivity detection without sacrificing practicality. However, barriers to clinical translation persist. Protein-based capture motifs like antibodies still face shelf-life stability and batch-to-batch variability challenges. At the same time, temperature sensitivity during storage and operation can compromise performance in resource-limited settings. Biofouling in complex matrices like blood or saliva also remains a critical limitation, often requiring surface passivation or frequent recalibration [81].

Additionally, many systems still rely on amplification steps or redox labels to achieve clinically relevant sensitivities, adding complexity to workflows. Despite these hurdles, disposable sensing surfaces and scalable production methods (like screen-printed graphene electrodes [82]) underscore the potential for widespread adoption. As electrochemical sensors continue to evolve—addressing stability, standardization, and environmental robustness—their integration into Internet of Things (IoT) ecosystems and clinical workflows shows promise in transforming both point-of-care diagnostics and population-scale surveillance, bringing precision medicine closer to patients and clinicians.

2.5. QCM

Gravimetric sensing provides another potential avenue for the clinical detection of viruses, offering operational simplicity by not requiring labeling steps, which may be present in optical sensing methods. Quartz crystal microbalance (QCM) instruments utilize the piezoelectric effect, passing an electric potential through a transducer surface to induce a vibrational oscillation through the contraction and expansion of the quartz crystal lattice [83]. The addition of mass to the surface can be tracked through changes in the transducer's resonant frequency [84]. If a viscoelastic material is added to the surface, like biological materials, then the dissipation of the oscillations can also be tracked [83]. Like other surface-based sensing methods, QCM can be adapted for viral diagnostics by immobilizing viral capture ligands to the surface. While QCM has generally seen less use in clinical diagnostics, it presents a promising alternative to traditional sensing methods. This is mainly due to the instruments' low cost, compact size, and operational simplicity.

Like other surface-based methods such as SPR and BLI, it is possible to immobilize capture ligands, such as antigens and antibodies, on the surface for viral detection. The antigen-capture approach has been used for various viruses, including SARS-CoV-2 [85] and African swine fever (ASF) [84]. Antibodies have also been used for viral antigen capture, with studies using this QCM approach to detect HIV-1 [86], bovine leukemia virus [87], and influenza-A [88]. While these approaches were practical, they are not directly conducive to clinical use. The immobilization of proteins or antibodies for capture would need to be performed in the clinic, adding costly and fragile reagents, more complicated assay administration, and more opportunities for procedural error. An alternative approach is likely required to bring QCM into the clinic.

Molecularly imprinted polymers (MIPs) are a potential alternative to protein-based capture on QCM systems. They offer the ability to cheaply functionalize surfaces without the complications of temperature stability that protein capture ligands incur. This means that surfaces can be premade and sent to clinics without storage concerns, and that sensor surfaces would be ready to accept diagnostic samples with no additional workup. MIP surfaces are often washable as well, which makes these surfaces more affordable for clinics by lowering the cost per run [11]. However, molecularly imprinted polymer QCM sensors are not a recent approach in viral diagnostics. For instance, in 2012, Jenik et al. reported the development of an assay to bind different picornaviruses, such as human rhinovirus (HRV) and foot-and-mouth disease virus (FMDV), using stamp imprinting with template viruses [89]. Being able to differentiate between different viral serotypes with signals multiple orders of magnitude stronger than non-specific binding, the process shows how a multi-virus sensor could be built on a QCM platform using MIPs for capture. In 2005, Tai et al. demonstrated a similar molecularly imprinted polymer process, demonstrating that smaller peptides from the linear epitope of the Dengue virus NS1 protein could also be used for detecting viral samples [90]. They achieved the quantitative detection of the NS1 protein in crude and purified samples. Despite these studies not being relatively recent, we have yet to see a mass-produced, readily available MIP solution for clinical adoption. For this QCM approach to reach the clinic, extra development work will need to be performed to bridge this gap.

Generally, QCM presents itself as a promising potential technique for clinical biosensing. It offers high sensitivity and real-time sensing while maintaining relatively simple operation. Although standard QCM can run into limitations when non-rigid analytes are added to the surface, this can be addressed through QCM-D sensing. QCM, like other methods such as SPR, BLI, and electrochemical processes, requires a capture ligand for viral sensing. However, the ability to use more straightforward, more durable capture methods like molecularly imprinted polymers can improve accessibility by simplifying the viral sensor system and by reducing the number of steps to be performed in the clinic itself. While multi-region QCM sensors exist [91], the more affordable instrumentation will likely be limited to individual channels, limiting its throughput.

3. Comparing Approaches

Optical biosensing methods carry several advantages that could make them well-suited for clinical use. They are often extremely sensitive, which is beneficial for detecting low viral loads in the early stages of infection. Further, many optical methods like SPR and BLI do not rely on labeling for detection, which reduces the complexity of assay preparation and the possible interfering effects that labels may have on the native interaction being measured for detection. The real-time capabilities of optical methods are a significant strength, enabling the rapid detection of relatively transient interactions. When combined with high-throughput and multiplexed detection strategies, optical biosensors hold the potential for high-throughput mass testing. This does not mean that optical detection is a silver bullet approach to viral detection, though. Often, optical instruments can be costly, and the bulky nature of many optical instruments can make them challenging to deploy in small clinics and for field testing.

A key weakness of many optical instruments is their cost. Both in terms of their consumables—like SPR chips, single-use BLI probes—and the instrumentation itself, their significantly higher prices can make clinical adoption much more difficult. Particularly in resource-limited regions, this factor alone could be enough to eliminate these instruments from contention. In this regard, movements towards affordable instrumentation could help bring these highly sensitive instruments into the hands of more clinics. A barrier to clinical adoption beyond cost is the additional considerations that must be taken into account when using optical instrumentation, such as the effects of temperature, buffer mismatches, and pH. While these factors can be accounted for in a research setting, they add further complexity and training for clinicians. Assay interpretation and false positives can also be a risk for the point-of-care deployment of these instruments, since non-specific binding could be interpreted as a detected infection in untrained hands. Conversely, colorimetric sensors offer far greater simplicity regarding administration and result interpretation but sacrifice some sensitivity and granularity.

Similarly, non-optical methods also have several benefits and drawbacks. Often, instruments like electrochemical sensors are more accessible to clinics than their optical counterparts due to their relatively low cost and size. Their compact nature makes them more feasible for various situations, from smaller clinics to field testing, providing strength in their suitability for both lab and field work. This is further bolstered by advancements in manufacturing capabilities, like printable electrodes and pre-functionalized surfaces, which can provide out-of-the-box functionality for rapid test administration. Their portability means they can also be scaled with less of a need for multiplexed testing, and the simplicity of electrochemical sensors means they could be integrated into portable solutions for community-level testing in the field. While these biosensors are less susceptible to environmental factors like refractive index mismatches, electrochemical sensors can suffer from weakness in their strong susceptibility to biofouling effects, particularly when sensing in complex matrices like serum. This drawback could be improved through prefunctionalizations geared towards surface passivation rather than capture motif immobilization. Further, signal interference can be possible in electrochemical sensors due to other materials in the sample that may be, for instance, redox-reactive. While QCM and electrochemical sensing hardware can be designed with reusability in mind, there is still a cost associated with using the instrumentation, and the degradation of electrodes over time is, to an extent, an unavoidable factor as the electrodes age.

When directly comparing these approaches (Table 1), none are an ideal technique for clinical deployment. While the portability and lower cost of electrochemical sensors may seem appealing, the sensitivity of optical sensors is also important. This means that clinics must weigh the benefits and drawbacks of each diagnostic method, determining which concessions are acceptable while meeting their needs. The innovations we see in recent diagnostic publications—both in optical and non-optical viral biosensing—have aimed to minimize the drawbacks of their respective approaches and minimize the concessions clinics must make. Logistics can also play a significant role in the feasibility of applying novel diagnostics to the point-of-care; as such, the development and production of biosensing surfaces, sensing hardware, and software also significantly determine whether a clinic could feasibly adopt new technologies.

Table 1. Comparative analysis of viral biosensing methods. (++++ = Excellent, +++ = Favorable, ++ = Moderate, + = Unfavorable).

Method	Sensitivity	Cost	Complexity	Portability	Clinical/POC Applicability	References
Lateral Flow Assay	+	++++	++++	++++	++++	[92,93]
RT-PCR	+++	+	+	+	++++	[92,93]
ELISA	++	++	++	+	++++	[92,93]
RT-LAMP	++	+++	+++	+++	+++	[93–95]
NGS	++++	+	+	+	+++	[96]
Plasmonic/SPR	++++	+	+	+	++	[97,98]
BLI	+++	++	+	+	++	[59,99]
QCM	++	++	++	+++	+	[100]
Electrochemical	++	++	++	+++	++	[74,77,98]

4. Diagnostic Development—Hardware and Software

Advancements in viral biosensor technology are essential, but ensuring ease of use and accessibility is crucial for widespread adoption at the point of care. Developing more straightforward instruments and protocols reduces the training required for operators, and lowering the costs of these instruments allows more clinics to utilize state-of-the-art sensing technologies.

Improvements in accessibility for viral diagnostics can be achieved in various ways, including providing pre-made sensing surfaces that offer out-of-the-box functionality, designing instruments that can be easily assembled within the clinic, and creating software that simplifies readouts for healthcare providers and patients. Reducing the number of preparation steps simplifies the administration of diagnostics and minimizes the risk of human error.

Additionally, pre-manufacturing essential diagnostic components, such as microfluidics and functionalized sensor chips, increases reliability and practicality in the field by reducing sample handling and instrumental setup steps for clinic employees. For instance, mass-producing chips with pre-immobilized capture ligands streamlines the setup process, reduces user error, and requires less skilled labor. Simplifying manufacturing processes can also lower costs through economies of scale, making innovative viral diagnostics more accessible.

4.1. Three-Dimensional Printing

Three-dimensional printing has emerged as a pivotal tool in improving the accessibility of diagnostic technology. With the versatility of additive manufacturing and the growing library of 3D-modeled components online, components ranging from instrumental housings and consumables to replacement parts can all be made with commercially, non-clinically oriented hardware (Figure 3A). With the costs of 3D printers falling and the library of open-source designs growing, it is now possible to repair, modify, or even outright assemble instruments by printing parts. In conjunction with the previously discussed improvements in biosensing technology, it is more feasible than ever to build diagnostic systems.

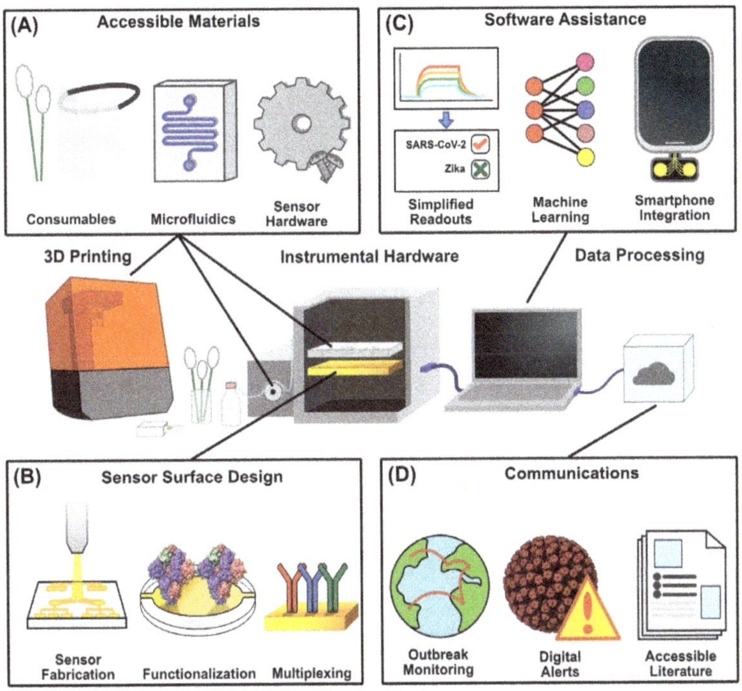

Figure 3. Aspects of POC Accessibility to Novel Diagnostics. (**A**) Additive manufacturing enables the increasingly affordable production of a range of physical diagnostic components, ranging from consumables like swabs and PPE to instrumental hardware like housings and microfluidic systems, as well as replacement internal components for instrumental repair. (**B**) Advancements in sensor surface production can lower the barrier to entry for administering viral assays either through scalable, automated manufacturing, pre-functionalization with immobilized capture motifs, or the utilization of multiple capture motifs for multiplexed sensing (PDB ID 6VXX). (**C**) Software assistance also lowers the barrier for data interpretation through simplified results for complex readouts, the machine learning-assisted processing of increasingly large datasets, and the incorporation of smartphones as processing devices to reduce cost. (**D**) Contact tracing drives outbreak monitoring and digital exposure alerts, easing the burden on clinicians and enabling the preemptive stocking of assay reagents. Additionally, accessible literature can equip both clinicians and the public with a greater understanding of viral infections and the administration of their respective assays (PDB ID 4V7Q)—structures were rendered in UCSF ChimeraX 1.9 [39] and figures were generated in Affinity Designer 2.

Several studies have demonstrated the ability of 3D printers to be used to make components of diagnostic systems, including microfluidic systems, at high resolutions. Standard methods like extrusion, stereolithography, and material jetting-based 3D printing have demonstrated microfluidic feature resolutions as high as single micrometers, with processes like direct laser writing coupled with photoresists even reaching the nanometer scale [101]. These processes can also be connected with biocompatible materials like polyethylene glycol diacrylate (PEGda) [102] and silicone [103] (among others [104]) to minimize the leaching of material and even enable wearable diagnostic technology. In a more indirect sense, low-cost printers can make molds to cast microfluidics from materials like PDMS [105]. Fluidics are not the only diagnostic components that 3D printing aims to democratize; materials like liquid silica resins and germanate glass have been successfully used to print optical components like prisms and lenses [106], as well as supporting mounts for hardware like smartphones [107] in do-it-yourself (DIY) detectors.

These 3D-printed components have already significantly impacted the realm of viral diagnostics. For instance, aerosol-jet 3D printing was used by Ali et al. to print 3D micropillar arrays that were coated in graphene and functionalized with viral antigens [108]. These 3D-printed sensors were used as electrodes to detect SARS-CoV-2 spike protein and nucleocapsid-specific antibodies in patient sera. Similarly, 3D-printed electrode cassettes were recently developed by Sharma et al., which were combined with paper electrodes to detect dengue and chikungunya antigens in serum through cyclic voltammetry [109]. Krejcova et al. developed a 3D-printable, bead-based microfluidic chip system that isolated paramagnetic beads, followed by voltammetry-based detection [110]. Printed fluidic systems have even been designed to automate ELISA reagent delivery, as demonstrated by Karamzadeh et al., in order to successfully detect the SARS-CoV-2 N-protein in a largely automated format [111]. Extending beyond sensing interfaces themselves, the COVID-19 pandemic saw the widespread use of 3D printing for manufacturing face shields [112], plastic supports for lateral flow assays [113], nasopharyngeal swabs [114], housings for respirator filters [115], and pumps [116]. While some 3D-printable technologies have been directly demonstrated in viral diagnostics, numerous other 3D-printed instruments that could be adapted for detecting viral infections have been developed, demonstrating the continued progress in this field. Examples include microspectroscopy imaging devices [117], PCR thermal cyclers [118,119], photosensitive drug delivery systems [120], quartz crystal microbalances, and electrochemical sensors [121,122].

While the capabilities of 3D printing have been shown across many methods of viral biosensing, continued development will still be needed to lower the barrier of technical knowledge to utilize these tools properly. This is partly aided by the tendency for 3D-printed, open-sourced electronic devices to be made with off-the-shelf components. This aspect is critical in bringing novel diagnostic technology to logistically constrained and underfunded areas. Utilizing generic components helps eliminate the need for niche or proprietary parts, a common hurdle regarding traditional laboratory equipment. Further, the upkeep of these instruments designed with this approach will be far more manageable, as replacement parts and instrument schematics will be more accessible to the end user than what may be seen for commercial biosensing technologies. While publications touting novel sensing methodologies initially seem accessible to clinics, they have little use if the instrumentation required to adopt the methods is prohibitively expensive. Further refinements beyond these initial studies need to be achieved to design easy-to-assemble, printable components to minimize the hardware limitations that clinics may often face.

4.2. Sensor Surface Fabrication

Another aspect of bringing more advanced bioanalytical techniques into the clinic for viral biosensing is the ability to produce highly consistent, functionalized, and ready-to-use sensing surfaces. Functionalizing sensing substrates in manufacturing reduces the potential for human error before being sent to a clinic, providing a more 'plug and play' process in diagnostics. Further, chip robustness and reusability reduce the cost per run, helping clinics digest the cost of expensive analytical techniques. Pre-functionalizations can range from self-assembled monolayers to polymer and nanomaterial coatings to surfaces pre-functionalized with viral capture motifs [123] (Figure 3B). Metallic sensor surfaces, such as those used in SPR, QCM, and electrochemical diagnostics, can be manufactured in bulk in a clean room using physical deposition methods [124]. This extends beyond metallic surfaces, and thermally grown silicon chips have also been used in viral biosensing [62,63]. Specific surface designs like microarrays, wells, and layered materials can be achieved by coupling physical deposition with photolithography [125,126]. These chips can then be connected to capture elements by carefully selecting the surface chemistry. Generic capture methods are also appealing for clinical use since a single capture protocol could be applied for the simple production of a multi-readout sensor. For instance, gold–thiol chemistry can immobilize thiol-modified capture motifs such as antigen-specific nucleic acids [127]. Materials like paper have even been shown to be an affordable diagnostic substrate. For example, freeze-dried, toehold switch-containing paper substrates have been combined with isothermal amplification and colorimetric readouts to detect the Zika virus [128].

In electrochemical sensing, screen printing offers a more accessible alternative to deposition-based electrode fabrication. Generally, screen-printed electrodes are cheaper and more disposable, and their compatibility with bulk production is conducive to scaled manufacturing [129]. Further, they lack the requirement for a high vacuum and a cleanroom environment for manufacturing, eliminating much of the training required for cleanroom manufacturing. They can enable printing on flexible materials like paper and plastic, whereas other sensor surfaces are constrained to glass or metal substrates [130]. Many printable electrode materials, such as conductive inks containing graphene, polymers like PEDOT: PSS, and nanoparticle-based solutions [131–133], are available. Screen-printed electrodes have already been used in viral diagnostics, including the detection of Dengue, HBV, HCV, Zika virus and H1N1, and even in variant-specific SARS-CoV-2 detection [78,134]. Due to the production cost of metallic surfaces like electrodes, screen printing presents an affordable, scalable alternative. Further, quickly disposable electrodes develop single-use, ready-to-run sensing surfaces.

While cheap, single-use sensing surfaces represent one approach to lowering the cost of viral diagnostics, while regenerable functionalized surfaces offer another approach to sustainability. The ability to cleanly strip analytes from a surface enables the reuse of sensing surfaces, which lowers the per-run cost of costly sensing surfaces that cannot be improved by cheaper alternatives (such as gold-coated chips in SPR and QCM). This has been shown using molecularly imprinted polymers [135] and surface-exposed NTA groups to capture histidine-tagged proteins [58,136,137]. Because approaches such as these also lack functional proteins for capture, they benefit from avoiding sensitive storage conditions. This holds the potential for cheaper running costs and more straightforward manufacturing scalability, which could further lower costs.

5. Software

Coupled with the development of accessible diagnostic hardware, novel software packages have also been developed to streamline the process of viral diagnostics. The overall goal of pushing new technology into the clinic can be aided by software in several ways (Figure 3C). From broad-scale infection reporting and tracking software to smartphone apps coupled with DIY hardware or even machine learning algorithms built to process complex samples, software advancements go hand in hand with improving the adoption of new diagnostic technology. By streamlining result delivery, simplifying wide-scale testing, and automating the processing of diagnostic data, software tools reduce the administrative burden on clinics, making it easier and more cost-effective to incorporate innovative diagnostics into routine care.

Machine learning is drastically changing the scientific world, and while we find ourselves in the early stages of its advancement into clinical settings, its potential can already be seen. Since machine learning and artificial intelligence tools are well-suited to processing enormous amounts of data, information sources like viral sequences, patient symptoms, and epidemiological trends could all be used to achieve higher-level analysis that individual diagnostic assays cannot attain. Often, machine learning tools are associated with the underlying analysis backing diagnostic development, such as the high-throughput screening of diagnostic target pairs [138]. However, recent advancements in machine learning extend beyond academic labs, with the benefits of their computational power now reaching the clinic. For instance, Khan et al. proposed a tunicate swarm machine learning algorithm for efficient data processing and the transmission of biosensor data between patients and clinicians [139]. The development of such data pipelines is vital for active symptom monitoring and testing recommendations, as it allows for the integration of a range of Internet of Things (IoT) sensors, including those for electrocardiograms (ECGs), body temperature, glucose, and electromyography (EMG). Beyond backend data management, machine learning has also been deployed for clinician and patient use. Machine learning tools can be used to track multiple parameters like symptoms, CD4 counts, and adherence to treatment regimens, which could provide clinicians with strong tools to predict infections and follow through with treatments [140]. Further, broad-scale infection predictions can be made through mass surveillance and publicly available information like search engine trends, where increases in searches related to symptoms could help anticipate waves of infections [141]. Such tools could drastically help preemptively stock assay reagents rather than chasing strong demand, as seen in the COVID-19 pandemic.

With the mass implementation of artificial intelligence tools in every realm of science and technology, it is hard not to see its promise in viral diagnostics. Particularly following the COVID-19 pandemic, we have seen the strong need for infrastructure that can handle diagnostics en masse. The mass data processing capabilities and ease of use of machine learning tools present a novel intersection between accessibility and technical strength, which could show substantial benefits in viral diagnostics. Nonetheless, there still needs to be a gradual, intentional implementation of such tools in clinical testing to ensure their proper use and an understanding of how these tools operate. The tendency for AI tools to 'hallucinate,' confidently giving incorrect results, carries much more dire consequences when lives are at stake [142]. In this regard, machine learning tools must be implemented cautiously to ensure that their convenience does not enable blind trust in the readouts given by the diagnostics they may use.

6. Smartphone Apps

Mainly due to isolation protocols from the COVID-19 pandemic, the coupling of viral diagnostics and smartphone apps became highly prevalent. Depending on the app, their purpose can range from symptom monitoring to contact tracing, or they can even be used in a smartphone in a home-built diagnostic device. Many apps that ask users to answer symptom questionnaires regularly and alert patients and healthcare providers when symptoms warrant testing have been published [143,144]. The pandemic also saw telemedicine's widespread adoption and expansion, which gave patients and clinicians a more convenient option [145]. Such types of app can also be coupled with methods to improve diagnostic throughput—like sample barcoding and automated alert systems—to rapidly diagnose and alert patients to positive results without additional visits to the clinic [146,147]. Alongside symptom-monitoring apps, the pandemic also saw the broad adoption of contact-tracing apps, which use the proximity monitoring of other signed-up users to anonymously alert potentially exposed individuals. These apps were mainly popularized due to their adoption by both Apple and Google, as well as their adoption by state-sponsored solutions like TraceTogether and COVIDSafe (from Singapore and Australia, respectively) [146]. Other apps that use smartphones as key components of assays have been developed, serving as detectors. For instance, smartphones have been used in bioluminescent RT-LAMP to detect the Zika virus in urine and saliva and HIV in blood [25]. Similarly, numerous colorimetric COVID-19 tests use smartphones for test readouts in colorimetric and fluorescent tests, taking advantage of the high-quality cameras present in modern smartphones [49,148,149]. Smartphone apps add further improve the usability of modern diagnostics by reducing our reliance on proprietary or complex systems for diagnostic readouts. Given the commonality of smartphones today, a well-designed smartphone app can eliminate the requirement for a clinician to process a readout, leaving them freer to run the assays themselves.

Pre-built diagnostic solutions appeal to clinics due to the simplicity of their setup and learning processes [12,150,151]. Still, the increase in home-built alternatives has also presented a more grassroots approach. Publicly available hardware and software have proven that there are significant cost and logistical benefits associated with the end user taking the burden of assembly upon themselves [152–154]. With the barrier to entry for processes like 3D printing continuing to become more accessible, it is becoming far more feasible to take this approach to bring new diagnostic technologies into the clinic [155]. An ideal home-built clinical solution for viral diagnostics could incorporate every aspect of these recent democratization efforts, coupling freely available hardware to open-sourced software packages and using regenerable sensing surfaces to maximize the number of assays per purchase.

7. Democratizing Scientific Communications

When discussing advancements that enable clinics to adopt new diagnostic technologies, it is also essential to acknowledge the role of accessible and understandable scientific information, including the pivotal role of scientific journals in disseminating information to clinicians (Figure 3D). In the same way that new technologies have accessibility concerns, particularly in terms of cost and training requirements, there is often a perceived gap between the academic community and the general public. This disparity is mainly present in the sciences that drive diagnostic development, like chemistry, material sciences, and informatics [156]. In recent years, and particularly following the broad impact of the COVID-19 pandemic on the public, many journals have made significant moves to improve the accessibility of their publications. New pilot programs have been adopted, such as translating abstracts from English to other languages like Japanese, Danish, Thai, and Arabic [157]. Recent years have also seen movements that promote disability accessibility in articles. Font sizes, acronym use, and design choices like color and contrast in figures are some considerations that can improve the accessibility of scientific literature for readers [158,159].

Regarding greater public access, journals have increasingly adopted plain-language summaries of their published papers [160,161]. For example, replacing the term 'hypoxia' with 'lack of oxygen' construes the same idea in a format that a wider audience can understand. There have also been additions like definitions of terms, lists of acronyms, bulleted key points, and video or audio abstracts to accompany paper graphics [162–164]. This is a subtle but important aspect in adopting new diagnostics, since accessible scientific literature equips clinicians with the knowledge necessary to administer and understand the diagnostics they may be adopting in their workflows. Improving patient access to scientific literature also helps them understand the procedures they may undergo in a clinic. There is no doubt that in the coming years, artificial intelligence will also see increasing use by clinicians and the public to summarize scientific information. With these tools often being trained by scrubbing vast amounts of data from online sources, it is even more relevant to make scientific details digestible, succinct, and easily accessible to ensure that their responses are grounded in truth.

The push for greater accessibility in scientific publishing mirrors the broader goal of democratizing viral diagnostics. By adopting plain-language summaries, multilingual abstracts, and inclusive design, journals are bridging the gap between academic researchers and clinicians by equipping them with the information necessary to adopt and administer state-of-the-art diagnostic techniques. These efforts also extend to patients whose understanding and receptivity of new diagnostic procedures are enhanced through digestible formats like bulleted key points or audio abstracts. As AI-driven tools increasingly facilitate access to scientific knowledge, ensuring clarity and standardization in publications will grow even more critical to prevent misinformation and align innovation with real-world needs. Accessible literature is supplemental to diagnostic advancement and a foundational step toward equitable adoption across diverse clinical and public health settings.

8. Conclusions

The drive to develop faster, more sensitive, and accessible viral diagnostics underscores the clinical world's commitment to early disease detection. While the COVID-19 pandemic focused on this need, emerging threats like avian flu and monkeypox reinforce the strong demand for innovation. Significant progress has been made in biosensing technologies, enhancing sensitivity, reproducibility, and the detection of novel analytes. However, transitioning from promising research to practical, point-of-care diagnostics still requires overcoming cost, scalability, and ease-of-use challenges. Techniques such as SPR, BLI, and electrochemical sensing offer remarkable sensitivity. However, their adoption in clinical settings hinges on reducing instrument costs, minimizing device footprints, and simplifying operation and data interpretation. In these sensing approaches, democratization efforts have focused on surface pre-functionalizations for ease of use, multiplexing, and physical improvements like cost reduction and a reduction in instrumental footprints. By reducing the cost and barriers to understanding these methods, SPR and BLI present themselves as promising tools for clinical adoption, particularly given their strong sensitivity and label-free sensing capabilities. Non-optical sensing methods have seen similar improvements, with multiplexed sensors and open-source sensing hardware offering affordability benefits in both assays and in the sensing hardware itself. The ability to mass manufacture relatively small electrochemical biosensors makes them a promising option for field testing, particularly regarding their clinical adoption in resource-limited and geographically constrained regions. This is further aided by robust capture motifs like nucleic acids and MIPs, since their stability strongly favors pre-fabrication, shipping, and storage. While these approaches also suffer from biofouling and hardware degradation over time, their relatively cheap cost can overcome these limitations using disposable sensing hardware.

Three-dimensional printing and automated, scalable manufacturing underlie the improvements in sensor technology by reducing the cost and logistical constraints experienced by disenfranchised clinics, allowing them to access otherwise unattainable sensing hardware. Further, the ability to print consumables and instrumental hardware for repairs strongly boosts clinics' sustainability. With the growing library of open-source hardware, many home-built biosensors are being developed and iterated upon. Self-assembled hardware and democratized biosensing approaches also benefit from a strong software backing. Smartphone-based analysis and simple-to-understand diagnostic readouts lower the training requirement for clinicians while making diagnoses easier for patients to understand. With underlying frameworks for contact tracing, infection reporting, and symptom monitoring, POC clinics can be given strong tools for anticipating and containing wide-scale infection events.

Finally, scientific research reporting works in tandem with innovations in sensing methodologies, hardware, and software provides the public with the information necessary to comprehend the results of diagnostic assays and the phenomena underlying the infections being diagnosed. Initiatives like multilingual plain-language summaries and AI-assisted knowledge translation ensure that advancements reach beyond academic circles to frontline healthcare workers and policymakers. In effect, the democratization of viral diagnostics does not solely lie in the sciences developing the assays. It requires a multidisciplinary effort between scientists, engineers, and journals to not only create new sensing technologies but to put them in clinicians' hands and equip them with the knowledge needed to utilize them effectively. The future of diagnostics lies not just in detecting pathogens, but in building ecosystems where technology, education, and accessibility converge to empower every community, regardless of resources, to act swiftly against emerging viral threats.

Author Contributions: Conceptualization, W.V.Z. and P.R.; methodology, W.V.Z. and P.R.; software, W.V.Z.; validation, W.V.Z.; formal analysis, W.V.Z.; investigation, W.V.Z. and P.R.; resources, W.V.Z. and P.R.; data curation, W.V.Z.; writing—original draft preparation, W.V.Z. and P.R.; writing—review and editing, W.V.Z. and P.R.; visualization, W.V.Z.; supervision, P.R.; project administration, P.R.; funding acquisition, P.R. All authors have read and agreed to the published version of the manuscript.

Funding: This research received no external funding.

Institutional Review Board Statement: Not applicable.

Informed Consent Statement: Not applicable.

Data Availability Statement: No new data were created or analyzed in this study. Data sharing is not applicable to this article.

Conflicts of Interest: The authors declare no conflict of interest.

References

1. Krammer, F.; Hermann, E.; Rasmussen, A.L. Highly Pathogenic Avian Influenza H5N1: History, Current Situation, and Outlook. *J. Virol.* **2025**, *99*, e02209-24. [CrossRef] [PubMed]
2. Mylne, A.; Brady, O.J.; Huang, Z.; Pigott, D.M.; Golding, N.; Kraemer, M.U.G.; Hay, S.I. A Comprehensive Database of the Geographic Spread of Past Human Ebola Outbreaks. *Sci. Data* **2014**, *1*, 140042. [CrossRef] [PubMed]
3. Chang, C.; Ortiz, K.; Ansari, A.; Gershwin, M.E. The Zika Outbreak of the 21st Century. *J. Autoimmun.* **2016**, *68*, 1–13. [CrossRef]
4. Larremore, D.B.; Wilder, B.; Lester, E.; Shehata, S.; Burke, J.M.; Hay, J.A.; Milind, T.; Mina, M.J.; Parker, R. Test Sensitivity Is Secondary to Frequency and Turnaround Time for COVID-19 Surveillance. *medRxiv* **2020**. [CrossRef]
5. Jean, S.; Burnham, C.-A.D.; Chapin, K.; Garner, O.B.; Pant Pai, N.; Turabelidze, G.; Butler-Wu, S. At-Home Testing for Infectious Diseases: The Laboratory Where You Live. *Clin. Chem.* **2022**, *68*, 19–26. [CrossRef]
6. Marrazza, G.; Ramalingam, M.; Jaisankar, A.; Cheng, L.; Selvolini, G.; Vitale, I.A. Advancements and Emerging Technologies in Biosensors for Rapid and Accurate Virus Detection. *TrAC Trends Anal. Chem.* **2024**, *172*, 117609. [CrossRef]
7. Otoo, J.A.; Schlappi, T.S. REASSURED Multiplex Diagnostics: A Critical Review and Forecast. *Biosensors* **2022**, *12*, 124. [CrossRef] [PubMed]
8. Chambers, J.P.; Arulanandam, B.P.; Matta, L.L.; Weis, A.; Valdes, J.J. Biosensor Recognition Elements. *Curr. Issues Mol. Biol.* **2008**, *10*, 1–12.
9. Thiviyanathan, V.; Gorenstein, D.G. Aptamers and the Next Generation of Diagnostic Reagents. *Proteom. Clin. Appl.* **2012**, *6*, 563–573. [CrossRef] [PubMed]
10. Karunakaran, R.; Keskin, M. Chapter 11—Biosensors: Components, Mechanisms, and Applications. In *Analytical Techniques in Biosciences*; Egbuna, C., Patrick-Iwuanyanwu, K.C., Shah, M.A., Ifemeje, J.C., Rasul, A., Eds.; Academic Press: Cambridge, MA, USA, 2022; pp. 179–190. [CrossRef]

11. Sajini, T.; Mathew, B. A Brief Overview of Molecularly Imprinted Polymers: Highlighting Computational Design, Nano and Photo-Responsive Imprinting. *Talanta Open* **2021**, *4*, 100072. [CrossRef]
12. Wallace, S.; Kartau, M.; Kakkar, T.; Davis, C.; Szemiel, A.; Samardzhieva, I.; Vijayakrishnan, S.; Cole, S.; De Lorenzo, G.; Maillart, E.; et al. Multiplexed Biosensing of Proteins and Virions with Disposable Plasmonic Assays. *ACS Sens.* **2023**, *8*, 3338–3348. [CrossRef] [PubMed]
13. Bhalla, N.; Jolly, P.; Formisano, N.; Estrela, P. Introduction to Biosensors. *Essays Biochem.* **2016**, *60*, 1–8. [CrossRef] [PubMed]
14. Damborský, P.; Švitel, J.; Katrlík, J. Optical Biosensors. *Essays Biochem.* **2016**, *60*, 91–100. [CrossRef]
15. Grieshaber, D.; MacKenzie, R.; Vörös, J.; Reimhult, E. Electrochemical Biosensors—Sensor Principles and Architectures. *Sensors* **2008**, *8*, 1400–1458. [CrossRef] [PubMed]
16. Marx, K.A. Quartz Crystal Microbalance: A Useful Tool for Studying Thin Polymer Films and Complex Biomolecular Systems at the Solution−Surface Interface. *Biomacromolecules* **2003**, *4*, 1099–1120. [CrossRef]
17. Mouro, J.; Pinto, R.; Paoletti, P.; Tiribilli, B. Microcantilever: Dynamical Response for Mass Sensing and Fluid Characterization. *Sensors* **2020**, *21*, 115. [CrossRef]
18. Ramanathan, K.; Danielsson, B. Principles and Applications of Thermal Biosensors. *Biosens. Bioelectron.* **2001**, *16*, 417–423. [CrossRef]
19. Chen, Y.-T.; Kolhatkar, A.G.; Zenasni, O.; Xu, S.; Lee, T.R. Biosensing Using Magnetic Particle Detection Techniques. *Sensors* **2017**, *17*, 2300. [CrossRef]
20. Yang, C.-T.; Wu, L.; Bai, P.; Thierry, B. Investigation of Plasmonic Signal Enhancement Based on Long Range Surface Plasmon Resonance with Gold Nanoparticle Tags. *J. Mater. Chem. C* **2016**, *4*, 9897–9904. [CrossRef]
21. Morales-Narváez, E.; Montón, H.; Fomicheva, A.; Merkoçi, A. Signal Enhancement in Antibody Microarrays Using Quantum Dots Nanocrystals: Application to Potential Alzheimer's Disease Biomarker Screening. *Anal. Chem.* **2012**, *84*, 6821–6827. [CrossRef]
22. Zhu, Z. An Overview of Carbon Nanotubes and Graphene for Biosensing Applications. *Nanomicro Lett.* **2017**, *9*, 25. [CrossRef] [PubMed]
23. Chowdhury, P.; Lawrance, R.; Lu, Z.-Y.; Lin, H.-C.; Chan, Y.-H. Recent Progress in Dual/Multi-Modal Detection Modes for Improving Sensitivity and Specificity of Lateral Flow Immunoassays Applied for Point-of-Care Diagnostics. *TrAC Trends Anal. Chem.* **2024**, *177*, 117798. [CrossRef]
24. Forssén, P.; Multia, E.; Samuelsson, J.; Andersson, M.; Aastrup, T.; Altun, S.; Wallinder, D.; Wallbing, L.; Liangsupree, T.; Riekkola, M.-L.; et al. Reliable Strategy for Analysis of Complex Biosensor Data. *Anal. Chem.* **2018**, *90*, 5366–5374. [CrossRef] [PubMed]
25. Song, J.; Pandian, V.; Mauk, M.G.; Bau, H.H.; Cherry, S.; Tisi, L.C.; Liu, C. Smartphone-Based Mobile Detection Platform for Molecular Diagnostics and Spatiotemporal Disease Mapping. *Anal. Chem.* **2018**, *90*, 4823–4831. [CrossRef]
26. Kadian, S.; Kumari, P.; Shukla, S.; Narayan, R. Recent Advancements in Machine Learning Enabled Portable and Wearable Biosensors. *Talanta Open* **2023**, *8*, 100267. [CrossRef]
27. Sekhwama, M.; Mpofu, K.; Sivarasu, S.; Mthunzi-Kufa, P. Applications of Microfluidics in Biosensing. *Discov. Appl. Sci.* **2024**, *6*, 303. [CrossRef]
28. Kulkarni, M.B.; Ayachit, N.H.; Aminabhavi, T.M. Biosensors and Microfluidic Biosensors: From Fabrication to Application. *Biosensors* **2022**, *12*, 543. [CrossRef]
29. Zhu, Y.; Zhang, M.; Jie, Z.; Tao, S. Nucleic Acid Testing of SARS-CoV-2: A Review of Current Methods, Challenges, and Prospects. *Front. Microbiol.* **2022**, *13*, 1074289. [CrossRef]
30. Zhao, Y.; Qiu, J.; Jiang, P.; Wang, M.; Sun, M.; Fan, G.; Yang, N.; Huang, N.; Han, Y.; Han, L.; et al. RNA Extraction-Free Reduced Graphene Oxide-Based RT-LAMP Fluorescence Assay for Highly Sensitive SARS-CoV-2 Detection. *Talanta* **2024**, *277*, 126413. [CrossRef]
31. Teymouri, M.; Mollazadeh, S.; Mortazavi, H.; Naderi Ghale-noie, Z.; Keyvani, V.; Aghababaei, F.; Hamblin, M.R.; Abbaszadeh-Goudarzi, G.; Pourghadamyari, H.; Hashemian, S.M.R.; et al. Recent Advances and Challenges of RT-PCR Tests for the Diagnosis of COVID-19. *Pathol. Res. Pract.* **2021**, *221*, 153443. [CrossRef]
32. Fowler, V.L.; Armson, B.; Gonzales, J.L.; Wise, E.L.; Howson, E.L.A.; Vincent-Mistiaen, Z.; Fouch, S.; Maltby, C.J.; Grippon, S.; Munro, S.; et al. A Highly Effective Reverse-Transcription Loop-Mediated Isothermal Amplification (RT-LAMP) Assay for the Rapid Detection of SARS-CoV-2 Infection. *J. Infect.* **2021**, *82*, 117–125. [CrossRef] [PubMed]
33. Hosseini, S.; Vázquez-Villegas, P.; Rito-Palomares, M.; Martinez-Chapa, S.O. Advantages, Disadvantages and Modifications of Conventional ELISA. In *Enzyme-Linked Immunosorbent Assay (ELISA): From A to Z*; Hosseini, S., Vázquez-Villegas, P., Rito-Palomares, M., Martinez-Chapa, S.O., Eds.; Springer: Singapore, 2018; pp. 67–115. [CrossRef]
34. Heidt, B.; Siqueira, W.F.; Eersels, K.; Diliën, H.; van Grinsven, B.; Fujiwara, R.T.; Cleij, T.J. Point of Care Diagnostics in Resource-Limited Settings: A Review of the Present and Future of PoC in Its Most Needed Environment. *Biosensors* **2020**, *10*, 133. [CrossRef] [PubMed]

35. Hsieh, K.; Melendez, J.H.; Gaydos, C.A.; Wang, T.-H. Bridging the gap between development of point-of-care nucleic acid testing and patient care for sexually transmitted infections. *Lab Chip* **2022**, *22*, 476–511. [CrossRef] [PubMed]
36. Ledgerwood-Lee, M.; Hubenko, A.; Ray, P.; Theriault, Y.; Brickner, H.; Vazquez, L.F.; Schooley, R.; Carlin, A.; Clark, A.; Garretson, A.; et al. A User-Centered Service Model for Accelerating COVID-19 Diagnostic Innovation: The RADx- Rad Dx Core Approach. *IEEE Open J. Eng. Med. Biol.* **2025**, *6*, 472–479. [CrossRef]
37. RADx® Tech | National Institute of Biomedical Imaging and Bioengineering. Available online: https://www.nibib.nih.gov/programs/radx-tech-program (accessed on 18 June 2025).
38. Baker, R.E.; Mahmud, A.S.; Miller, I.F.; Rajeev, M.; Rasambainarivo, F.; Rice, B.L.; Takahashi, S.; Tatem, A.J.; Wagner, C.E.; Wang, L.-F.; et al. Infectious Disease in an Era of Global Change. *Nat. Rev. Microbiol.* **2022**, *20*, 193–205. [CrossRef]
39. Meng, E.C.; Goddard, T.D.; Pettersen, E.F.; Couch, G.S.; Pearson, Z.J.; Morris, J.H.; Ferrin, T.E. UCSF ChimeraX: Tools for Structure Building and Analysis. *Protein Sci.* **2023**, *32*, e4792. [CrossRef]
40. Nguyen, H.H.; Park, J.; Kang, S.; Kim, M. Surface Plasmon Resonance: A Versatile Technique for Biosensor Applications. *Sensors* **2015**, *15*, 10481–10510. [CrossRef]
41. Abbas, A.; Linman, M.J.; Cheng, Q. Sensitivity Comparison of Surface Plasmon Resonance and Plasmon-Waveguide Resonance Biosensors. *Sens. Actuators B Chem.* **2011**, *156*, 169–175. [CrossRef] [PubMed]
42. Liu, B.; Liu, X.; Shi, S.; Huang, R.; Su, R.; Qi, W.; He, Z. Design and Mechanisms of Antifouling Materials for Surface Plasmon Resonance Sensors. *Acta Biomater.* **2016**, *40*, 100–118. [CrossRef]
43. Riedel, T.; Rodriguez-Emmenegger, C.; de los Santos Pereira, A.; Bědajánková, A.; Jinoch, P.; Boltovets, P.M.; Brynda, E. Diagnosis of Epstein–Barr Virus Infection in Clinical Serum Samples by an SPR Biosensor Assay. *Biosens. Bioelectron.* **2014**, *55*, 278–284. [CrossRef]
44. Florschütz, K.; Schröter, A.; Schmieder, S.; Chen, W.; Schweizer, P.; Sonntag, F.; Danz, N.; Baronian, K.; Kunze, G. 'Phytochip': On-Chip Detection of Phytopathogenic RNA Viruses by a New Surface Plasmon Resonance Platform. *J. Virol. Methods* **2013**, *189*, 80–86. [CrossRef] [PubMed]
45. Yano, T.; Kajisa, T.; Ono, M.; Miyasaka, Y.; Hasegawa, Y.; Saito, A.; Otsuka, K.; Sakane, A.; Sasaki, T.; Yasutomo, K.; et al. Ultrasensitive Detection of SARS-CoV-2 Nucleocapsid Protein Using Large Gold Nanoparticle-Enhanced Surface Plasmon Resonance. *Sci. Rep.* **2022**, *12*, 1060. [CrossRef] [PubMed]
46. Tombelli, S.; Minunni, M.; Luzi, E.; Mascini, M. Aptamer-Based Biosensors for the Detection of HIV-1 Tat Protein. *Bioelectrochemistry* **2005**, *67*, 135–141. [CrossRef] [PubMed]
47. Chen, R.; Kan, L.; Duan, F.; He, L.; Wang, M.; Cui, J.; Zhang, Z.; Zhang, Z. Surface Plasmon Resonance Aptasensor Based on Niobium Carbide MXene Quantum Dots for Nucleocapsid of SARS-CoV-2 Detection. *Microchim. Acta* **2021**, *188*, 316. [CrossRef]
48. P4SPR 2.0. Affinite Instruments. Available online: https://www.affiniteinstruments.com/p4spr-2-0 (accessed on 22 April 2025).
49. Dong, T.; Han, C.; Jiang, M.; Zhang, T.; Kang, Q.; Wang, P.; Zhou, F. A Four-Channel Surface Plasmon Resonance Sensor Functionalized Online for Simultaneous Detections of Anti-SARS-CoV-2 Antibody, Free Viral Particles, and Neutralized Viral Particles. *ACS Sens.* **2022**, *7*, 3560–3570. [CrossRef]
50. Meet OpenSPR | Benchtop SPR Instrument | Nicoya. Available online: https://nicoyalife.com/products/spr-instruments/openspr/ (accessed on 22 April 2025).
51. NanoSPR—Devices and Glass Plates for Surface Plasmon Resonance. Available online: https://nanospr.com/ (accessed on 22 April 2025).
52. Choudhary, S.; Altintas, Z. Development of a Point-of-Care SPR Sensor for the Diagnosis of Acute Myocardial Infarction. *Biosensors* **2023**, *13*, 229. [CrossRef]
53. Murali, S.; Rustandi, R.R.; Zheng, X.; Payne, A.; Shang, L. Applications of Surface Plasmon Resonance and Biolayer Interferometry for Virus–Ligand Binding. *Viruses* **2022**, *14*, 717. [CrossRef]
54. Jug, A.; Bratkovič, T.; Ilaš, J. Biolayer Interferometry and Its Applications in Drug Discovery and Development. *TrAC Trends Anal. Chem.* **2024**, *176*, 117741. [CrossRef]
55. Guo, Z.; Wilson, J.R.; York, I.A.; Stevens, J. Biosensor-Based Epitope Mapping of Antibodies Targeting the Hemagglutinin and Neuraminidase of Influenza A Virus. *J. Immunol. Methods* **2018**, *461*, 23–29. [CrossRef]
56. Lu, W.; Du, W.; Somovilla, V.J.; Yu, G.; Haksar, D.; de Vries, E.; Boons, G.-J.; de Vries, R.P.; de Haan, C.A.M.; Pieters, R.J. Enhanced Inhibition of Influenza A Virus Adhesion by Di- and Trivalent Hemagglutinin Inhibitors. *J. Med. Chem.* **2019**, *62*, 6398–6404. [CrossRef]
57. Wei, X.; Du, W.; Duca, M.; Yu, G.; de Vries, E.; de Haan, C.A.M.; Pieters, R.J. Preventing Influenza A Virus Infection by Mixed Inhibition of Neuraminidase and Hemagglutinin by Divalent Inhibitors. *J. Med. Chem.* **2022**, *65*, 7312–7323. [CrossRef] [PubMed]
58. Poolsup, S.; Zaripov, E.; Hüttmann, N.; Minic, Z.; Artyushenko, P.V.; Shchugoreva, I.A.; Tomilin, F.N.; Kichkailo, A.S.; Berezovski, M.V. Discovery of DNA Aptamers Targeting SARS-CoV-2 Nucleocapsid Protein and Protein-Binding Epitopes for Label-Free COVID-19 Diagnostics. *Mol. Ther. Nucleic Acids* **2023**, *31*, 731–743. [CrossRef] [PubMed]

59. Dzimianski, J.V.; Lorig-Roach, N.; O'Rourke, S.M.; Alexander, D.L.; Kimmey, J.M.; DuBois, R.M. Rapid and Sensitive Detection of SARS-CoV-2 Antibodies by Biolayer Interferometry. *Sci. Rep.* **2020**, *10*, 21738. [CrossRef] [PubMed]
60. Daaboul, G.G.; Freedman, D.S.; Scherr, S.M.; Carter, E.; Rosca, A.; Bernstein, D.; Mire, C.E.; Agans, K.N.; Hoenen, T.; Geisbert, T.W.; et al. Enhanced Light Microscopy Visualization of Virus Particles from Zika Virus to Filamentous Ebolaviruses. *PLoS ONE* **2017**, *12*, e0179728. [CrossRef]
61. Daaboul, G.G.; Lopez, C.A.; Chinnala, J.; Goldberg, B.B.; Connor, J.H.; Ünlü, M.S. Digital Sensing and Sizing of Vesicular Stomatitis Virus Pseudotypes in Complex Media: A Model for Ebola and Marburg Detection. *ACS Nano* **2014**, *8*, 6047–6055. [CrossRef]
62. Daaboul, G.G.; Yurt, A.; Zhang, X.; Hwang, G.M.; Goldberg, B.B.; Ünlü, M.S. High-Throughput Detection and Sizing of Individual Low-Index Nanoparticles and Viruses for Pathogen Identification. *Nano Lett.* **2010**, *10*, 4727–4731. [CrossRef]
63. Aslan, M.; Seymour, E.; Brickner, H.; Clark, A.E.; Celebi, I.; Townsend, M.B.; Satheshkumar, P.S.; Riley, M.; Carlin, A.F.; Ünlü, M.S.; et al. A Label-Free Optical Biosensor-Based Point-of-Care Test for the Rapid Detection of Monkeypox Virus. *Biosens. Bioelectron.* **2025**, *269*, 116932. [CrossRef]
64. Wang, X.; Shang, X.; Huang, X. Next-Generation Pathogen Diagnosis with CRISPR/Cas-Based Detection Methods. *Emerg. Microbes Infect.* **2020**, *9*, 1682–1691. [CrossRef]
65. Kaminski, M.M.; Abudayyeh, O.O.; Gootenberg, J.S.; Zhang, F.; Collins, J.J. CRISPR-Based Diagnostics. *Nat. Biomed. Eng.* **2021**, *5*, 643–656. [CrossRef]
66. Wang, X.; Xiong, E.; Tian, T.; Cheng, M.; Lin, W.; Wang, H.; Zhang, G.; Sun, J.; Zhou, X. Clustered Regularly Interspaced Short Palindromic Repeats/Cas9-Mediated Lateral Flow Nucleic Acid Assay. *ACS Nano* **2020**, *14*, 2497–2508. [CrossRef]
67. Weerathunge, P.; Ramanathan, R.; Torok, V.A.; Hodgson, K.; Xu, Y.; Goodacre, R.; Behera, B.K.; Bansal, V. Ultrasensitive Colorimetric Detection of Murine Norovirus Using NanoZyme Aptasensor. *Anal. Chem.* **2019**, *91*, 3270–3276. [CrossRef] [PubMed]
68. Gupta, R.; Gupta, P.; Wang, S.; Melnykov, A.; Jiang, Q.; Seth, A.; Wang, Z.; Morrissey, J.J.; George, I.; Gandra, S.; et al. Ultrasensitive Lateral-Flow Assays via Plasmonically Active Antibody-Conjugated Fluorescent Nanoparticles. *Nat. Biomed. Eng.* **2023**, *7*, 1556–1570. [CrossRef] [PubMed]
69. Gugliandolo, G.; Pilato, G.; Donato, N. Compact System for Colorimetric Sensor Arrays Characterization Based on Computer Screen Photo-Assisted Technology. *Electronics* **2021**, *10*, 2587. [CrossRef]
70. Kim, H.; Jang, H.; Song, J.; Lee, S.M.; Lee, S.; Kwon, H.-J.; Kim, S.; Kang, T.; Park, H.G. A CRISPR/Cas12 Trans-Cleavage Reporter Enabling Label-Free Colorimetric Detection of SARS-CoV-2 and Its Variants. *Biosens. Bioelectron.* **2024**, *251*, 116102. [CrossRef] [PubMed]
71. Lopez-Tellez, J.; Ramirez-Montes, S.; Ferreira, T.A.; Santos, E.M.; Rodriguez, J.A. Application of Voltammetric Sensors for Pathogen Bacteria Detection: A Review. *Chemosensors* **2022**, *10*, 424. [CrossRef]
72. Hammond, J.L.; Formisano, N.; Estrela, P.; Carrara, S.; Tkac, J. Electrochemical Biosensors and Nanobiosensors. *Essays Biochem.* **2016**, *60*, 69–80. [CrossRef]
73. Jarosińska, E.; Zambrowska, Z.; Witkowska Nery, E. Methods of Protection of Electrochemical Sensors against Biofouling in Cell Culture Applications. *ACS Omega* **2024**, *9*, 4572–4580. [CrossRef]
74. Zamani, M.; Robson, J.M.; Fan, A.; Bono, M.S., Jr.; Furst, A.L.; Klapperich, C.M. Electrochemical Strategy for Low-Cost Viral Detection. *ACS Cent. Sci.* **2021**, *7*, 963–972. [CrossRef]
75. Huang, K.-J.; Li, J.; Liu, Y.-M.; Cao, X.; Yu, S.; Yu, M. Disposable Immunoassay for Hepatitis B Surface Antigen Based on a Graphene Paste Electrode Functionalized with Gold Nanoparticles and a Nafion-Cysteine Conjugate. *Microchim. Acta* **2012**, *177*, 419–426. [CrossRef]
76. Salahandish, R.; Jalali, P.; Tabrizi, H.O.; Hyun, J.E.; Haghayegh, F.; Khalghollah, M.; Zare, A.; Berenger, B.M.; Niu, Y.D.; Ghafar-Zadeh, E.; et al. A Compact, Low-Cost, and Binary Sensing (BiSense) Platform for Noise-Free and Self-Validated Impedimetric Detection of COVID-19 Infected Patients. *Biosens. Bioelectron.* **2022**, *213*, 114459. [CrossRef]
77. de Lima, L.F.; Ferreira, A.L.; Ranjan, I.; Collman, R.G.; de Araujo, W.R.; de la Fuente-Nunez, C. A Bacterial Cellulose-Based and Low-Cost Electrochemical Biosensor for Ultrasensitive Detection of SARS-CoV-2. *Cell Rep. Phys. Sci.* **2023**, *4*, 101476. [CrossRef] [PubMed]
78. Szydlowska, B.M.; Pola, C.C.; Cai, Z.; Chaney, L.E.; Hui, J.; Sheets, R.; Carpenter, J.; Dean, D.; Claussen, J.C.; Gomes, C.L.; et al. Biolayer-Interferometry-Guided Functionalization of Screen-Printed Graphene for Label-Free Electrochemical Virus Detection. *ACS Appl. Mater. Interfaces* **2024**, *16*, 25169–25180. [CrossRef] [PubMed]
79. Zhao, H.; Liu, F.; Xie, W.; Zhou, T.-C.; OuYang, J.; Jin, L.; Li, H.; Zhao, C.-Y.; Zhang, L.; Wei, J.; et al. Ultrasensitive Supersandwich-Type Electrochemical Sensor for SARS-CoV-2 from the Infected COVID-19 Patients Using a Smartphone. *Sens. Actuators B Chem.* **2021**, *327*, 128899. [CrossRef]

80. Torrente-Rodríguez, R.M.; Lukas, H.; Tu, J.; Min, J.; Yang, Y.; Xu, C.; Rossiter, H.B.; Gao, W. SARS-CoV-2 RapidPlex: A Graphene-Based Multiplexed Telemedicine Platform for Rapid and Low-Cost COVID-19 Diagnosis and Monitoring. *Matter* **2020**, *3*, 1981–1998. [CrossRef] [PubMed]
81. Russo, M.J.; Han, M.; Desroches, P.E.; Manasa, C.S.; Dennaoui, J.; Quigley, A.F.; Kapsa, R.M.I.; Moulton, S.E.; Guijt, R.M.; Greene, G.W.; et al. Antifouling Strategies for Electrochemical Biosensing: Mechanisms and Performance toward Point of Care Based Diagnostic Applications. *ACS Sens.* **2021**, *6*, 1482–1507. [CrossRef]
82. Kamsong, W.; Primpray, V.; Pasakon, P.; Sriprachuabwong, C.; Pakapongpan, S.; Mensing, J.P.; Wisitsoraat, A.; Tuantranont, A.; Karuwan, C. Highly Sensitive and Disposable Screen-Printed Ionic Liquid/Graphene Based Electrochemical Sensors. *Electrochem. Commun.* **2022**, *135*, 107209. [CrossRef]
83. Dixon, M.C. Quartz Crystal Microbalance with Dissipation Monitoring: Enabling Real-Time Characterization of Biological Materials and Their Interactions. *J. Biomol. Tech.* **2008**, *19*, 151–158.
84. Uttenthaler, E.; Kößlinger, C.; Drost, S. Quartz Crystal Biosensor for Detection of the African Swine Fever Disease. *Anal. Chim. Acta* **1998**, *362*, 91–100. [CrossRef]
85. Nilsson, P.H.; Al-Majdoub, M.; Ibrahim, A.; Aseel, O.; Suriyanarayanan, S.; Andersson, L.; Fostock, S.; Aastrup, T.; Tjernberg, I.; Rydén, I.; et al. Quartz Crystal Microbalance Platform for SARS-CoV-2 Immuno-Diagnostics. *Int. J. Mol. Sci.* **2023**, *24*, 16705. [CrossRef]
86. Ly, T.N.; Park, S.; Park, S.J. Detection of HIV-1 Antigen by Quartz Crystal Microbalance Using Gold Nanoparticles. *Sens. Actuators B Chem.* **2016**, *237*, 452–458. [CrossRef]
87. Makaraviciute, A.; Ruzgas, T.; Ramanavicius, A.; Ramanaviciene, A. A QCM-D Study of Reduced Antibody Fragments Immobilized on Planar Gold and Gold Nanoparticle Modified Sensor Surfaces. *Key Eng. Mater.* **2014**, *605*, 340–343. [CrossRef]
88. Owen, T.W.; Al-Kaysi, R.O.; Bardeen, C.J.; Cheng, Q. Microgravimetric Immunosensor for Direct Detection of Aerosolized Influenza A Virus Particles. *Sens. Actuators B Chem.* **2007**, *126*, 691–699. [CrossRef] [PubMed]
89. Jenik, M.; Schirhagl, R.; Schirk, C.; Hayden, O.; Lieberzeit, P.; Blaas, D.; Paul, G.; Dickert, F.L. Sensing Picornaviruses Using Molecular Imprinting Techniques on a Quartz Crystal Microbalance. *Anal. Chem.* **2009**, *81*, 5320–5326. [CrossRef]
90. Tai, D.-F.; Lin, C.-Y.; Wu, T.-Z.; Chen, L.-K. Recognition of Dengue Virus Protein Using Epitope-Mediated Molecularly Imprinted Film. *Anal. Chem.* **2005**, *77*, 5140–5143. [CrossRef]
91. Shen, H.; Zhou, T.; Hu, J. A High-Throughput QCM Chip Configuration for the Study of Living Cells and Cell-Drug Interactions. *Anal. Bioanal. Chem.* **2017**, *409*, 6463–6473. [CrossRef]
92. Liu, Y.; Zhan, L.; Qin, Z.; Sackrison, J.; Bischof, J.C. Ultrasensitive and Highly Specific Lateral Flow Assays for Point-of-Care Diagnosis. *ACS Nano* **2021**, *15*, 3593–3611. [CrossRef]
93. Bukasov, R.; Dossym, D.; Filchakova, O. Detection of RNA Viruses from Influenza and HIV to Ebola and SARS-CoV-2: A Review. *Anal. Methods* **2021**, *13*, 34–55. [CrossRef]
94. Iqbal, B.N.; Arunasalam, S.; Divarathna, M.V.M.; Jabeer, A.; Sirisena, P.; Senaratne, T.; Muthugala, R.; Noordeen, F. Diagnostic Utility and Validation of a Newly Developed Real Time Loop Mediated Isothermal Amplification Method for the Detection of SARS CoV-2 Infection. *J. Clin. Virol. Plus* **2022**, *2*, 100081. [CrossRef]
95. Hoffmann, E.d.R.; Balzan, L.d.R.; Inamine, E.; Pancotto, L.R.; Gaboardi, G.; Cantarelli, V.V. Performance of Reverse Transcription Loop-Mediated Isothermal Amplification (RT-LAMP) Targeting the RNA Polymerase Gene for the Direct Detection of SARS-CoV2 in Nasopharyngeal Swabs. *Int. J. Mol. Sci.* **2023**, *24*, 13056. [CrossRef]
96. Kuang, Y.; Tan, W.; Hu, C.; Dai, Z.; Bai, L.; Wang, J.; Liao, H.; Chen, H.; He, R.; Zhu, P.; et al. Diagnosis Value of Targeted and Metagenomic Sequencing in Respiratory Tract Infection. *Front. Cell. Infect. Microbiol.* **2024**, *14*, 1498512. [CrossRef]
97. SPR Instruments. Available online: https://www.sprpages.nl/spr-instruments (accessed on 18 June 2025).
98. Aziz, A.; Asif, M.; Ashraf, G.; Farooq, U.; Yang, Q.; Wang, S. Trends in Biosensing Platforms for SARS-CoV-2 Detection: A Critical Appraisal against Standard Detection Tools. *Curr. Opin. Colloid. Interface Sci.* **2021**, *52*, 101418. [CrossRef] [PubMed]
99. Wilson, J.L.; Scott, I.M.; McMurry, J.L. Optical Biosensing: Kinetics of Protein A-IGG Binding Using Biolayer Interferometry. *Biochem. Mol. Biol. Educ.* **2010**, *38*, 400–407. [CrossRef] [PubMed]
100. Afzal, A.; Mujahid, A.; Schirhagl, R.; Bajwa, S.Z.; Latif, U.; Feroz, S. Gravimetric Viral Diagnostics: QCM Based Biosensors for Early Detection of Viruses. *Chemosensors* **2017**, *5*, 7. [CrossRef]
101. Su, R.; Wang, F.; McAlpine, M.C. 3D Printed Microfluidics: Advances in Strategies, Integration, and Applications. *Lab Chip* **2023**, *23*, 1279–1299. [CrossRef]
102. Three-Dimensional Direct Laser Writing of PEGda Hydrogel Microstructures with Low Threshold Power Using a Green Laser Beam. Available online: https://www.light-am.com/en/article/doi/10.37188/lam.2021.003 (accessed on 22 April 2025).
103. Jin, Y.; Xiong, P.; Xu, T.; Wang, J. Time-Efficient Fabrication Method for 3D-Printed Microfluidic Devices. *Sci. Rep.* **2022**, *12*, 1233. [CrossRef]

104. Guttridge, C.; Shannon, A.; O'Sullivan, A.; O'Sullivan, K.J.; O'Sullivan, L.W. Biocompatible 3D Printing Resins for Medical Applications: A Review of Marketed Intended Use, Biocompatibility Certification, and Post-Processing Guidance. *Ann. 3D Print. Med.* **2022**, *5*, 100044. [CrossRef]
105. Poskus, M.D.; Wang, T.; Deng, Y.; Borcherding, S.; Atkinson, J.; Zervantonakis, I.K. Fabrication of 3D-Printed Molds for Polydimethylsiloxane-Based Microfluidic Devices Using a Liquid Crystal Display-Based Vat Photopolymerization Process: Printing Quality, Drug Response and 3D Invasion Cell Culture Assays. *Microsyst. Nanoeng.* **2023**, *9*, 140. [CrossRef]
106. Vaidya, N.; Solgaard, O. 3D Printed Optics with Nanometer Scale Surface Roughness. *Microsyst. Nanoeng.* **2018**, *4*, 18. [CrossRef]
107. Shanmugam, A.; Usmani, M.; Mayberry, A.; Perkins, D.L.; Holcomb, D.E. Imaging Systems and Algorithms to Analyze Biological Samples in Real-Time Using Mobile Phone Microscopy. *PLoS ONE* **2018**, *13*, e0193797. [CrossRef]
108. Ali, M.A.; Zhang, G.F.; Hu, C.; Yuan, B.; Jahan, S.; Kitsios, G.D.; Morris, A.; Gao, S.-J.; Panat, R. Ultrarapid and Ultrasensitive Detection of SARS-CoV-2 Antibodies in COVID-19 Patients via a 3D-printed Nanomaterial-based Biosensing Platform. *J. Med. Virol.* **2022**, *94*, 5808–5826. [CrossRef]
109. Sharma, P.; Hasan Mohd, R.; Yardim, A.; Celik, E.G.; Timur, S.; Abdin, M.Z.; Narang, J. Next-Generation Point-of-Care Diagnostics: Silver Nanoparticle-Enhanced 3D-Printed Multiplex Electrochemical Biosensor for Detecting Dengue and Chikungunya Viruses. *ACS Appl. Bio Mater.* **2025**, *8*, 1251–1260. [CrossRef] [PubMed]
110. Krejcova, L.; Nejdl, L.; Rodrigo, M.A.M.; Zurek, M.; Matousek, M.; Hynek, D.; Zitka, O.; Kopel, P.; Adam, V.; Kizek, R. 3D Printed Chip for Electrochemical Detection of Influenza Virus Labeled with CdS Quantum Dots. *Biosens. Bioelectron.* **2014**, *54*, 421–427. [CrossRef] [PubMed]
111. Karamzadeh, V.; Sohrabi-Kashani, A.; Shen, M.; Juncker, D. Digital Manufacturing of Functional Ready-to-Use Microfluidic Systems. *Adv. Mater.* **2023**, *35*, 2303867. [CrossRef] [PubMed]
112. Armijo, P.R.; Markin, N.W.; Nguyen, S.; Ho, D.H.; Horseman, T.S.; Lisco, S.J.; Schiller, A.M. 3D Printing of Face Shields to Meet the Immediate Need for PPE in an Anesthesiology Department during the COVID-19 Pandemic. *Am. J. Infect. Control* **2021**, *49*, 302–308. [CrossRef]
113. Prototyping Rapid Diagnostics for COVID-19 with 3D Printing. Formlabs. Available online: https://formlabs.com/covid-19-response/covid-19-rapid-diagnostics/?srsltid=AfmBOoq2JMJEfdWqop-UvUqu4hrgqFwz8Ngiu-GrmlGGl162dFemATCz (accessed on 22 April 2025).
114. Ford, J.; Goldstein, T.; Trahan, S.; Neuwirth, A.; Tatoris, K.; Decker, S. A 3D-Printed Nasopharyngeal Swab for COVID-19 Diagnostic Testing. *3D Print. Med.* **2020**, *6*, 21. [CrossRef]
115. Goswami, M.; Yadav, A.K.; Chauhan, V.; Singh, N.; Kumar, S.; Das, A.; Yadav, V.; Mandal, A.; Tiwari, J.K.; Siddiqui, H.; et al. Facile Development of Graphene-Based Air Filters Mounted on a 3D Printed Mask for COVID-19. *J. Sci. Adv. Mater. Devices* **2021**, *6*, 407–414. [CrossRef]
116. Naz, S.A.; Huynh, V.T.; Doeven, E.H.; Adams, S.; Kouzani, A.; Guijt, R.M. Closed-Loop Control Systems for Pumps Used in Portable Analytical Systems. *J. Chromatogr. A* **2023**, *1695*, 463931. [CrossRef]
117. Moya Muñoz, G.G.; Brix, O.; Klocke, P.; Harris, P.D.; Luna Piedra, J.R.; Wendler, N.D.; Lerner, E.; Zijlstra, N.; Cordes, T. Single-Molecule Detection and Super-Resolution Imaging with a Portable and Adaptable 3D-Printed Microscopy Platform (Brick-MIC). *Sci. Adv.* **2024**, *10*, eado3427. [CrossRef]
118. Mendoza-Gallegos, R.A.; Rios, A.; Garcia-Cordero, J.L. An Affordable and Portable Thermocycler for Real-Time PCR Made of 3D-Printed Parts and Off-the-Shelf Electronics. *Anal. Chem.* **2018**, *90*, 5563–5568. [CrossRef]
119. Sheu, S.C.; Kijamnajsuk, P.; Chan, C.K.; Chen, J.J. A 3D-Printed Oscillatory Polymerase Chain Reaction System Using a Single Heater. *Appl. Therm. Eng.* **2024**, *239*, 122201. [CrossRef]
120. Torres Fredes, I.P.; Cortés-Adasme, E.N.; Barrientos, B.A.; Real, J.P.; Gomez, C.G.; Palma, S.D.; Kogan, M.J.; Real, D.A. 3D-Printed Plasmonic Nanocomposites: VAT Photopolymerization for Photothermal-Controlled Drug Release. *Pharmaceuticals* **2024**, *17*, 1453. [CrossRef] [PubMed]
121. Horst, R.J.; Katzourakis, A.; Mei, B.T.; de Beer, S. Design and Validation of a Low-Cost Open-Source Impedance Based Quartz Crystal Microbalance for Electrochemical Research. *HardwareX* **2022**, *12*, e00374. [CrossRef] [PubMed]
122. Glasco, D.L.; Elhassan, M.M.; McLeod, W.T.; Bell, J.G. Nonenzymatic Detection of Glucose Using 3D Printed Carbon Electrodes in Human Saliva. *ECS Sens. Plus* **2024**, *3*, 020602. [CrossRef]
123. Lee, J.W.; Sim, S.J.; Cho, S.M.; Lee, J. Characterization of a Self-Assembled Monolayer of Thiol on a Gold Surface and the Fabrication of a Biosensor Chip Based on Surface Plasmon Resonance for Detecting Anti-GAD Antibody. *Biosens. Bioelectron.* **2005**, *20*, 1422–1427. [CrossRef]
124. Nakamoto, K.; Kurita, R.; Niwa, O.; Fujii, T.; Nishida, M. Development of a Mass-Producible on-Chip Plasmonic Nanohole Array Biosensor. *Nanoscale* **2011**, *3*, 5067–5075. [CrossRef]
125. Wang, Z.; Wilkop, T.; Han, J.H.; Dong, Y.; Linman, M.J.; Cheng, Q. Development of Air-Stable, Supported Membrane Arrays with Photolithography for Study of Phosphoinositide–Protein Interactions Using Surface Plasmon Resonance Imaging. *Anal. Chem.* **2008**, *80*, 6397–6404. [CrossRef]

126. Liu, Y.; Zhang, X. Microfluidics-Based Plasmonic Biosensing System Based on Patterned Plasmonic Nanostructure Arrays. *Micromachines* **2021**, *12*, 826. [CrossRef]
127. Sun, S.; Zhao, Y.; Hao, Y.; Xue, P.; Guo, X.; Zhang, W.; Zhang, C.; Zhang, Q.; Zhou, W.; Ong, H.C.; et al. Rapid, Sensitive and Multiplexed Detection of SARS-CoV-2 Viral Nucleic Acids Enabled by Phase-Based Surface Plasmon Resonance of Metallic Gratings. *Biomed. Opt. Express* **2024**, *15*, 5215–5226. [CrossRef]
128. Pardee, K.; Green, A.A.; Takahashi, M.K.; Braff, D.; Lambert, G.; Lee, J.W.; Ferrante, T.; Ma, D.; Donghia, N.; Fan, M.; et al. Rapid, Low-Cost Detection of Zika Virus Using Programmable Biomolecular Components. *Cell* **2016**, *165*, 1255–1266. [CrossRef]
129. Crapnell, R.D.; Banks, C.E. Electroanalytical Overview: Screen-Printed Electrochemical Sensing Platforms. *ChemElectroChem* **2024**, *11*, e202400370. [CrossRef]
130. Inam, A.K.M.S.; Costa Angeli, M.A.; Shkodra, B.; Douaki, A.; Avancini, E.; Magagnin, L.; Petti, L.; Lugli, P. Flexible Screen-Printed Electrochemical Sensors Functionalized with Electrodeposited Copper for Nitrate Detection in Water. *ACS Omega* **2021**, *6*, 33523–33532. [CrossRef] [PubMed]
131. He, P.; Cao, J.; Ding, H.; Liu, C.; Neilson, J.; Li, Z.; Kinloch, I.A.; Derby, B. Screen-Printing of a Highly Conductive Graphene Ink for Flexible Printed Electronics. *ACS Appl. Mater. Interfaces* **2019**, *11*, 32225–32234. [CrossRef] [PubMed]
132. Wu, K.; Kim, K.-W.; Kwon, J.H.; Kim, J.K.; Kim, S.H.; Moon, H.C. Direct Ink Writing of PEDOT:PSS Inks for Flexible Micro-Supercapacitors. *J. Ind. Eng. Chem.* **2023**, *123*, 272–277. [CrossRef]
133. Fernandes, I.J.; Aroche, A.F.; Schuck, A.; Lamberty, P.; Peter, C.R.; Hasenkamp, W.; Rocha, T.L.A.C. Silver Nanoparticle Conductive Inks: Synthesis, Characterization, and Fabrication of Inkjet-Printed Flexible Electrodes. *Sci. Rep.* **2020**, *10*, 8878. [CrossRef]
134. Sher, M.; Faheem, A.; Asghar, W.; Cinti, S. Nano-Engineered Screen-Printed Electrodes: A Dynamic Tool for Detection of Viruses. *TrAC Trends Anal. Chem.* **2021**, *143*, 116374. [CrossRef]
135. Sestaioni, D.; Giurlani, W.; Ciacci, G.; Camagni, V.; Palladino, P.; Barucci, A.; Scarano, S. Sustainable and Effective Reconditioning of SPR Gold Chips Functionalized with Molecularly Imprinted Polynorepinephrine. *Anal. Chim. Acta* **2024**, *1321*, 343037. [CrossRef]
136. Kimple, A.J.; Muller, R.E.; Siderovski, D.P.; Willard, F.S. A Capture Coupling Method for the Covalent Immobilization of Hexahistidine Tagged Proteins for Surface Plasmon Resonance. *Methods Mol. Biol.* **2010**, *627*, 91–100. [CrossRef]
137. Auer, S.; Azizi, L.; Faschinger, F.; Blazevic, V.; Vesikari, T.; Gruber, H.J.; Hytönen, V.P. Stable Immobilisation of His-Tagged Proteins on BLI Biosensor Surface Using Cobalt. *Sens. Actuators B Chem.* **2017**, *243*, 104–113. [CrossRef]
138. Mantena, S.; Pillai, P.P.; Petros, B.A.; Welch, N.L.; Myhrvold, C.; Sabeti, P.C.; Metsky, H.C. Model-Directed Generation of Artificial CRISPR–Cas13a Guide RNA Sequences Improves Nucleic Acid Detection. *Nat. Biotechnol.* **2024**, 1–8. [CrossRef]
139. Khan, S.; Singh, Y.V.; Singh, P.; Singh, R.S. An Optimized Artificial Intelligence System Using IoT Biosensors Networking for Healthcare Problems. *Comput. Intell. Neurosci.* **2022**, *2022*, 2206573. [CrossRef]
140. Padhi, A.; Agarwal, A.; Saxena, S.K.; Katoch, C.D.S. Transforming Clinical Virology with AI, Machine Learning and Deep Learning: A Comprehensive Review and Outlook. *Virusdisease* **2023**, *34*, 345–355. [CrossRef] [PubMed]
141. Ginsberg, J.; Mohebbi, M.H.; Patel, R.S.; Brammer, L.; Smolinski, M.S.; Brilliant, L. Detecting Influenza Epidemics Using Search Engine Query Data. *Nature* **2009**, *457*, 1012–1014. [CrossRef] [PubMed]
142. Sun, Y.; Sheng, D.; Zhou, Z.; Wu, Y. AI Hallucination: Towards a Comprehensive Classification of Distorted Information in Artificial Intelligence-Generated Content. *Humanit. Soc. Sci. Commun.* **2024**, *11*, 1278. [CrossRef]
143. Zens, M.; Brammertz, A.; Herpich, J.; Südkamp, N.; Hinterseer, M. App-Based Tracking of Self-Reported COVID-19 Symptoms: Analysis of Questionnaire Data. *J. Med. Internet Res.* **2020**, *22*, e21956. [CrossRef]
144. Çelik Ertuğrul, D.; Çelik Ulusoy, D. A Knowledge-Based Self-Pre-Diagnosis System to Predict Covid-19 in Smartphone Users Using Personal Data and Observed Symptoms. *Expert. Syst.* **2022**, *39*, e12716. [CrossRef]
145. Aborode, A.T.; Adesola, R.O.; Scott, G.Y.; Arthur-Hayford, E.; Otorkpa, O.J.; Kwaku, S.D.; Elebesunu, E.E.; Nibokun, E.O.; Aruorivwooghene, I.J.; Bakre, A.A.; et al. Bringing Lab to the Field: Exploring Innovations in Point-of-Care Diagnostics for the Rapid Detection and Management of Tropical Diseases in Resource-Limited Settings. *Adv. Biomark. Sci. Technol.* **2025**, *7*, 28–43. [CrossRef]
146. Chen, H.-C. A Systematic Review of the Barcoding Strategy That Contributes to COVID-19 Diagnostics at a Population Level. *Front. Mol. Biosci.* **2023**, *10*, 1141534. [CrossRef]
147. Farrell, L. Automated Digital Notification of COVID-19 Diagnoses Through Text and Email Messaging—North Carolina, December 2020–January 2021. *MMWR Morb. Mortal. Wkly. Rep.* **2021**, *70*, 1603–1607. [CrossRef]
148. Ding, X.; Yin, K.; Li, Z.; Lalla, R.V.; Ballesteros, E.; Sfeir, M.M.; Liu, C. Ultrasensitive and Visual Detection of SARS-CoV-2 Using All-in-One Dual CRISPR-Cas12a Assay. *Nat. Commun.* **2020**, *11*, 4711. [CrossRef]
149. Martín-Sierra, C.; Chavez, M.T.; Escobedo, P.; García-Cabrera, V.; López-Delgado, F.J.; Guardia-Monteagudo, J.J.; Ruiz-García, I.; Erenas, M.M.; Sanchez-Martin, R.M.; Capitán-Vallvey, L.F.; et al. SARS-CoV-2 Viral RNA Detection Using the Novel CoVradar Device Associated with the CoVreader Smartphone App. *Biosens. Bioelectron.* **2023**, *230*, 115268. [CrossRef]

150. Wu, K.; Chugh, V.K.; Girolamo, A.d.; Liu, J.; Saha, R.; Su, D.; Krishna, V.D.; Nair, A.; Davies, W.; Wang, A.Y.; et al. Portable Magnetic Particle Spectrometer (MPS) for Future Rapid and Wash-Free Bioassays. *ACS Appl. Mater. Interfaces* **2021**, *13*, 7966–7976. [CrossRef] [PubMed]
151. Pennisi, I.; Cavuto, M.L.; Miglietta, L.; Malpartida-Cardenas, K.; Stringer, O.W.; Mantikas, K.-T.; Reid, R.; Frise, R.; Moser, N.; Randell, P.; et al. Rapid, Portable, and Electricity-Free Sample Extraction Method for Enhanced Molecular Diagnostics in Resource-Limited Settings. *Anal. Chem.* **2024**, *96*, 11181–11188. [CrossRef] [PubMed]
152. Xie, Y.; Sha, Z.; Huang, S.; Yin, C.; Wan, L.; Li, J.; Ling, J.; Wu, C.; Dai, L. An Open Source, PCR Based, Point-of-Care Testing Platform. *Sci. Rep.* **2025**, *15*, 12025. [CrossRef] [PubMed]
153. Ramos-Mandujano, G.; Grünberg, R.; Zhang, Y.; Bi, C.; Guzmán-Vega, F.J.; Shuaib, M.; Gorchakov, R.V.; Xu, J.; Tehseen, M.; Takahashi, M.; et al. An Open-Source, Automated, and Cost-Effective Platform for COVID-19 Diagnosis and Rapid Portable Genomic Surveillance Using Nanopore Sequencing. *Sci. Rep.* **2023**, *13*, 20349. [CrossRef]
154. Moukarzel, G.; Wang, Y.; Xin, W.; Hofmann, C.; Joshi, A.; Loughney, J.W.; Bowman, A. Automation of Biochemical Assays Using an Open-Sourced, Inexpensive Robotic Liquid Handler. *SLAS Technol.* **2024**, *29*, 100205. [CrossRef]
155. 3D Printing: Overview, Impacts, and the Federal Role. Available online: https://www.congress.gov/crs-product/R45852 (accessed on 18 June 2025).
156. Peters, H.P. Gap between Science and Media Revisited: Scientists as Public Communicators. *Proc. Natl. Acad. Sci. USA* **2013**, *110* (Suppl. 3), 14102–14109. [CrossRef]
157. Dankel, M.; Lambert, M. Abstracts for a Wider Audience: Promoting Linguistic Diversity in JBI Evidence Synthesis. *JBI Evid. Synth.* **2023**, *21*, 833. [CrossRef]
158. How to Make Your Research Accessible. Author Services. Available online: https://authorservices.taylorandfrancis.com/publishing-your-research/writing-your-paper/how-to-make-your-research-accessible/ (accessed on 23 April 2025).
159. Stack Whitney, K.; Perrone, J.; Bahlai, C.A. Open Access Journals Lack Image Accessibility Guidelines. *Quant. Sci. Stud.* **2025**, *6*, 46–62. [CrossRef]
160. Shailes, S. Plain-Language Summaries of Research: Something for Everyone. *eLife* **2017**, *6*, e25411. [CrossRef]
161. Sedgwick, C.; Belmonte, L.; Margolis, A.; Shafer, P.O.; Pitterle, J.; Gidal, B.E. Extending the Reach of Science—Talk in Plain Language. *Epilepsy Behav. Rep.* **2021**, *16*, 100493. [CrossRef]
162. Viruses: Instructions for Authors. Available online: https://www.mdpi.com/journal/viruses/instructions (accessed on 23 April 2025).
163. Highlights of Journal Articles | Elsevier. Available online: https://www.elsevier.com/researcher/author/tools-and-resources/highlights (accessed on 23 April 2025).
164. Bredbenner, K.; Simon, S.M. Video Abstracts and Plain Language Summaries Are More Effective than Graphical Abstracts and Published Abstracts. *PLoS ONE* **2019**, *14*, e0224697. [CrossRef] [PubMed]

biosensors

MDPI

Review

Quantum Dot Applications Using Kinetic Data: A Promising Approach for Enhanced Analytical Determinations

Rafael C. Castro, Ricardo N. M. J. Páscoa *, David S. M. Ribeiro * and João L. M. Santos

LAQV/REQUIMTE—Laboratório Associado Para a Química Verde da Rede de Química e Tecnologia, Laboratory of Applied Chemistry, Department of Chemical Sciences, Faculty of Pharmacy, University of Porto, Rua de Jorge Viterbo Ferreira n° 228, 4050-313 Porto, Portugal; rafael.castro.cl@hotmail.com (R.C.C.); joaolms@ff.up.pt (J.L.M.S.)

* Correspondence: rnpascoa@ff.up.pt (R.N.M.J.P.); dsmribeiro@gmail.com (D.S.M.R.)

Abstract: The acquisition of kinetic data in QD-based PL sensing methodologies has been revealed to be an auspicious alternative in applying these nanomaterials in analytical chemistry, enabling enhanced discrimination and quantification of analytes, even in complex sample matrices. The accessibility of kinetic measurements, which use routine laboratory instrumentation, is a significant advantage that increases the practicality of this methodology. The simple acquisition of these kinds of second-order data combined with chemometric analysis can ensure accurate results in environmental, biomedical, and food monitoring applications. These developments emphasize the vital importance of kinetic approaches in increasing sensitivity, improving analyte discrimination, and making the application of QDs in complex samples possible.

Keywords: quantum dots; kinetics; second-order advantage; fluorescence; chemometrics

1. Introduction

Quantum dots (QDs) have unusual properties, making them one of the most promising nanomaterials in nanoscience and nanotechnology, with a panoply of applications, including bioimaging/biosensing [1–4], photocatalysis [5–7], chemical analysis [8–10], and electronics, especially in their use in light-emitting diodes (LEDs) or light-harvesting systems [11,12]. Additionally, QDs can be arranged in single-electron semiconductor quantum dot arrays, which can be applied in ultra-sensitive detection and quantum computing [13–15]. There has been a growing interest in the scientific community in applying these nanomaterials in analytical chemistry due to their exceptional optical properties. For example, the simultaneous excitation of multiple QDs with distinct sizes and/or compositions using a single excitation wavelength for multiplexed detections is possible due to their high molar attenuation coefficients and broad excitation bands [16–18]. Additionally, the combination of multiple fluorophores within the same probe without significant spectral overlap due to their narrow emission bands allows for obtaining enhanced detection capabilities, generating specific analyte response profiles [19]. It is also important to emphasize the possibility of obtaining nanocrystals that emit light across the entire UV/visible-to-infrared electromagnetic spectrum only by adjusting the size of these nanomaterials during the synthesis [16–18]. In addition, the ability to detect a multitude of analytical targets with greater selectivity is provided by the enhanced specific recognition capacity of QDs, which could be obtained because of their highly tunable surface chemistry, enabling their conjugation with numerous biomolecules and polymers, such as antibodies, oligonucleotides, proteins, and molecularly imprinted polymers (MIPs) [20,21].

Finally, QDs can provide enhanced sensitivity for sensing applications as they possess high photoluminescent (PL) quantum yield (QY) [16–18].

QDs have two essential characteristics that make them valuable sensing platforms for implementing kinetic assays. The first one is their resistance to photobleaching, which is much higher than that of organic fluorophores, permitting prolonged irradiation time during time-based measurements. The other one is their long PL lifetimes, ranging from tens to hundreds of nanoseconds, making them ideal nanomaterials for tracking dynamic processes over time [10].

Despite all the aforementioned advantages, broader applications in chemical analysis, particularly in samples with complex matrices, could be a significant drawback because of the QDs' limited selectivity, which occurs when their surfaces are not functionalized with specific recognition elements [9,10]. The analysis of PL data using chemometric models has emerged as a useful strategy to circumvent this issue [8,10]. In this context, a more efficient description of the relationships between samples and analyzed variables can be obtained, which allows for a better understanding of complex systems and improves the analytical results.

Three-dimensional datasets, including Excitation–Emission Matrix (EEM) and kinetic data, also called second-order data, can be processed by appropriate chemometric models, which provide the advantage of circumventing the occurrence of unexpected sample components not considered during the calibration step (second-order advantage). This can be extremely important for distinguishing between different components or predicting potential interferences that impact analytical results, particularly when analyzing complex samples, such as food, environmental, or biological samples [8,10]. So, in these cases, acquiring more complex data such as second- or even higher-order data can successfully mitigate the effects of these interferents, which allows for obtaining more accurate and reliable analytical results as well as better discrimination in the qualitative analysis compared to what would be obtained with zeroth- or first-order data.

Despite being the most explored second-order data in QD-based methodologies, acquiring EEM data requires expensive and specific instrumentation. A simpler and alternative way to acquire second-order data involves recording over time the QD PL spectra modulation in the presence of increasing analyte concentrations, which can be easily obtained using a common fluorimeter [8]. Nevertheless, despite its simplicity, only a very restricted number of analytical methodologies have explored this possibility [8].

Kinetics is essential in chemistry, biology, and environmental science because it enables an enhanced understanding of how and why reactions occur at certain rates and under specific conditions, providing relevant information about mechanisms and dynamics [22]. Consequently, kinetic approaches are frequently employed in enzyme studies, quality control processes, environmental monitoring, and drug development [22].

Kinetic determinations in analytical chemistry involve using techniques and methods to measure the variations in the concentration of reactants or products over time and determine the rate at which a given reaction or chemical process occurs. Kinetic assays are mostly interesting for quantitative analysis as relating the reaction rate to the concentration of the analytes in a sample provides diverse advantages regarding fixed-time assays, where measurements are performed at a single time point, usually after the reaction development. On the contrary, the kinetic approach enables the continuous measurement of reaction rates, which provides valuable insights into reaction mechanisms and consequently facilitates more accurate quantification of analyte concentrations in complex samples [22,23].

Additionally, kinetic assays usually exhibit a wider dynamic range than fixed-time assays, allowing for the accurate determination of analytes across a broader concentration range. Effectively, kinetic assays can measure reaction rates at multiple time points, accommodating a wide range of analyte concentrations. In contrast, in fixed-time assays, the measurements may be limited by the linear range of detection. This wider dynamic range allows for the accurate quantification of analytes in diverse sample matrices, demonstrating the high versatility and applicability of kinetic assays.

In summary, the ability of kinetic assays to provide real-time monitoring, higher accuracy, and a wider dynamic range makes them indispensable analytical methods for quantitative analysis in the fields of analytical chemistry and biochemistry [23,24].

Kinetic determinations using quantum dots as PL nanoprobes represent an interesting challenge in analytical chemistry. By taking advantage of the abovementioned properties of QDs, researchers have opened new avenues for studying real-time reaction kinetics with improved sensitivity and accuracy. This review provides a comprehensive overview of recent advances, applications, and future perspectives of kinetic determinations using quantum dots as PL probes. Despite the potential of this symbiosis, the number of works on this topic is relatively low and can be divided into two main areas of focus: seeking selectivity enhancement or analytical performance improvement.

2. Kinetic Strategies in Quantum Dot Applications

2.1. Selectivity Enhancement by Coupling with Chemometrics

QD-based PL sensing platforms have been widely used in analytical applications; however, their poor selectivity is the largest issue that remains to be solved. Effectively, QDs lack molecular recognition abilities without appropriate surface functionalization, becoming highly reactive sensing elements because of their susceptibility to interact with various chemical species in a non-selective way. This limitation makes it difficult to detect multiple analytes in a single analysis or to monitor them in the presence of interfering substances [9,10].

The capability of QDs to serve as sensing elements is attributed to the modulation of their optical properties upon interaction with the target molecule, with the nanoparticle surface playing a crucial role [8,10,16,18]. The influence of the QDs' surface on their optical properties can be understood in terms of trap states arising from structural defects, such as dangling bonds or adsorbates present at the surface, among others. Depending on its properties, the analyte can induce changes in the intensity of the QDs' PL emission, either enhancing or impairing the efficiency of electron-hole recombination, leading to either a photoluminescence-enhancing or -quenching effect, respectively [9,10].

Consequently, research efforts have been focused on developing strategies to overcome this challenge. These efforts seek to enable accurately determining one or more analytes in a single analysis and monitoring analytes in complex sample matrices containing interfering species. Some strategies can be highlighted to circumvent the reduced selectivity of QDs, including the following:

(i) separation techniques to separate the analyte from the sample matrix, reducing the presence of potential interferents. However, these strategies involve labour-intensive and time-consuming procedures, which frequently require using strong acids, heating, auxiliary oxidants, and organic solvents [25]; (ii) tuning the surface chemistry of QDs through the functionalization of their surface with appropriate molecular recognition elements, namely oligonucleotides, polymers, biomolecules, or enzymes. This allows QDs to obtain specific recognition capabilities to detect target analytes. Surface functionalization is essential for multiplexed detection approaches or analyte quantification, especially in complex samples. Thus, the enhanced recognition capabilities ensure that the PL modulation of the functionalized sensing platform results specifically from the interaction with the target analyte. The success of this strategy depends on the efficiency of the recognition element, which guarantees selectivity and consequently allows multiplexed detection. However, this approach is usually elaborate, time-consuming, and expensive [10,20,26]; (iii) exploring Förster Resonance Energy Transfer (FRET) sensing schemes can help overcome selectivity issues in QD-based assays by taking advantage of the distance-dependent energy transfer process between a donor fluorophore and an acceptor. In FRET-based assays, energy is transferred from the donor to the acceptor when they are close enough (less than 10 nm), reducing the donor's fluorescence. This process can be used successfully in the specific detection of molecular interactions or binding events due to its high sensitivity to changes in distance. This allows for specifically detecting interactions between the QDs and target analytes, minimizing interference from non-target molecules. These strategies have several drawbacks such as the need for advanced materials, complex experimental setups, sophisticated instrumentation, restrictions on the number of fluorophores that can be combined in a given experiment, and complex data interpretation procedures. Furthermore, the selectivity of FRET-based assays is still poor without efficient functionalization of surface chemistry [27–29]; (iv) using chemometric models to analyze and process the acquired PL data. This approach enables the identification and quantification of each analyte in complex sample matrices or mixtures, even in the presence of interferents. Chemometric models that employ statistical and mathematical methods allow for describing the relationship between PL data and analytical parameters, facilitating a deeper understanding of the dataset. Extracting more valuable insights from gathered data through chemometric analysis facilitates the selective analysis of an analyte, even without any sample purification steps or improvements in QDs' selectivity in a FRET process or QD surface functionalization. This selectivity and the accuracy of the methodology can be further improved depending on the complexity of the acquired data and the selection of the most appropriate chemometric models. Therefore, the chemometric analysis of PL data represents a simpler and more economical strategy to overcome selectivity problems [30–32].

In the following sections, the use of chemometric models to circumvent selectivity issues related to QD-based sensing platforms will be discussed in detail.

2.2. *Relating Data Complexity and Their Processing Using Chemometric Models*

Chemometrics, a term first introduced by Svante Wold in the 1970s [33], refers to the science of extracting valuable chemical information from complex experimental systems and transforming the acquired data into meaningful insights [33,34]. While univariate methods analyze one variable independently of others, they fail to recognize the intercorrelation among multiple variables. In contrast, multivariate strategies can consider various variables, ensuring a more comprehensive interpretation of the available data [34].

The instrumental data obtained for QD-based PL methodologies can be categorized based on their complexity, each type offering distinct advantages. Analytical data are classified in terms of increasing complexity as zeroth-order, first-order, second-order, and higher-order data, depending on the characteristics of the instrument or method used to acquire it. It is anticipated that as the complexity of the collected data per sample analysis increases, more refined quantitative or qualitative analytical estimations can be achieved, along with greater selectivity [35,36].

Zeroth-order data are acquired when the instrument generates a single output per sample. An example is the acquisition of a fluorescence intensity, at a specific wavelength, for each analyte concentration (Figure 1a). This implies that each sample is defined by a single numerical value. In analytical contexts, it is crucial to emphasize that zeroth-order data, employed in univariate calibrations, necessitate complete selectivity towards the analyte under investigation and are applicable only to samples with a known composition and that lack interfering species [34,36].

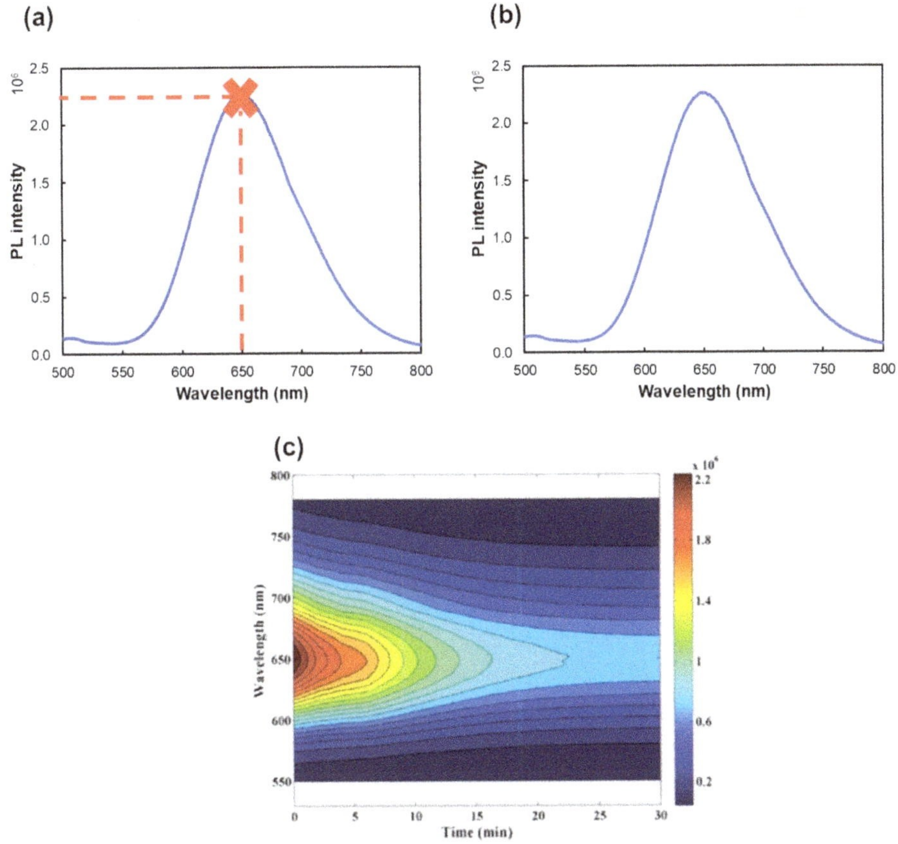

Figure 1. Summary of the different data structures that can be obtained for a sample using PL-based methodologies. (**a**) Zeroth-order data, representing fluorescence intensity at a single wavelength as a function of analyte concentration (red cross); (**b**) First-order data, corresponding to the fluorescence emission spectrum at a fixed excitation wavelength; and (**c**) Second-order data, related to an Excitation-emission matrix or the evolution of the sample's PL spectrum over time at a fixed excitation wavelength. Adapted with permission from [37]. Copyright 2021 Elsevier.

First-order data consist of vector data for each sample, such as acquiring a PL emission spectrum at a fixed excitation wavelength (Figure 1b). This spectrum includes a series of intensity values at various wavelengths, forming a data vector. Analyzing these first-order data with appropriate first-order multivariate calibration methods enables the circumvention of selectivity limitations through the application of efficient mathematical algorithms. This is known as the first-order advantage, which allows for the quantification of the analyte in samples containing known interferents, if these interferents are included in the standard solutions used during calibration. However, unexpected constituents can make the quantification of the target analyte impossible [34,36].

In second-order data, a data matrix is obtained for each sample (Figure 1c). These complex data contain extractible complementary information, which, through second-order advantage, enables the monitoring of the analyte, even when unexpected interferences are present. This can be especially relevant in complex samples (such as biological, environmental, or food samples), where the sample composition is unknown and, therefore, unexpected interferents may be present [34,36].

Second-order data in QD-based PL analytical methodologies can be acquired by two main approaches: (i) employing a spectrofluorometer able to record an EEM (Figure 2a) or (ii) monitoring the kinetics of the QDs/sample interaction, where the evolution of the emission spectrum over time at a fixed excitation wavelength is assessed (Figure 2b). Another alternative, but unusual, strategy involves the use of hyphenated instruments with multiblock data analysis techniques. In this approach, data of diverse natures are collected from different instrumental modes, extracting complementary and useful information from the sample [8,10].

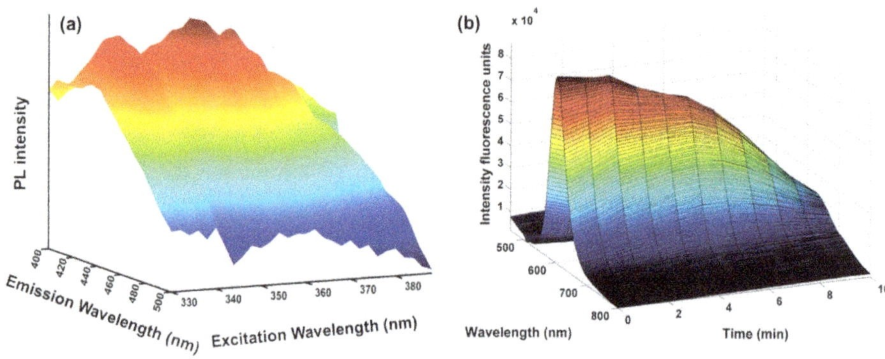

Figure 2. Second-order data using QD-based methodologies: (**a**) excitation–emission spectra (λ_{em} ranging from 330 nm to 380 nm and λ_{ex} ranging from 400 to 500 nm) of CDs and (**b**) PL kinetics data of the combined nanoprobe encompassing CdTe/AgInS_2 QDs upon interaction with acetylsalicylic acid over 10 min. The colors represent the intensity of the emission, with red indicating higher intensity and blue indicating lower intensity. Adapted with permission from [38,39]. Copyright 2024 Elsevier and MDPI 2023.

In fact, either the EEM or the kinetics of the QDs/sample interaction ensures ample spectral information essential for circumventing the presence of unexpected interfering compounds in complex samples. On one hand, the EEM can enhance the specificity of fluorescence spectroscopy. By collecting emission spectra at various excitation wavelengths, EEMs generate a matrix of PL intensities. This matrix provides a comprehensive overview of the sample's fluorescence behavior across different excitation and emission wavelengths, allowing for a more detailed analysis and improved specificity in identifying and characterizing fluorescent compounds or substances present in the sample. On the other hand, the collection of the emission spectra at different times of the QDs/analyte interaction allows for obtaining more information and better understanding the interaction between them. However, when quantum dots (QDs) are utilized as a sensing platform, the complexity of the matrices and the data they generate requires the use of sophisticated mathematical models and computational techniques for accurate decoding. Therefore, effectively handling the complexity of data matrices in QD-based methodologies necessitates a comprehensive approach. This involves combining advanced experimental procedures with robust data analysis strategies to extract meaningful insights [8,10].

Within the two main alternatives discussed to obtain second-order data using QDs as fluorescent probes, EEM data acquisition demands costly and specialized instrumentation. Alternatively, the time-based recording of QD PL spectra modulation in the presence of increasing analyte concentrations, easily monitored using a common fluorimeter, offers a straightforward method to gather second-order data, without requiring an investment in specialized equipment. Nevertheless, a very restricted number of analytical methodologies have explored this possibility [8]. It must be considered that the utilization of kinetic data carries analytical significance only if consistent variations in the optical properties of the QDs are observed over time. However, in certain instances, when QDs are mixed with the analyte, their optical properties are swiftly altered, thereafter remaining unchanged for extended periods of time, rendering them analytically irrelevant.

There are several algorithms available for second-order data analysis, but only a few have been used in QDs applications (namely, multivariate curve resolution, unfolded partial least squares, multiway partial least squares, and artificial neural networks). These algorithms can be categorized into three different groups: noniterative (e.g., direct trilinear decomposition), iterative (e.g., parallel factor analysis and multivariate curve resolution with alternating least squares), and residual bilinearization (e.g., unfolded partial least squares and multiway partial least squares, both followed by residual bilinearization) [40]. It is beyond the scope of this manuscript to conduct an extensive and detailed analysis of all available algorithms for second-order data analysis. Instead, only a brief overview of the algorithms described in this manuscript used in QD applications will be made. For more detailed information about these algorithms, we recommend reading the reference [40]. The multivariate curve resolution with alternating least squares (MCR-ALS) extracts meaningful information from a system through bilinear model decomposition. The application of constraints (e.g., unimodality and non-negativity, among others) during ALS optimization helps reduce the problem of rotational ambiguity, thereby improving model accuracy and enabling the identification of components in the system. Additionally, applying the correlation constraint enables quantitative measurements of the components present in the system. Thus, this model allows handling second-order data that do not follow a trilinear structure, featuring overlapping signals, background contributions, and noise, while obtaining the second-order advantage [41,42]. Unfolded partial least squares (U-PLS) is capable of handling second-order data by unfolding them into vectors and performing a bilinear model decomposition [40,43]. Calibration then begins using these vectors (X data) against the concentration values of the parameters to be analyzed (Y data), aiming to explain the most variance contained in the X data and maximize its covariance with Y data. During this process, it is important to select the optimal number of latent variables to avoid overfitting. The most common method for estimating the best number of latent variables is the leave-one-out cross-validation method. After optimizing using the calibration set, the validation samples are projected to assess the accuracy of the developed model. However, in the presence of unexpected constituents in the validation samples, the U-PLS method may not be able to fully mitigate their influence, but it can identify them as outliers because their residuals will be significantly higher. Therefore, a complementary algorithm named residual bilinearization (RBL) can be employed to reduce the residual errors of these samples to values similar to the noise of the instrument used. Multiway partial least squares (N-PLS) operates under the same principles as U-PLS but without the need to unfold the second-order data and maintain covariance maximization [44]. This method is often considered superior to unfolding methods such as U-PLS due to its stable multiway decomposition, which makes it less susceptible to handling noisy signals and easier to interpret without losing information through unfolding [44]. Artificial neural network (ANN) models are capable of modeling complex nonlinear data and consist of three layers (input, hidden, and output layers), also known as architecture, which are interconnected in multiple ways with the next layer. These models can "learn by example" through iterative training processes such as backpropagation, where the weights of the neurons within each layer are adjusted to minimize the error using only the calibration set [45,46]. During this process, it is crucial to adjust several parameters properly to avoid overfitting of the model [46]. The primary difference between multilayer feed-forward (MLF) and radial basis function (RBF) NNs lies in the activation function used, where MLF typically employs sigmoid functions to introduce nonlinearity, while RBF uses radial basis functions such as the Gaussian function. After optimization, validation samples are projected to assess the accuracy of the developed ANN models [45,46].

It should be highlighted that the MCR-ALS, U-PLS/RBL, and N-PLS/RBL models are more suitable for linear data, although U-PLS/RBL and N-PLS/RBL can handle mild nonlinear data [47]. When dealing with nonlinear data, ANNs are often the best option.

2.3. *Kinetic Determinations to Enhance Analytical Performance Using QDs as PL Nanoprobes*

Acquiring the time-based evolution of QD PL spectra resulting from the interaction between QDs and the target analyte could be regarded as having great analytical value for analyte concentration quantification. Given the complex structure of second-order data, these data should be processed using suitable second-order multivariate calibration methodologies. In this way, not only is analyte quantification simplified but it also allows for mitigating the presence of uncalibrated interfering species, whose effects are thus excluded from the quantification process. However, using QD-based PL kinetics data processed by chemometric models for analytical purposes remains relatively limited among the methodologies proposed in the scientific literature.

In the first proposed methodology from 2009 (Table 1), which involved studying the interaction between multiple analytes and QDs over time, it was demonstrated that the investigation of kinetics could effectively discern between different metals [48]. This research focused on the quenching kinetics induced by Fe^{2+} and Fe^{3+} ions on glutathione (GSH) capped-CdTe QDs. Both Fe^{2+} and Fe^{3+} can quench the fluorescence of GSH-CdTe QDs, but with different kinetics. Upon the addition of Fe^{3+}, the fluorescence of GSH-CdTe QDs was rapidly quenched by approximately 18% within one minute, subsequently reaching equilibrium. Contrastingly, the addition of Fe^{2+} resulted in a gradual quenching of the fluorescence intensity of Cd-based QDs. The fluorescence decreased by approximately 65% within the first 5 min, followed by a slower decrease of 15% over the next 25 min. This distinct kinetic behavior suggests differences in the interaction between Fe^{2+} and Fe^{3+} with GSH-CdTe QDs (Figure 3a). Additionally, it was observed that different metals also exhibited distinct kinetic behaviors (Figure 3b). The observed variations in quenching behaviors among different metal ions for CdTe QDs can be attributed to differences in their electronic structures and dissimilar redox potentials. This understanding of the kinetics and mechanisms involved in the interaction between metal ions and CdTe QDs is crucial for the development of sensitive detection methods and the optimization of QD-based sensing platforms [48]. This work was pioneering in the utilization of the kinetic behavior of the interaction between QDs and target compounds. However, this approach was only applied to iron speciation in that study, without the use of chemometric models.

Choosing the optimal chemometric model and modulation strategy is crucial for ensuring the desired outcomes of the analysis when employing QDs as fluorescence probes. In this regard, various approaches have been investigated to develop a reliable sensing platform based on QDs:

(i) Using single-emitter QDs involves their interaction with the analyte, resulting in the modulation of their optical properties over time (Figure 4a). This interaction, which leads to various PL responses, is mainly influenced by the surface chemistry of the QDs as well as the stability of QDs upon the interaction with the target analytes. Indeed, the nanocrystal's reactivity towards the analyte is determined by the distinct terminal functional groups of the capping ligand employed to stabilize the QD's surface. Consequently, changes in optical properties can occur through alterations in PL intensity or shifts in the maximum emission wavelength.

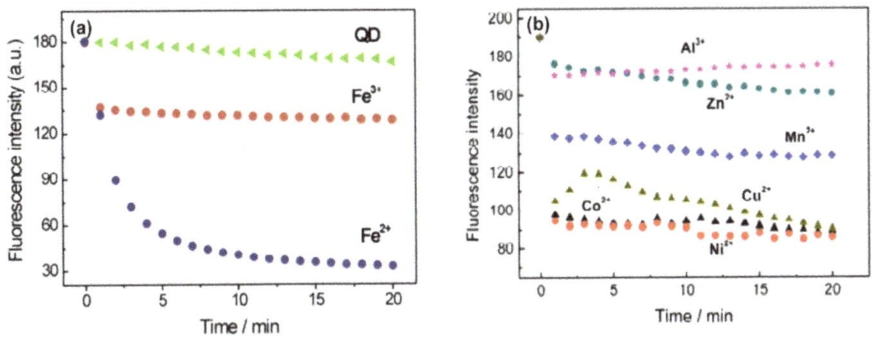

Figure 3. (**a**) Variation in the PL signal of GSH-CdTe QDs during 20 min without and with Fe^{2+} and Fe^{3+}. (**b**) Kinetic behavior of GSH-CdTe in the presence of different transition metal ions. Adapted with permission from [48]. Copyright 2009 American Chemical Society.

(ii) Using multi-emitter probes involves combining various multi-colored fluorophores (including QDs and/or other molecules) with different affinities towards the analyte. This enables the gathering of complementary information, which is crucial for obtaining a distinctive profile of analyte responses. Indeed, through individual interactions with each single-emitter probe, targets can be more easily quantified and discriminated based on the cumulative nonspecific responses. This capability is achieved with multi-emitter nanoprobes designed for multipoint detection. The presence of multiple responses in the same probe significantly enhances the amount of available information, providing a more comprehensive and accurate understanding of the analyzed sample. In such multi-emitter probes, two fluorophores can be strategically used, obtaining an emission spectrum of the combined probe with well-separated emission bands, which allow for distinct signal detection and analysis (Figure 4b). Alternatively, the spectral overlap between the emission bands of the two fluorophores from the combined probe can be exploited (Figure 4c).

Abdollahi et al. [49] used three-dimensional kinetic data to determine copper ions. In the study, the authors investigated the modulation of the PL of L-cysteine (CYS)-capped CdS QDs in the presence of Cu^{2+} over time. The data were analyzed using various chemometric models, including multivariate curve resolution with alternating least squares (MCR-ALS) and partial least squares (PLS), after row-wise augmentation of the second-order data. MCR-ALS was employed to extract spectral profiles of all chemical species present, while PLS was used for quantifying the metal ions. Additionally, the authors examined the kinetic interaction of the nanoprobe with other ions (Ag^{+}, Ni^{2+}, and Hg^{2+}), and they observed that each tested metal ion produced a distinctive emission spectra profile, indicating unique kinetic behaviors in the CdS QD PL. This observation suggests potential applications for accurately quantifying Cu^{2+} in the presence of interfering species and simultaneously determining co-existing metal ions that affect the QD PL.

Our research group also developed a methodology using single-emitter QDs, thiomalic acid (TMA)—capped-$AgInS_2$ (AIS) QDs. Initially, a ratiometric sensing platform comprising two dynamic fluorophores was developed for analytical purposes. In this platform, oxytetracycline (OTC) served as both the analyte and fluorophore, while the QDs dynamically responded to the presence of the target molecule. Consequently, as the concentration of OTC increased, the corresponding emission band also increased, while the emission of AIS QDs was gradually suppressed. The proposed ratiometric sensing approach was tested for determining OTC in a commercially available veterinary pharmaceutical formulation. The results showed that the excipient, lactose monohydrate (94.5% of the sample content), significantly influenced the ratiometric measurements. Although the inhibition of AIS QDs'

PL emission remained unaffected, the presence of the excipient hindered the emergence of the OTC emission band. Due to this excipient effect, the feasibility of using the proposed ratiometric sensing approach for OTC determination in pharmaceutical formulations was impracticable. To circumvent the presence of the excipient in pharmaceutical formulations, the kinetic behavior of AIS QDs' PL quenching in the presence of OTC was evaluated. Immediately after the addition of OTC (0 min), a slight quenching of the PL of AIS QDs was observed, which intensified with increasing antibiotic concentration over time. This gradual enhancement of the quenching effect on QDs' PL emission with reaction time emphasized the potential of this kinetic approach for OTC determination (Figure 5a). The acquisition of kinetic data not only reduced the limit of detection (LOD) but also enhanced sensitivity and selectivity. Therefore, the corresponding kinetic data were analyzed using unfolded partial least squares (U-PLS) to overcome selectivity issues. This was achieved through the acquisition of second-order data, which proved beneficial for quantifying OTC in the presence of uncalibrated lactose monohydrate.

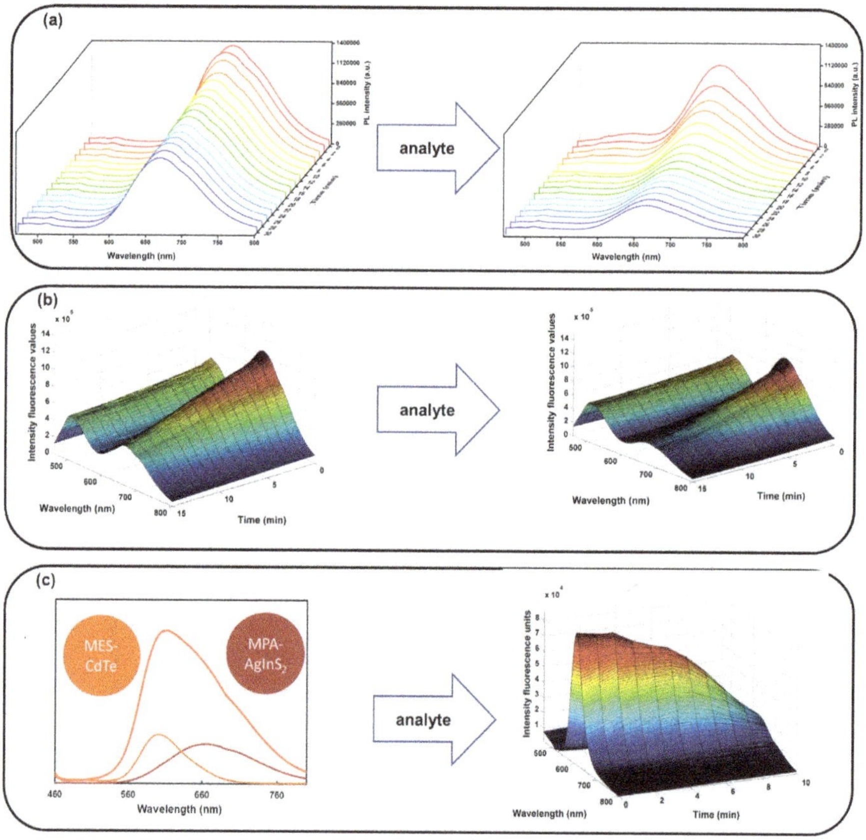

Figure 4. Alternative approaches for achieving effective kinetics-based sensing platforms using QDs: (**a**) nanoprobe with a single emitter for oxytetracycline detection; (**b**) multi-emission nanoprobe with distinct emission spectra in a ratiometric probe for the detection of histamine in foodstuffs; and (**c**) PL spectra of the nanohybrid probe composed of MES-CdTe/MPA-AgInS_2 with the overlap of both individual nanoparticle emission bands by the determination of acetylsalicylic acid in pharmaceutical formulations. The colors represent the intensity of the emission, with red indicating higher intensity and blue indicating lower intensity.Adapted with permission from [37,39,50]. Copyright 2021 Elsevier, 2023 Elsevier, 2023 MDPI.

Despite their effective use as PL sensors, QDs also exhibit excellent photocatalytic properties [6,51]. Effectively, their ability to generate reactive oxygen species (ROS) in aqueous media upon irradiation with ultraviolet (UV) and visible light has been exploited for several applications. This photocatalytic capacity has led to diverse applications in environmental remediation [52–54], microbiological interventions [55,56], and analytical chemistry [7,57,58]. Notably, QDs have been utilized to eliminate persistent pollutants in aquatic environments, serve as antimicrobial nanomaterials, and facilitate the determination of analytes, namely, analytes with antioxidant properties.

In another work, the ability of Cd-free AIS ternary QDs to generate ROS, upon visible light irradiation, was used for Aflatoxin B1 (AFB1) determination [59]. The photogenerated ROS induced the formation of highly fluorescent AFB1 degradation products, which were used for mycotoxin monitoring. This work demonstrated that although semiconductor QDs were not suitable for direct use as PL sensing platforms for AFB1 determination, AIS QDs showed potential as a photocatalytic agent, enabling mycotoxin detection. The gradual increase in PL emission (more accentuated in the first 15 min) can be attributed to the increased fluorescence of the photodegradation products compared to the native fluorescence of AFB1. Effectively, the photodegradation products resulting from the photocatalytic activity of QDs increased in solution over time (Figure 5b). In contrast, the fluorescence of the AFB1 solution showed minimal change without QDs, showing the central role of QDs as photocatalytic agents. Subsequently, the use of a flexible multivariate calibration model like unfolded partial least squares followed by residual bilinearization procedure (U-PLS/RBL) allowed for the extraction of useful information from the PL dataset, making the accurate determination of AFB1 in some foodstuffs possible, even in the presence of unknown interfering species not included in the calibration set [59].

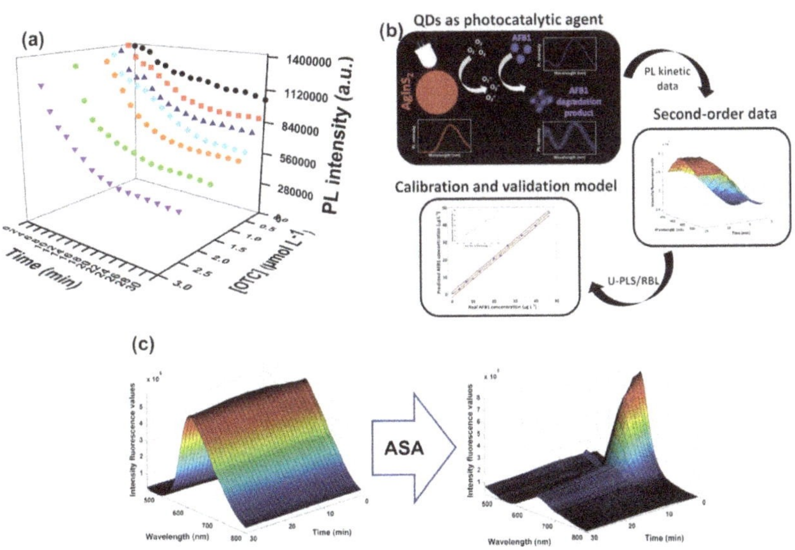

Figure 5. (**a**) Progression of the PL emission intensity of the AIS QDs at the maximum emission wavelength over time, both in the absence of OTC and upon the addition of increasing concentrations of OTC. (**b**) Schematic representation summarizing the detection principle of AFB1 via the photocatalytic process involving the mycotoxin and AIS QDs and AFB1 determination using the U-PLS model. (**c**) Second-order data PL spectra of the combined nanoprobe before and after the interaction with 35.6 mg L^{-1} of acetylsalicylic acid over 30 min. The colors represent the intensity of the emission, with red indicating higher intensity and blue indicating lower intensity. Adapted with permission from [37,39,59]. Copyright 2021 Elsevier, 2023 Elsevier, 2023 MDPI.

The use of nanoprobes encompassing multiple emitters with multipoint detection, combined with the analysis of PL data using appropriate chemometric models, presents a viable strategy for quantifying analytes in complicated samples. This approach provides complementary and valuable information, allowing for an analytical methodology with enhanced accuracy and selectivity. Among the different types of multi-emitter nanoprobes, ratiometric sensing platforms, involving the combination of QDs of different sizes or nature, each emitting light at specific wavelengths, have been most commonly explored. Our research group combined CDs (non-reactive reference fluorophores) and 3-mercaptopropionic acid (MPA)-capped AIS QDs (reactive fluorophores whose PL intensity is modulated by the target analyte) to determine histamine in foodstuffs [50]. In this study, two different ratios of AIS QDs and CDs were explored (different CD dilutions) to assess the effectiveness of the multi-emitter ratiometric sensing platform in detecting histamine. The results indicated that the ratiometric probe's performance varied in terms of the determination coefficient (R_2C) and root mean square error of calibration (RMSEC) for different combination ratios. This demonstrates that the selection of the sensing platform is essential, even in cases where the fluorophore is inert to the presence of the target analyte. Moreover, the assessment of two chemometric models, specifically N-way partial least squares (N-PLS) and U-PLS, involves evaluating them using validation samples across different time intervals. This evaluation occurred within the first 15 min. Although both chemometric models proved capable of accurately determining histamine in complex matrix samples, the N-PLS model outperformed the U-PLS model in terms of prediction accuracy. Additionally, accurate results were attained with just 5 min of kinetic spectral acquisition. Therefore, by making a better compromise between time consumed per sample and the results' accuracy, the shorter time was selected, thus allowing an increase in the sampling rate. The developed methodology demonstrated accuracy, reliability, and simplicity, and at the same time, it was less labour-intensive and more environmentally friendly compared to the reference procedures, namely high-performance liquid chromatography or capillary electrochromatography. Effectively, utilizing nanoprobes with multiple emitters and multipoint detection, along with employing suitable chemometric models to analyze PL data, provides a practical method for quantifying analytes in complex samples such as tuna, tomato, and hake fish [50].

Lastly, an interesting approach was developed using a dual-emitter fluorescent probe that combines binary CdTe and ternary AIS QDs for the determination of acetylsalicylic acid (ASA) [39]. Matrices were acquired using a time-based approach (kinetic method) to ensure the gathering of second-order data capable of exploiting the second-order advantage. Contrarily to the abovementioned work, the PL spectrum of the multi-emitter nanoprobe displayed a significantly broad band, with undefined maximum emission wavelengths. This was attributed to the overlap of the individual nanoparticle emission bands, and this overlapping phenomenon led to an enhancement in the PL intensity of both emission bands. The presence of ASA caused a gradual PL quenching of the combined nanoprobe, which was more pronounced in the first 10 min of the interaction (Figure 5c). Additionally, four distinct chemometric models, namely U-PLS, N-PLS, multilayer feed-forward neural networks (MLF-NNs), and radial basis function neural networks (RBF-NNs), were evaluated to determine the most suitable chemometric approach. While some minor differences were observed during calibration, the overall accuracy did not significantly differ between the models tested. The best results were obtained when fluorescence signals were mean-centered. In the validation phase, all models demonstrated high accuracy, confirming the reliability of the proposed methodology for ASA quantification in pharmaceutical samples [39].

This study conducted an important investigation to evaluate the stability of various synthesized QDs. Effectively, the stability of QDs is of paramount importance in kinetic assays since ensuring the stability of QDs guarantees consistent and reliable performance throughout the assay, minimizing fluctuations in signal intensity or emission spectra. Unfortunately, some QDs exhibited a significant inhibition of PL emission (~70%) within the first 15 min [39]. The predominant method for optimizing synthetic routes for QDs remains univariate approaches. Furthermore, this synthesis is generally optimized considering the maximum emission wavelength or their quantum yield, almost always neglecting their stability. Since the stability of nanomaterials is vital for implementing kinetic assays, this parameter must receive greater consideration in future research.

Table 1. Analytical approaches exploring the use of kinetic data using QDs as a sensing platform.

Analytes	Sensing Platform	Chemometric Tool	Time Acquisition	LOD	Ref.
Fe^{2+} and Fe^{3+}	GSH-CdTe	n.a.	30 min.	5 nmol L^{-1}	[48]
Cu^{2+}	CYS-CdS	MCR-ALS PLS	30 min.	13 nmol L^{-1}	[49]
OTC	TMA-AIS	U-PLS	30 min.	0.144 µmol L^{-1}	[37]
AFB1	MPA-AIS	U-PLS	15 min.	1.2 µg L^{-1}	[59]
Histamine	CDs@MPA-AIS	U-PLS N-PLS	5 min.	1.26 mg L^{-1}	[50]
ASA	MES-CdTe@MPA-AIS	U-PLS N-PLS MLF-NNs RBF-NNs	10 min.	2.82 mg L^{-1} 3.10 mg L^{-1} 3.38 mg L^{-1} 3.26 mg L^{-1}	[39]

n.a. means not applicable.

3. Conclusions and Prospects

Using kinetic data in QD-based PL sensing platforms is a promising strategy for improving the results obtained in the discrimination and quantification of analytes, which is crucial in analytical chemistry. The accurate quantification of the concentration of the analytes, even at minimal levels and in the presence of interfering species, is important for multiple applications, such as environmental monitoring and biomedical diagnostics. Kinetic approaches can provide comprehensive knowledge about complex systems by collecting a higher amount of data, which allows, for example, the quantification of analytes in the presence of uncalibrated species.

Additionally, obtaining kinetic data is accessible since routine laboratory instrumentation is available. Employing kinetic assays provides an effective solution for circumventing interference issues that commonly occur in complex sample matrices. Using second-order data with appropriate chemometric analysis allows for obtaining second-order advantage, which enables the accurate quantification of analytes, even in the presence of uncalibrated interferents. Therefore, this approach can successfully distinguish analytes from interfering and uncalibrated substances, especially in samples with unknown and/or variable compositions, such as environmental, food, or biological samples. This ability is particularly important for the application of QDs in real-world samples. Unfortunately, only a small number of scientific studies have explored the use of kinetic data employing QDs as PL emission nanoplatforms.

Moreover, kinetic analysis using quantum dots as PL probes can benefit from developments in other scientific areas. An example is an improvement in nanomaterial stability, which, in several cases, loses its stability in a few hours, impairing the application of kinetic measurements. In this sense, the development of synthesis strategies should prioritize the stability of QDs to foster the use of this kinetic approach in various analytical applications. Furthermore, silica coating has been explored as a strategy to enhance the stability of QDs, with several reported applications in some areas, namely, chemical analysis and biocatalysis. However, coating the surface of QDs with silica can compromise the QD's reactivity, making them less prone to establishing chemical interactions with the target analyte, which may be a significant limitation depending on the intended analytical application. Indeed, more studies may be performed to obtain the best compromise between the stability and reactivity of the nanoparticles. Additionally, the increased interest in synthesizing eco-friendly alternatives for conventional Cd-based QDs is expected to contribute to the development of greener analytical methodologies with minimal ecological impact. These nanoparticles should provide similar performance in terms of sensitivity for kinetic analysis.

Another important improvement in kinetic methodologies is their portability, which ensures a broader accessibility. A promising alternative is the use of smartphone-based analytical platforms, employing the capabilities of mobile devices for kinetic data collection and analysis. As smartphones are used extensively in our daily lives, researchers can enable their worldwide use to perform advanced kinetic analyses more simply. Visual detection techniques allow for the identification and/or quantification of analytes without requiring complex external instruments, using color changes detected during interactions with PL QDs. Although QD-based visual sensing methods allow for qualitative or semi-quantitative evaluations through visual inspection, the use of smartphones presents an interesting solution for immediate and on-site quantitative measurements without depending on complex laboratory equipment. High-quality imaging capabilities available in most modern smartphones, combined with proper software applications that can convert the acquired color data into RGB values, emerge as efficient sensing platforms for monitoring various target species. To the best of our knowledge, all the studies that use smartphones for the capture of RGB data focus on single-point measurements. However, continuous advancements in smartphone camera resolution and image processing capabilities indicate their potential application in kinetic measurements in the future. Exploring this potential could open new perspectives for real-time portable kinetic analyses using smartphone-based platforms.

As final remarks, the use of kinetic data for QD-based analytical nanoplatforms promises to foster the development of analytical methodologies through increased sensitivity, which will improve the discrimination capacity, as well as the accurate quantification of analytes in the presence of uncalibrated interfering species. These features are crucial for providing comprehensive knowledge about complex systems and developing reliable analytical techniques.

Author Contributions: R.C.C.: Conceptualization, Investigation, Methodology, Visualization, Validation, and Writing—original draft. R.N.M.J.P.: Conceptualization, Validation, Methodology, Writing—original draft, Visualization, Investigation, Writing—review and editing, and Supervision. D.S.M.R.: Conceptualization, Validation, Methodology, Writing—original draft, Visualization, Investigation, Writing—review and editing, and Supervision. J.L.M.S.: Funding acquisition, Validation, Writing—review and editing, and Supervision. All authors have read and agreed to the published version of the manuscript.

Funding: This work received financial support from FCT/MCTES (UIDB/50006/2020 DOI 10.54499/UIDB/50006/2020) through national funds.

Institutional Review Board Statement: Not applicable.

Informed Consent Statement: Not applicable.

Data Availability Statement: The raw data supporting the conclusions of this article will be made available by the authors upon request.

Acknowledgments: David S. M. Ribeiro (DL 57/2016/CP1346/CT0033, DOI 10.54499/DL57/2016/CP1346/CT0033) thanks FCT (Fundação para a Ciência e Tecnologia) for funding through program DL 57/2016—Norma transitória. This work received support and help from FCT/MCTES (LA/P/0008/2020 DOI 10.54499/LA/P/0008/2020, UIDP/50006/2020 DOI 10.54499/UIDP/50006/2020 and UIDB/50006/2020 DOI 10.54499/UIDB/50006/2020) through national funds.

Conflicts of Interest: The authors declare no conflicts of interest.

Abbreviations

The following abbreviations are used in this manuscript:

AFB1	Aflatoxin B1
AIS	AgInS2
ASA	acetylsalicylic acid
CDs	carbon dots
CYS	L-cysteine
EEM	Excitation–Emission Matrix
FRET	Förster Resonance Energy Transfer
GSH	glutathione
LEDs	light-emitting diodes
LOD	limit of detection
MCR-ALS	multivariate curve resolution with alternating least squares
MIPs	molecularly imprinted polymers
MLF-NNs	multilayer feed-forward neural networks
MPA	3-mercaptopropionic acid
N-PLS	N-way partial least squares
OTC	oxytetracycline
PL	photoluminescent
PLS	partial least squares
QDs	quantum dots
QY	quantum yield
R2C	determination coefficient
RBF-NNs	radial basis function neural networks
RGB	red, green, and blue
RMSEC	root mean square error of calibration
ROS	reactive oxygen species
TMA	thiomalic acid
U-PLS	unfolded partial least squares
U-PLS/RBL	unfolded partial least squares/residual bilinearization
UV	ultraviolet

References

1. Daoudi, W.; Tiwari, A.; Tyagi, M.; Singh, P.; Saxena, A.; Verma, D.K.; Dagdag, O.; Sharma, H.K.; Fuertes, P.O.; El Aatiaoui, A. Carbon dots nanoparticles: A promising breakthrough in biosensing, catalysis, biomedical and authers applications. *Nano-Struct. Nano-Objects* **2024**, *37*, 101074. [CrossRef]
2. Pandey, S.; Bodas, D. High-quality quantum dots for multiplexed bioimaging: A critical review. *Adv. Colloid Interface Sci.* **2020**, *278*, 102137. [CrossRef] [PubMed]
3. Sengupta, S.; Pal, S.; Pal, A.; Maity, S.; Sarkar, K.; Das, M. A review on synthesis, toxicity profile and biomedical applications of graphene quantum dots (GQDs). *Inorganica Chim. Acta* **2023**, *557*, 121677. [CrossRef]
4. Sobhanan, J.; Rival, J.V.; Anas, A.; Sidharth Shibu, E.; Takano, Y.; Biju, V. Luminescent quantum dots: Synthesis, optical properties, bioimaging and toxicity. *Adv. Drug Deliv. Rev.* **2023**, *197*, 114830. [CrossRef]
5. Gao, W.; Zhang, S.; Wang, G.; Cui, J.; Lu, Y.; Rong, X.; Gao, C. A review on mechanism, applications and influencing factors of carbon quantum dots based photocatalysis. *Ceram. Int.* **2022**, *48*, 35986–35999. [CrossRef]

6. Sun, P.; Xing, Z.; Li, Z.; Zhou, W. Recent advances in quantum dots photocatalysts. *Chem. Eng. J.* **2023**, *458*, 141399. [CrossRef]
7. Ribeiro, D.S.M.; Castro, R.C.; Soares, J.X.; Santos, J.L.M. Photocatalytic activity of $AgInS_2$ quantum dots upon visible light irradiation for melatonin determination through its reactive oxygen species scavenging effect. *Microchem. J.* **2020**, *155*, 104728. [CrossRef]
8. Castro, R.C.; Páscoa, R.N.M.J.; Saraiva, M.L.M.F.S.; Santos, J.L.M.; Ribeiro, D.S.M. Chemometric models for data processing in quantum dots-based photoluminescence methodologies. *Coord. Chem. Rev.* **2024**, *502*, 215605. [CrossRef]
9. Castro, R.C.; Ribeiro, D.S.M.; Santos, J.L.M. Visual detection using quantum dots sensing platforms. *Coord. Chem. Rev.* **2021**, *429*, 213637. [CrossRef]
10. Castro, R.C.; Saraiva, M.L.M.F.S.; Santos, J.L.M.; Ribeiro, D.S.M. Multiplexed detection using quantum dots as photoluminescent sensing elements or optical labels. *Coord. Chem. Rev.* **2021**, *448*, 214181. [CrossRef]
11. Liu, C.; Huang, W.-T.; Liu, R.-S. Stable glass-protected $CsPbX_3$ (X = Cl, Br, and I) perovskite quantum dots and their applications in backlight LED. *Prog. Mater. Sci.* **2024**, *143*, 101243. [CrossRef]
12. Fan, X.; Yang, X.; Kong, X.; Zhang, T.; Wang, S.; Lin, Y.; Chen, Z. Recent progresses on perovskite quantum dots patterning techniques for color conversion layer in micro-LED displays. *Next Nanotechnol.* **2024**, *5*, 100045. [CrossRef]
13. Luo, Y.; Liu, X.-F.; Liu, Z.-H.; Li, W.; Yan, S.; Gao, H.; Su, H.; Pan, D.; Zhao, J.; Wang, J.-Y.; et al. One-Dimensional Quantum Dot Array Integrated with Charge Sensors in an InAs Nanowire. *Nano Lett.* **2024**, *24*, 14012–14019. [CrossRef] [PubMed]
14. Pomorski, K. Analytical View on Non-Invasive Measurement of Moving Charge by Position Dependent Semiconductor Qubit. In Proceedings of the Future Technologies Conference (FTC) 2020, Vancouver, BC, Canada, 5–6 November 2020; Springer: Cham, Switzerland, 2021; Volume 2, pp. 31–53.
15. Xia, H.; Hu, J.; Tang, J.; Xu, K.; Hou, X.; Wu, P. A RGB-Type Quantum Dot-based Sensor Array for Sensitive Visual Detection of Trace Formaldehyde in Air. *Sci. Rep.* **2016**, *6*, 36794. [CrossRef]
16. Castro, R.C.; Ribeiro, D.S.M.; Páscoa, R.N.M.J.; Soares, J.X.; Mazivila, S.J.; Santos, J.L.M. Dual-emission CdTe/$AgInS_2$ photoluminescence probe coupled to neural network data processing for the simultaneous determination of folic acid and iron (II). *Anal. Chim. Acta* **2020**, *1114*, 29–41. [CrossRef]
17. Ribeiro, D.S.M.; Castro, R.C.; Páscoa, R.N.M.J.; Soares, J.X.; Rodrigues, S.S.M.; Santos, J.L.M. Tuning CdTe quantum dots reactivity for multipoint detection of mercury(II), silver(I) and copper(II). *J. Lumin.* **2019**, *207*, 386–396. [CrossRef]
18. Bittar, D.B.; Ribeiro, D.S.M.; Páscoa, R.N.M.J.; Soares, J.X.; Rodrigues, S.S.M.; Castro, R.C.; Pezza, L.; Pezza, H.R.; Santos, J.L.M. Multiplexed analysis combining distinctly-sized CdTe-MPA quantum dots and chemometrics for multiple mutually interfering analyte determination. *Talanta* **2017**, *174*, 572–580. [CrossRef]
19. Shen, Y.; Wei, Y.; Zhu, C.; Cao, J.; Han, D.-M. Ratiometric fluorescent signals-driven smartphone-based portable sensors for onsite visual detection of food contaminants. *Coord. Chem. Rev.* **2022**, *458*, 214442. [CrossRef]
20. Foubert, A.; Beloglazova, N.V.; Rajkovic, A.; Sas, B.; Madder, A.; Goryacheva, I.Y.; De Saeger, S. Bioconjugation of quantum dots: Review & impact on future application. *TrAC Trends Anal. Chem.* **2016**, *83*, 31–48.
21. Keçili, R.; Hussain, C.G.; Hussain, C.M. Fluorescent nanosensors based on green carbon dots (CDs) and molecularly imprinted polymers (MIPs) for environmental pollutants: Emerging trends and future prospects. *Trends Environ. Anal. Chem.* **2023**, *40*, e00213. [CrossRef]
22. Pardue, H.L. Kinetic aspects of analytical chemistry. *Anal. Chim. Acta* **1989**, *216*, 69–107. [CrossRef]
23. Valcárcel, M. Selectivity and kinetics in analytical chemistry. Plenary lecture. *Analyst* **1987**, *112*, 729–737. [CrossRef]
24. Marangoni, A.G. *Kinetic Analysis of Food Systems*; Springer Nature: Cham, Switzerland, 2017.
25. Hyötyläinen, T. Critical evaluation of sample pretreatment techniques. *Anal. Bioanal. Chem.* **2009**, *394*, 743–758. [CrossRef] [PubMed]
26. Díaz-Álvarez, M.; Martín-Esteban, A. Molecularly Imprinted Polymer-Quantum Dot Materials in Optical Sensors: An Overview of Their Synthesis and Applications. *Biosensors* **2021**, *11*, 79. [CrossRef]
27. Cardoso Dos Santos, M.; Algar, W.R.; Medintz, I.L.; Hildebrandt, N. Quantum dots for Förster Resonance Energy Transfer (FRET). *TrAC Trends Anal. Chem.* **2020**, *125*, 115819. [CrossRef]
28. Snee, P.T. Semiconductor quantum dot FRET: Untangling energy transfer mechanisms in bioanalytical assays. *TrAC Trends Anal. Chem.* **2020**, *123*, 115750. [CrossRef]
29. Kaur, A.; Dhakal, S. Recent applications of FRET-based multiplexed techniques. *TrAC Trends Anal. Chem.* **2020**, *123*, 115777. [CrossRef]
30. Monakhova, Y.B.; Goryacheva, I.Y. Chemometric analysis of luminescent quantum dots systems: Long way to go but first steps taken. *TrAC Trends Anal. Chem.* **2016**, *82*, 164–174. [CrossRef]
31. Mazivila, S.J.; Soares, J.X.; Santos, J.L.M. A tutorial on multi-way data processing of excitation-emission fluorescence matrices acquired from semiconductor quantum dots sensing platforms. *Anal. Chim. Acta* **2022**, *1211*, 339216. [CrossRef]

32. Kumar, N.; Bansal, A.; Sarma, G.S.; Rawal, R.K. Chemometrics tools used in analytical chemistry: An overview. *Talanta* **2014**, *123*, 186–199. [CrossRef]
33. Wold, S. Spline functions in data analysis. *Technometrics* **1974**, *16*, 1–11. [CrossRef]
34. Oliveri, P.; Forina, M. Data Analysis and Chemometrics. In *Chemical Analysis of Food: Techniques and Applications*; Elsevier: Waltham, MA, USA, 2012; pp. 25–57.
35. Mas, S.; de Juan, A.; Tauler, R.; Olivieri, A.C.; Escandar, G.M. Application of chemometric methods to environmental analysis of organic pollutants: A review. *Talanta* **2010**, *80*, 1052–1067. [CrossRef] [PubMed]
36. Olivieri, A.C. Analytical advantages of multivariate data processing. One, two, three, infinity? *Anal. Chem.* **2008**, *80*, 5713–5720. [CrossRef]
37. Castro, R.C.; Páscoa, R.N.M.J.; Saraiva, M.L.M.F.S.; Santos, J.L.M.; Ribeiro, D.S.M. Chemometric-assisted kinetic determination of oxytetracycline using $AgInS_2$ quantum dots as PL sensing platforms. *Anal. Chim. Acta* **2021**, *1188*, 339174. [CrossRef]
38. Castro, R.C.; Páscoa, R.N.M.J.; Saraiva, M.L.M.F.S.; Santos, J.L.M.; Ribeiro, D.S.M. Selective determination of Fe (III) with carbon dots as photoluminescence Probes: Chemometric analysis using Excitation-Emission matrices. *Microchem. J.* **2024**, *205*, 111207. [CrossRef]
39. Castro, R.C.; Páscoa, R.N.M.J.; Saraiva, M.L.M.F.S.; Santos, J.L.M.; Ribeiro, D.S.M. Kinetic Determination of Acetylsalicylic Acid Using a $CdTe/AgInS_2$ Photoluminescence Probe and Different Chemometric Models. *Biosensors* **2023**, *13*, 437. [CrossRef]
40. Olivieri, A.C.; Escandar, G.M.; Goicoechea, H.C.; de la Peña, A.M. *Fundamentals and Analytical Applications of Multiway Calibration*; Elsevier: Waltham, MA, USA, 2015.
41. Tauler, R. Multivariate curve resolution applied to second order data. *Chemom. Intell. Lab. Syst.* **1995**, *30*, 133–146. [CrossRef]
42. Jaumot, J.; de Juan, A.; Tauler, R. MCR-ALS GUI 2.0: New features and applications. *Chemom. Intell. Lab. Syst.* **2015**, *140*, 1–12. [CrossRef]
43. Wold, S.; Geladi, P.; Esbensen, K.; Öhman, J. Multi-way principal components-and PLS-analysis. *J. Chemom.* **1987**, *1*, 41–56. [CrossRef]
44. Bro, R. Multiway calibration. multilinear pls. *J. Chemom.* **1996**, *10*, 47–61. [CrossRef]
45. Olivieri, A.C.; Olivieri, A.C. Non-linearity and Artificial Neural Networks. In *Introduction to Multivariate Calibration: A Practical Approach*; Springer Nature: Cham, Switzerland, 2018; pp. 207–226.
46. Marini, F. 3.24-Non-linear Modeling: Neural Networks. In *Comprehensive Chemometrics: Chemical and Biochemical Data Analysis, Second Edition: Four Volume Set*; Elsevier: Waltham, MA, USA, 2020; pp. 519–541.
47. Culzoni, M.J.; Damiani, P.C.; García-Reiriz, A.; Goicoechea, H.C.; Olivieri, A.C. Experimental study of non-linear second-order analytical data with focus on the second-order advantage. *Analyst* **2007**, *132*, 654–663. [CrossRef] [PubMed]
48. Wu, P.; Li, Y.; Yan, X.-P. CdTe Quantum Dots (QDs) Based Kinetic Discrimination of Fe^{2+} and Fe^{3+}, and CdTe QDs-Fenton Hybrid System for Sensitive Photoluminescent Detection of Fe^{2+}. *Anal. Chem.* **2009**, *81*, 6252–6257. [CrossRef]
49. Abdollahi, H.; Shamsipur, M.; Barati, A. Kinetic fluorescence quenching of CdS quantum dots in the presence of Cu(II): Chemometrics-assisted resolving of the kinetic data and quantitative analysis of Cu(II). *Spectrochim. Acta Part A Mol. Biomol. Spectrosc.* **2014**, *127*, 137–143. [CrossRef] [PubMed]
50. Castro, R.C.; Páscoa, R.N.M.J.; Saraiva, M.L.M.F.S.; Cunha, S.C.; Santos, J.L.M.; Ribeiro, D.S.M. A ratiometric carbon dots/$AgInS_2$ quantum dots sensing probe for the kinetic determination of histamine resorting to N-PLS chemometric model. *Food Control* **2023**, *153*, 109934. [CrossRef]
51. Wen, J.; Zhou, L.; Tang, Q.; Xiao, X.; Sun, S. Photocatalytic degradation of organic pollutants by carbon quantum dots functionalized g-C_3N_4: A review. *Ecotoxicol. Environ. Saf.* **2023**, *262*, 115133. [CrossRef]
52. Meng, F.; Hao, T.; Tian, W.; Zhao, J.; Wang, S.; Zhang, H. Photocatalytic aqueous environmental remediation via two-dimensional carbon nitride nanosheets. *Surf. Interfaces* **2024**, *44*, 103717. [CrossRef]
53. Tian, M.; Hu, C.; Yu, J.; Chen, L. Carbon quantum dots (CQDs) mediated Z-scheme g–C_3N_4–CQDs/$BiVO_4$ heterojunction with enhanced visible light photocatalytic degradation of Paraben. *Chemosphere* **2023**, *323*, 138248. [CrossRef]
54. Mei, A.; Xu, Z.; Wang, X.; Liu, Y.; Chen, J.; Fan, J.; Shi, Q. Photocatalytic materials modified with carbon quantum dots for the degradation of organic pollutants under visible light: A review. *Environ. Res.* **2022**, *214*, 114160. [CrossRef]
55. Rajendiran, K.; Zhao, Z.; Pei, D.-S.; Fu, A. Antimicrobial Activity and Mechanism of Functionalized Quantum Dots. *Polymers* **2019**, *11*, 1670. [CrossRef]
56. Ghirardello, M.; Ramos-Soriano, J.; Galan, M.C. Carbon Dots as an Emergent Class of Antimicrobial Agents. *Nanomaterials* **2021**, *11*, 1877. [CrossRef]
57. Chen, H.; Lin, L.; Li, H.; Lin, J.-M. Quantum dots-enhanced chemiluminescence: Mechanism and application. *Coord. Chem. Rev.* **2014**, *263–264*, 86–100. [CrossRef]

58. Song, H.; Su, Y.; Zhang, L.; Lv, Y. Quantum dots-based chemiluminescence probes: An overview. *Luminescence* **2019**, *34*, 530–543. [CrossRef] [PubMed]
59. Castro, R.C.; Páscoa, R.N.M.J.; Saraiva, M.L.M.F.S.; Lapa, R.A.S.; Fernandes, J.O.; Cunha, S.C.; Santos, J.L.M.; Ribeiro, D.S.M. Fluorometric kinetic determination of aflatoxin B1 by combining Cd-free ternary quantum dots induced photocatalysis and chemometrics. *Microchem. J.* **2023**, *185*, 108300. [CrossRef]

biosensors

MDPI

Review

The Interpretation of Carbon Nanotubes' Electrochemistry: Electrocatalysis and Mass Transport Regime in the Apparent Promotion of Electron Transfer

Josipa Dugeč [1], Ivana Škugor Rončević [1], Nives Vladislavić [1], Josip Radić [2], Maša Buljac [2] and Marijo Buzuk [1,*]

[1] Department of General and Inorganic Chemistry, Faculty of Chemistry and Technology, University of Split, 21000 Split, Croatia; josipa.dugec@ktf-split.hr (J.D.); skugor@ktf-split.hr (I.Š.R.); nives@ktf-split.hr (N.V.)
[2] Department of Analytical and Environmental Chemistry, Faculty of Chemistry and Technology, University of Split, 21000 Split, Croatia; jradic@ktf-split.hr (J.R.); masa@ktf-split.hr (M.B.)
* Correspondence: buzuk@ktf-split.hr; Tel.: +385-21-329-474

Abstract: Carbon nanotubes (CNTs) are widely used in electrochemical sensors due to their significant impact on the electroanalytical signal. However, there remain significant doubts regarding the origin of the improved electroanalytical response observed in CNT-based sensors, particularly concerning the precise role of CNTs in these systems. In particular, the origin of the electrochemical response is controversial, as it may be due to either electrocatalytic or non-electrocatalytic processes. The latter implies that the electroanalytical response is mainly governed by the mass transport phenomena within the porous CNT layer. This article briefly reviews the several comprehensive models based on the role of porosity (diffusion in a 'thin-layer') on the electrochemical behavior as well as on the electrocatalytic properties of CNTs to resolve conflicts arising from misinterpretations of the electroanalytical response of CNT-based sensors. However, even though there are some explanations and conclusions on this topic, they seem to be valuable for specific electroactive species, the type of CNTs and/or electrode architecture, the electrode surface, etc. Accordingly, general theories and conclusions are not yet defined, so different approaches to this topic are still needed, since the main phenomenological effects responsible for the nature of the electrochemical response of the electrodes modified with CNTs need to be determined in a rational way.

Keywords: voltammetry; chronoamperometry; electrochemical impedance spectroscopy; (bio)sensors; carbon nanotubes; porosity; electrocatalysis; thin layer; diffusion

1. Introduction

The modified electrodes have been broadly applied in the development of electrochemical sensors and biosensors [1]. By simply entering the keyword ((bio)sensor, carbon, nanomaterial, nanotubes, graphene, yarn, etc.) in the Web of Knowledge, a large number of published articles from the last 5 years—about 10,419—is listed.

As an attractive material, CNTs have many benefits that make them suitable for the development of electrochemical (bio)sensors, due to their apparent ability in the promotion of heterogeneous electron transfer (i.e., to act as an 'electrocatalyst') [2]. Here, we have to specify that the term 'electrocatalytic' refers to an improved electroanalytical response (e.g., lower overpotential and higher current signal) in the case of the electrode modified with a CNT layer in comparison to an unmodified one. However, there are disagreements concerning the origin of the enhanced electrochemical activity of CNTs. For example, direct electron transfer between carbon nanotubes and electrochemically active species has been reported in the literature [3,4], leading to the simple explanation that the enhanced electrochemical response of the electrode modified with CNTs is the consequence of the highly specific electroactive surface area. Although pristine CNTs may show inherent electrochemical activity, the most notable explanation is that the active sites (prepared by oxidation) present on CNTs (hydroxyl, alcoholic, carbonyl, and carboxyl) are responsible for the electrochemical activity. In certain cases, for example, in sensing H_2O_2 with MWCNTs, the origin of the improved electrochemical signal is attributed to the presence of the residual metal impurities [5,6]. We can define all the above aspects with the 'kinetic' approach to elucidate the role of CNTs.

However, as the modification of the electrode by these materials usually involves the formation of multiple porous layers on the electrode surface [1], another approach can be used in elucidating the improved electrochemical response of the CNT-modified electrode. Basically, this means that the effects and experimental results can be analyzed in the light of mass transport in the pockets of solution present in the CNTs' porous layer. This means that the 'electrocatalytic' behavior of the pristine CNTs can be alternatively explained by the mass transport phenomena in the CNTs' porous layer for kinetic independent reaction.

Nevertheless, the boundary between the two approaches mentioned above is not sharp, as the phenomena of mass transport (Fick's laws) and the kinetics of heterogeneous electron transfer (HET) (Butler–Volmer kinetics) must be taken into account in both. As the 'kinetic' approach uses semi-infinite diffusion (Fick's law, Cottrell) in explanation of the electrochemical response; in the case when the response is mainly controlled by mass transport in the porous CNTs, the finite diffusion model is useful. The latter is often interrelated with the 'thin-layer' diffusion of the electroactive species that occurs in the porous CNTs. Despite this fact, 'thin-layer' diffusion is a phenomenon that is reserved for situations when mass transfer equilibrium is achieved rapidly, as electroactive species have immediate access to the electroactive sites of the electrode. This roughly means that the electroactive species are located within 0.001 cm of the electrode (working and auxiliary) surfaces (it should be noted that the thickness of the diffusion layer can even reach 500 μm [7]). In this case, this phenomena can be simply described by Faraday's and Nernst's laws [8].

Furthermore, the situation becomes complicated when electroactive species are present in the porous layer of the electrode and in the bulk of the electrolyte. An even more complicated situation can be expected when the electrochemical reversible reaction or even the slow kinetics of HET occur. Also, modified electrodes may exhibit a complex mass transport regime that is a combination of semi-infinite and 'thin-layer' diffusions, especially when the porous layer is electrochemically active, but the substrate (the electrode) is not [2]. In addition, many of the proposed explanations show variations depending on the electroactive species as well as the type of CNTs. For example, the electrochemical behavior of $Fe(CN)_6^{4-}$ on the SWCNT-modified glassy carbon electrode (GCE) or H_2O_2 on MWCNTs can be explained with the 'thin-layer' model, while this model shows many weaknesses in the case of etoposide and $Fe(CN)_6^{4-}$ on MWCNTs. Further problems can arise from the fact that a heterogeneous electrochemical process takes place in the porous layer, which can lead to an uneven distribution of the electrolyte and the electroactive species, which affects the kinetics of HET [4].

In summary, many details regarding the electrochemical properties of CNTs are still unclear and controversial and need to be clarified to gain a complete insight into the electrochemical processes occurring in these electrode systems.

The aim of this article is to provide an overview of the approaches used to elucidate the processes and phenomena that govern the electrochemical response of the electrode modified within a porous carbon layer. This means a comprehensive insight into theoretical models and simulations, not only specified for CNTs, but for general porous electrode systems: a conductive or non-conductive porous layer in combination with a conductive or non-conductive substrate. Although, at first glance, these models seem unrelated to CNTs, they can provide important insights that can be helpful in the assessment of the electrochemical properties of CNTs. In addition to this 'non-electrocatalytic' approach, here, we present an overview of recent efforts to explore the 'electrocatalytic' properties of CNTs.

2. Carbon Nanomaterials as 'Workhorse' in Electroanalysis

During the last 20 years, among all carbon nanomaterials, different forms of carbon nanotubes and graphene have emerged as the most promising materials in various electrochemical systems, especially in the various architecture of electrochemical sensors. The benefits of these materials as a 'workhorse' in the construction of electrochemical (bio)sensors arise from the fact that (a) they can act as a transducer; (b) they can serve as a support for a recognition element or/and mediator; and (c) they can act as a mediator themselves [9–15].

Despite the avalanche number of articles on the development of the various types of electrochemical (bio)sensors, there is still no complete understanding of the underlying physical and chemical phenomena that occur in the (bio)sensors based on different CNTs. This can be manifested in various explanations of the mechanism of the analyte recognition, even for very similar systems [16].

Accordingly, many electrochemists fall into the snare and usually explain the improved electrochemical behavior with these electrodes by its 'inherent' electric properties, while neglecting other phenomena. The latter arise from the morphology of these materials. Basically, carbon nanomaterials with a hollow structure have a large specific surface area and porosity, which contribute to the improved possibilities of electrochemical energy storage and conversion (supercapacitors and batteries). As CNTs have diameters on a nanoscale, the unexpected behavior of fluids and ions inside the CNT channel can be noticed, including enhanced ionic currents [17–21]. This can be applied in the development of the field-effect ion-transport devices [22–24]. Also, 2D carbon nanomaterials (such as graphene) show potential as useful electrode materials in the above context, due to possibility to be arranged in a layered structure. The above properties (porosity and layered structure) can be harnessed in the modification of these materials with various inorganic, organic, biological, and other recognition elements for (electro)chemical detection. This versatility is one of the main reasons why these materials are widely used in the development of (bio)sensors.

3. Origin of the Electrochemical Response

The origin of the electrochemical response of the pristine CNTs have occupied the attention of the many scientists and led to the publication of several papers with provocative titles [4,25–30]. As already mentioned, 'pristine' CNTs can exhibit 'electrocatalytic' behavior due to the presence of metallic impurities [5,6]. Procedures (chemical or electrochemical) for the purification of CNTs in order to obtain 'really pristine' materials involve exposure to the various oxidation agents, which can result in cutting of the CNTs' ends followed by the formation of the various morphology defects and oxygen functional group on the edges. Furthermore, the formation of the oxygen-containing functional group can also occur on the sidewall (basal side) of the CNTs. As a consequence, these changes have a significant influence on the Fermi state of the electrons in the CNTs; as the oxygen-containing functional groups increase, the gentle trend of decline in conductivity can be noticed [31–38]. In particular, these functional groups can increase the (pseudo)capacity of

calculated slope of 0.62 of the log i vs. log υ dependence was attributed to the possible mixed mass transport regime comprised of the thin-layer diffusion in MWCNTs and semi-infinite diffusion in the bulk. The contribution of the thin-layer diffusion was higher for a thicker MWCNT layer, as the slope obtained was 0.78. As a consequence of the rapid depletion of the MWCNT layer by nicotine, the cyclic voltammogram deviates significantly from what is expected for semi-infinite diffusion. For the further illustration of the contribution of the thin-layer phenomena, cyclic voltammetry simulations at various scan rates were performed. As a model, vertically oriented cylindrical electrodes, of which only the sides are considered conductive, on an insulating substrate was used. Although this model is far from the real situation where the MWCNTs are randomly oriented, the essence of the phenomena is retained. The simulation shows that the slope of log i vs. log σ (where σ is the normalized scan rate) varies from 0.5 at the lowest σ, since at this scan rate, the size of the thin layer becomes negligible compared to the diffuse layer, towards 1 at intermediate scan rates, where the situation is reversed. Finally, at high scan rates, a slope of 0.5 is reached again. From this, the authors conclude that the thin-layer effects depend on the relationship between the size of the diffuse layer and the separation of the cylinders as well as their height.

A simulation study on the influence of the mass transport regime on the electrochemical response obtained by linear sweep and various pulse voltammetry was reported by Laborda et al. [1]. First, the change in the current signal was analyzed by altering the kinetic parameter of the electrode (from a completely irreversible to a reversible process) and the pulse height for the semi-infinite diffusion model. Subsequently, the dependence of the shape and the current signal on the layer thickness, the kinetic parameter, the pulse height, and the pulse duration for the thin diffusion model was examined. As a model, the (hemi)spherical electrode (unmodified and modified with a conductive porous layer) was chosen, as it considers a simplified model due to a unique distance to the electrode surface. By decreasing the layer thickness, smaller current peaks, a shift in the peak potential towards the standard potential, and a decrease in the half-peak width were observed. This observation was interpreted as an increase in the HET kinetic. Indeed, the same effects were obtained for the semi-infinite model with an increasing heterogeneous rate constant. In the porous structure, however, all these observations were attributed to the presence and depletion of electroactive species in the porous layer. Accordingly, the experimental results for porous electrodes can be (mis)interpreted. In addition, two other phenomena were observed for thin-layer regimes: (a) a 'quasireversible maximum' as a consequence of a deviation of the current response at long pulse times, which is due to depletion in the thin-layer finite diffusion space and (b) splitting of the current signal, for fast kinetic upon large amplitudes. As will be emphasized later, these observations can be used to diagnose or determine the contribution of thin-layer diffusion to the overall signal.

5.1. *Cylindrical Electrodes and Arrays*

For the illustration of the thin-layer diffusion in the mixed transport regime, in the above reviewed articles, previously reported simulations of the cyclic voltammetry and chronoamperometric response for various geometry and electrical properties of the porous layers were used. These simulations include variation in the shapes/geometry (cylinders (pillar), cylindrical pores, and spheres) of the porous layers, their dimensions (depth/high, surface area, and separation between shapes), and their electrical properties (active (e.g., conductive), inactive (e.g., non-conductive), or partially active)) in conjunction with a conductive or non-conductive substrate (electrode). Different combinations of the above parameters were used in the simulation of the (cyclic) voltammetric and chronoamperometric response. For example, as a model for the simulation of the irreversible oxidation of nicotine onto MWCNT-modified basal plane pyrolytic graphite (MWCNT-BPPG) [2], they considered vertically oriented cylindrical (pillar) electrodes on the non-conductive support [73]. In the latter, they modeled chronoamperometric response and the steady-state concentration profile for various substrate conductivities and cylinders (full, top-only, and side-only) and various cylinder radii and heights (depth), as well as the chronoamperometric response for the arrays of the above-mentioned cylinders. For the insulated cylinder arrays at a small depth, they obtained a transition from the planar Cottrellian response at short times, along the non-planar diffusion effects caused by the situation, where the diffusion layers of the neighboring cylinders are separated, to the longer times in which planar diffusion is again achieved due to the overlapping of the diffusion layers of the neighboring cylinders, as the diffusion layer exceeds the top of the cylinders. The latest was not observed for higher cylinders in the time frame of the simulation. For the top-only conductive cylinder, non-planar diffusion is quickly reached, and it is followed by strong dependence of the depth and radius of the cylinders, as for shorter cylinders, the final planar diffusion is more quickly attained. For fully conductive cylinders, at short times, a greater current can be expected due to an increase in the electroactive area. This is more evident at a higher depth of the cylinders; however, planar Cottrellian response cannot be achieved at longer times for deeper cylinders.

The same model of cylindrical arrays was used to study the difference in the current signal between staircase and analog voltammetry in the case of a porous electrode [74]. They found that, in the case of staircase voltammetry and for reversible kinetics, the signal of the voltammogram depends on the point of the current measurements, as the peak current decreases with an increase in the time when the current is measured. This has a large implication when measurements were performed at the different scan rates for the staircase voltammetry measurements.

Besides the above simulation model, other authors have designed, fabricated, and characterized (theoretically and experimentally) a gold-based micropillar electrode. Sanchez et al. [75] discovered many similarities in the behavior between porous and micropillar electrodes, especially for dense micropillars, and concluded that hemispherical diffusion plays a minor role in mass transport. Thus, planar diffusion is a process that controls overall mass. The current response is also governed by planar diffusion at slow scan rates in cyclic voltammetry (CV) and at longer times in chronoamperometry (CA), as the depletion of electroactive species between micropillars becomes complete, and planar diffusion from the bulk to the electrode surfaces dominates. Interestingly, the authors found that certain phenomena, such as hindered degassing and issues with the full penetration of the electrolyte, can occur in high-density micropillars. This leads to the conclusion that conditions within dense porous structures can differ significantly from those in the bulk solution.

Chen et al. [76] used polydimethylsiloxane (PDMS) micropillars sputtered with gold in order to obtain conductive layers. The resulting arrays were modified with a Pd-MWCNT nanocomposite to obtain a layer sensitive to sarcosine. However, the authors did not comment whether any mass transfer phenomena could potentially occur in this system.

5.2. More Complex Cylindrical Arrays

In further theoretical investigations, the Compton group introduced a new diffusion indicator to characterize the nature of the diffusion (linear vs. convergent) in chronoamperometric measurements. It is derived from simulations of more complex cylinder-shaped geometries, which, in addition to the previously mentioned variations in the cylinder, included the 'annular band' and 'embedded ring' models (Figure 1) [77].

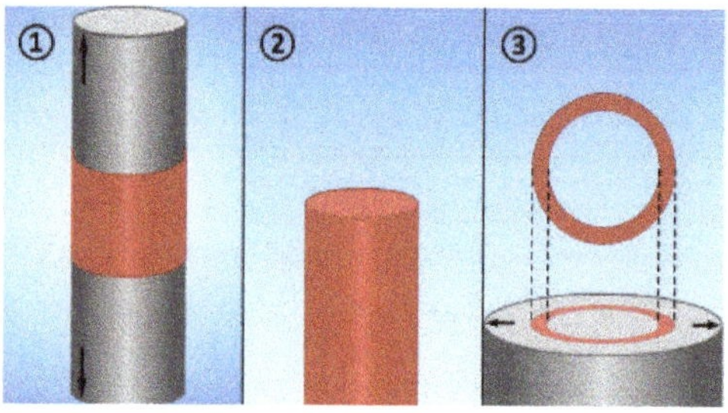

Figure 1. Illustrations of the models investigated: ① the 'annular band' electrode model, ② the cylinder electrode model, and ③ the 'embedded ring' electrode model. The black arrows in the figures represent infinitely expanding insulating areas [77]. Reproduced with permission from J. Electroanal. Chem.; published by Elsevier, 2019.

They determined that the geometrical parameters under which diffuse fluxes toward micropillar electrodes, as defined by Aoki et al. [78] and Szabo et al. [79], can be used to predict the chronoamperometric response for these types of electrodes.

Finally, the cylindrical model was used for the prediction of the splitting voltammetric signal into double peaks for the porous film electrode, where electrolysis occurs at the substrate, at the porous film surface, and within it [80]. The simulation of the porous film used in this model includes the influence of the cylinder separation and cylinder depth (film thickness) on the current and current peak, using linear sweep voltammetry. For reversible kinetics, under boundary conditions (very high and very low surface density of the cylinders), a planar Cottrell response is obtained, which corresponds to the Randles–Sevcik limit. In the moderate surface density of the cylinders (moderate separation), the current increases, due to the increased electroactive surface area, followed by a decrease in the current and by a further decrease in the cylinder density. This decrease is explained by the fast depletion of the electroactive substance in the thin layer (in the pores), suggesting that not only a large electroactive surface area is important but also an amount of the electroactive specie in the pores. For a very high density of the cylinders, the electrode behavior can be depicted by the Randles–Sevcik limit (for the planar Cottrellian response), since the electrode acts as a planar electrode (where the electrode surface is the top of the cylinders). This behavior becomes more pronounced with the increasing cylinder depth, due to the increased electroactive surface area as well as the pore volume (increased amount of electroactive species in the pores). Accordingly, the shift in the peak current potential towards a higher overpotential than for the planar macroelectrode is expected with an increase in the separation between cylinders as a consequence of the increased thin-layer diffusion. This indicates that mass transport is not a limiting factor, as is the case with the planar macroelectrode. On the contrary, by increasing the cylinder density, a lower overpotential is expected, as the volume of the pores decreases. Accordingly, thin-layer diffusion is observed to a lesser extent due to the reduced amount of electroactive species in the pores and the shorter diffusion path between cylinders. In moderate cylinder surface density, a transition is observed from a decrease in peak current potential with decreasing cylinder density to an increase in the peak current at a high cylinder density. This transition can be explained in the same manner as described earlier (see current dependence at high cylinder density). Of course, this behavior becomes more pronounced with an increase in the cylinder depth. In the case of irreversible kinetics, a similar pattern of current peak as in reversible kinetics can be expected. However, at a high cylinder density, the peak current can split into two peaks. One is a consequence of the thin-layer effect, while the other is related to planar diffusion. This splitting depends strongly on the cylinder density, especially in the low current density region, as the current peak from thin-layer diffusion becomes more prominent with an increase in pore size, until the second peak, which corresponds to planar diffusion, is hindered. Interestingly, this separation cannot be observed for reversible kinetics, as the difference in the peak potential (for a given time frame of the experiment, e.g., a scan rate of 10 mV s^{-1}) for thin-layer and planar diffusion, for such fast kinetics, cannot be observed (Figure 2).

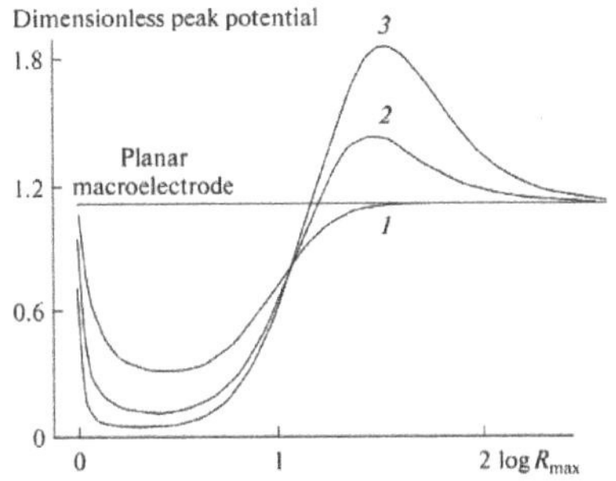

Figure 2. Plot of peak potential against R_{max} for a reversible system. Data are plotted for different values of film thickness with the peak potential for a planar macroelectrode plotted for comparison. Film thickness: (1) 30, (2) 100, and (3) 300 [80]. Reproduced with permission from Russ. J. Electrochem.; published by Springer Nature, 2012.

5.3. Cylindrical Pores

Moreover, the same group has considered another geometry for the porous layer's pores with cylindrical symmetry, known as the pinhole model [81,82] (Figure 3). To simplify the diffusion problem for this geometry, the diffusion domain was treated as cylindrical (with the pore as the axis of symmetry), and for the calculations, it was approximated as hexagonal with the same surface area to ensure that the entire domain was included 'without empty space' (Figure 4).

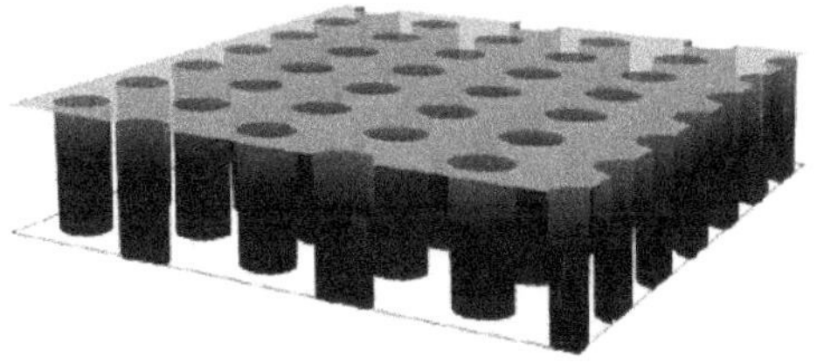

Figure 3. Model of the electrode surface [81]. Reproduced with permission from Electroanalysis; published by WILEY Analytical Science, 2008.

Simulations of the cyclic voltammetric response were performed at different pore depths, scan rates, and constants of heterogeneous electron transfer. It must be emphasized that the pores are considered electroinactive. Accordingly, the electrochemical reaction takes place on the underlying electrode surface. Regardless of the kinetics and the scan rate, the simulated voltammograms show a tendency to be similar to the voltammograms of the flat electrode (without pores) as the depth of the pores decreases. However, the current decreases with increasing pore depth at a constant scan rate. This was attributed to the hampered mass transfer from the bulk solution into the pores, which is obviously more pronounced at faster scan rates. Furthermore, this effect is more emphasized at faster scan rates, as the diffusion layer becomes thinner with an increase in the scan rate. Consequently, diffusion within the pores and diffusion from the bulk do not interfere with each other. This behavior has an influence on the peak-to-peak separation, since it is dependent on the mass transfer (with the same constant of the HET). Accordingly, the peak-to-peak separation becomes constant with an increase in the scan rate, as the electroactive species in the pores becomes depleted. Finally, this phenomenon is more pronounced in deeper pores (Figure 5).

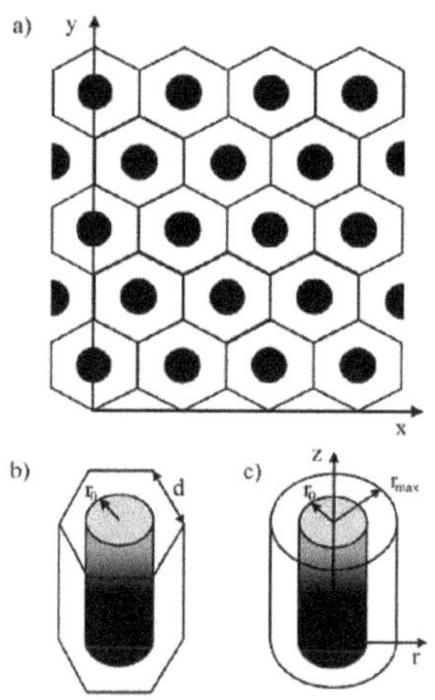

Figure 4. (**a**) Schematic diagram of the unit cell for an array of cylindrical pores; (**b**) single unit cell in Cartesian coordinates; (**c**) equivalent diffusion domain in cylindrical coordinates [81]. Reproduced with permission from Electroanalysis; published by WILEY Analytical Science, 2008.

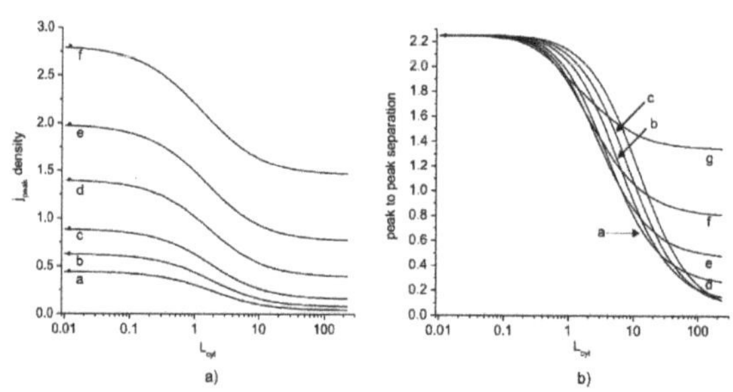

Figure 5. Peak current (**a**) and peak-to-peak separation (**b**) at different scan rates and cylinder depth values L_{cyl}; infinitely fast rate of electrode transfer; (a) σ = 0.1, (b) σ = 0.2, (c) σ = 0.4, (d) σ = 1, (e) σ = 2, (f) σ = 4, (g) σ = 10. Dots show the values calculated for the flat macroelectrode limit with DigiSim [81]. Reproduced with permission from Electroanalysis; published by WILEY Analytical Science, 2008.

This approach is successfully used to arbitrarily distinguish mass transport through a film (e.g., through a polymer membrane) from pore diffusion [83]. Depending on the various parameters, such as scan rate, pore depth, and diffusion coefficient, different types of simulated cyclic voltammograms were obtained, as follows, from steady-state (at slow scan rates), across Randles–Sevcik dependence, toward a thin-layer type of voltammograms for fast scan rates. Other parameters were also simulated for these two cases (film and pore), such as peak-to-peak separation, peak current, etc.

The behavior of the voltammetric signal versus the geometrical parameters of the conductive cylindrical pores (support is insulating), together with the determination of the extent of this effect, was considered by Ward et al. [84]. The simulations were performed for the different scan rates and for fully reversible and fully irreversible kinetics for the various pore radius, depths, and separation between pores (Figure 6).

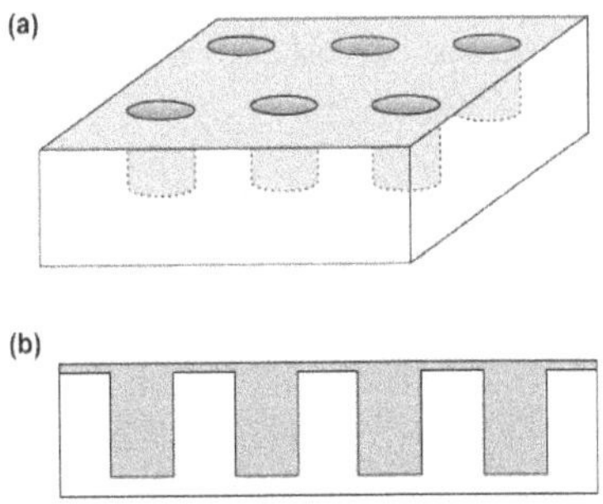

Figure 6. (**a**) Schematic of a porous surface; (**b**) side on view [84]. Reproduced with permission from J. Electroanal. Chem.; published by Elsevier, 2014.

For fully reversible kinetics at fast scan rates, the voltammetric signal is expected to resemble those observed in the thin layer. This phenomenon depends on the cylinder pore depth, as increasing the depth allows thin-layer diffusion to be observed even at slower scan rates. This is followed by the dependence of the current peak upon the Randles–Sevcik equation. Of course, there is a transition region in which the contributions of the two diffusions predominate. For irreversible kinetics (slow kinetics), there is a threshold value for the pore depth at which the voltammetric signal shows an 'apparent catalytic effect'. However, this model is not so important for the subject of this work, as it assumes that support is non-conductive.

The variation in the cylindrical pores was also considered for an infinite cylindrical tube with a partially electroactive inner surface (like a ring electrode), where the electroactive species are located in the cylinder interior surface (Figure 7) [85]. The variation in the height of the inner electroactive ring surface and the radius of the ring were investigated when changing the scan rate. The authors established four limiting cases. For an infinitely large cylinder radius, voltammetry governed by the Randles–Sevcik equation can be expected. As the radius of the cylinder decreases, thin-layer behavior occurs, because the length of the active area increases.

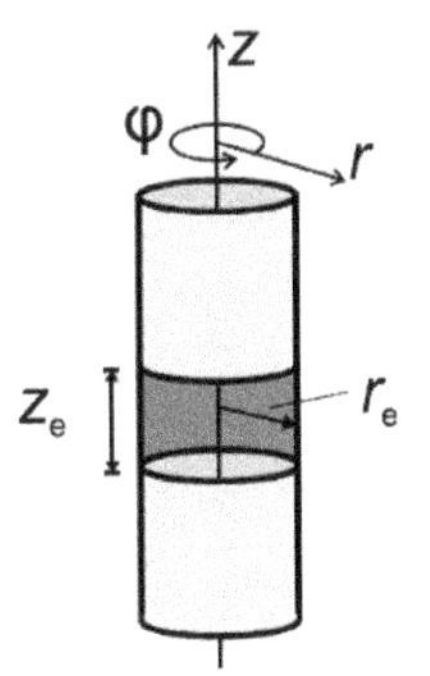

Figure 7. Band of conducting electrode of height z_e, embedded in a hollow insulating cylinder of radius r_e and the (r, z, φ) cylindrical polar coordinate system. The front section of the conducting band is drawn as transparent [85]. Reproduced with permission from J. Electroanal. Chem.; published by Elsevier, 2013.

Finally, for the situation in which the radius and length converge to 0 (zero), the current is equivalent to that of a macroelectrode with an electroactive area twice the top surface area (hole) of the cylinder.

5.4. Spheres

The same group has developed a theory for the porous electrode based on the hollow conductive spheres (Figure 8) [86].

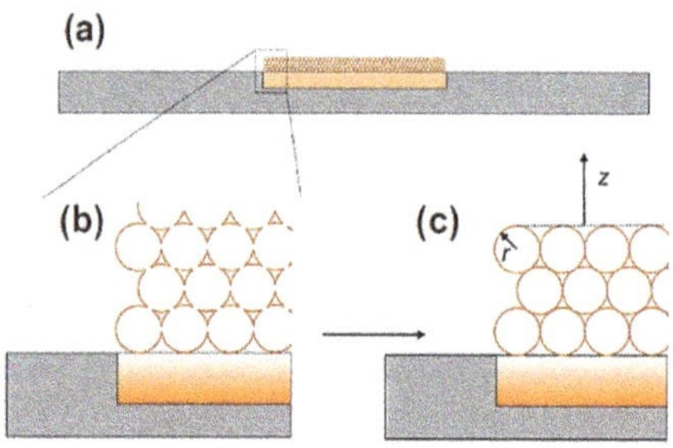

Figure 8. (**a**) Schematic cross-section of a porous electrode consisting of a series of hollow spheres supported on a disc electrode (the size of the sphere relative to the disc electrode is greatly exaggerated); (**b**) shows a zoomed in section, highlighting then interconnected nature of the spheres; and (**c**) shows a depiction of the model used to simulate electrochemistry at the electrode [86]. Reproduced with permission from J. Electroanal. Chem.; published by Elsevier, 2014.

Furthermore, they were guided by the assumption that, in the case of the similar potentials of two analytes, the thin-layer effect can be used to separate these two current signals, while simultaneously lowering the overpotential. For the purpose of the experiment, diffusion between the spheres and inside spheres and the bulk was neglected. This allows the assumption that the reaction can take place in the spheres, at the electrode (substrate) surface, and on the top layer of the spheres. In the case of a small sphere radius, the overall electrochemical behavior is, therefore, similar to that of the flat electrode. Accordingly, the authors try to separate these models using mass transport. This model fits the case of a porous electrode fabricated using the drop-casting method with carbon nanotubes. For the flat electrode (diffusion from the bulk to the surface of the spheres), the authors relied on the previous considerations on the electrochemical reaction at flat heterogeneous electrodes [87,88], partially blocked flat electrodes [89–91], and flat microelectrodes with random or regular arrays [92,93]. Accordingly, they used the Shoup–Szabo equation for the chronoamperometric simulations, while for cyclic voltammetry, they relied on Butler–Volmer kinetics. In the case of chronoamperometry within a sphere, they reduced the spherical (symmetric) space to a one-dimensional system, which allows them to express the dependence of the concentration of the electroactive species according to Fick's second law. This model is only modified in terms of the surface boundary condition, where Butler–Volmer kinetics is used.

Simulated chronoamperometric data show that, at short times, there is no difference between the current generated inside the sphere and the flat electrode. The response follows the Cottrell equation, as the diffusion layer is close to the electroactive surfaces. However, at longer electrolysis times, the current within the sphere decreases rapidly, indicating a depletion of the electroactive species in the sphere's volume (Figure 9). In contrast, at the same times ($\log t$), the current ($\log i$) of the flat electrode maintains a linear dependence, consistent with the Cottrell equation. This allows us, by simply adding the current response within the sphere (with the number of spheres known), including the parameter for the ratio of the flat electrode radius to the sphere radius, to approximate the total response of the porous electrode. Accordingly, the total chronoamperometric response at short times suggests a significant contribution of the thin-layer phenomenon when simulations are performed with a constant sphere radius. As expected, by increasing the number of spheres, the thin-layer response becomes more pronounced. Additionally, the thin-layer effect diminishes (within the time frame of the simulation) as the sphere radius decreases. Simulation data for cyclic voltammetry within the sphere reveal a strong dependence of the current on the scan rates. Since the entire electroactive substance can be consumed inside the sphere at low scan rates, a thin-layer response is observed, with the slope of the log υ vs. log i plot equal to 1, the same as for adsorption-controlled electrolysis. At fast scan rates, this depletion is not complete. Bearing in mind that at fast scan rates the diffusion layer is not as extended as at slow scan rates, diffusion occurs within the sphere. This has implications for the phenomenon of splitting the voltammetric signal (Figure 10), in the same manner as presented in previous work [80,86].

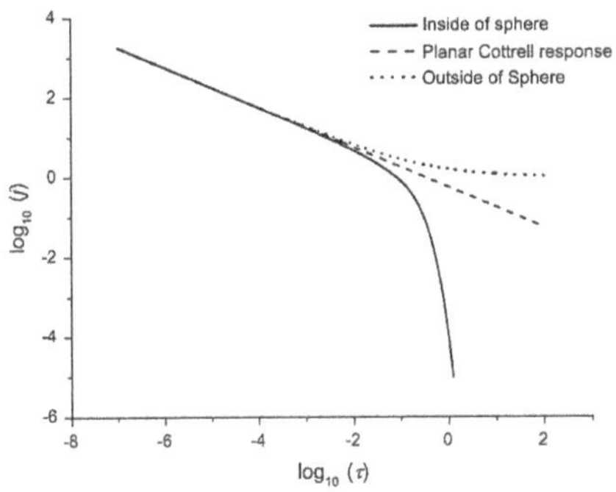

Figure 9. Chronoamperometric responses (in dimensionless parameters) of the inside of a sphere (solid line), a macrodisc (dashed line), and the outside of a sphere (dotted line) [86]. Reproduced with permission from J. Electroanal. Chem.; published by Elsevier, 2014.

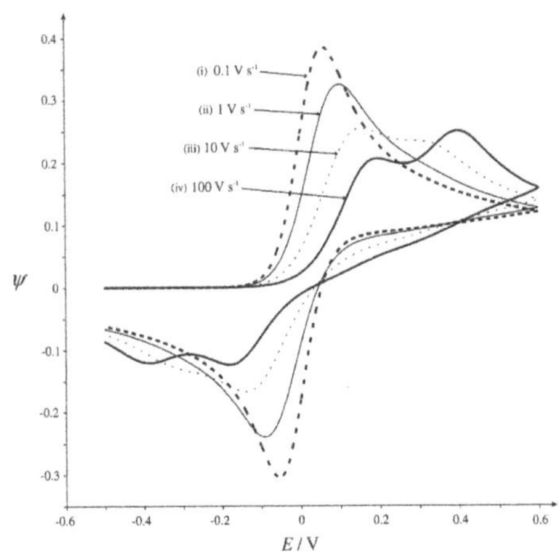

Figure 10. Simulated dimensionless current cyclic voltammograms for the R_0 = 10 μm diffusion domain where the scan rate is (i) 0.1 V s^{-1}, (ii) 1 V s^{-1}, (iii) 10 V s^{-1}, and (iv) 100 V s^{-1} [87]. Reproduced with permission from J. Electroanal. Chem.; published by Elsevier, 2004.

Generally, this splitting cannot be observed at fast scan rates (regardless of the kinetics) nor in systems with a large HET constant (reversible reactions?). This phenomenon is more prominent at slow kinetics; as the number of spheres increases, the radius of the spheres increases, while it is almost independent upon the surface of the flat electrode. The position of the peak potential is also the result of the interplay of electrochemical and mass transfer phenomena that occur both on the flat electrode surface and within the sphere at different electrochemical parameters. At fast scan rates, regardless of kinetics, splitting of the peaks cannot be noticed. At slow scan rates and fast kinetics, this splitting is less pronounced. The large splitting of the current peaks can be observed at slow scan rates and slow kinetics.

Elliot et al. [94] used the aforementioned diffusion indicator [77] to characterize the nature (convergent or linear) of the diffusion fluxes, qualitatively and quantitatively, toward a truncated electroactive sphere. By modeling 'the one sphere', they examined the different contributions, ranging from thin-layer to linear and convergent diffusion, depending on the extent of truncation of the sphere.

Also, simulation models were used in unscrambling the 'catalysis effect' on an electrode modified with electroinactive spheres, which were considered a special case of the so-called 'blocked electrode' [89–91] in the works of Kätelhön et al. [30] and Yang et al. [95]. As these particles have a crucial impact, through mass transport phenomena, on the electrode electrochemical behavior, this can be easily misinterpreted as electrocatalysis. Of course, phenomena of voltammetric signal splitting can also occur, as previously reported [80,86,87]. Although unexpected and unusual behavior was described in previous reports [80,86,96,97], Kätelhön et al. [30] provided a description of the phenomena that led to the unusual variation in the peak-to-peak separation with the scan rate, as well as the decrease in the peak height with the increasing scan rate. Their explanation is based on the extent of the decoupling of mass transfer (diffusion) in different volumes. One volume is between the conductive electrode and the inactive sphere, while the other is the bulk (Figure 11).

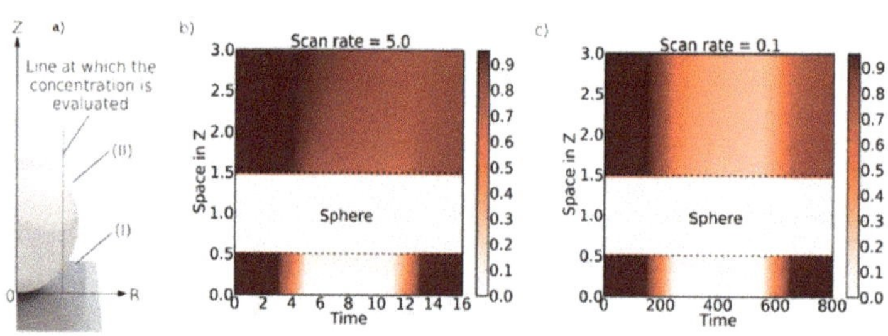

Figure 11. Concentration profiles as a function of Z and time evaluated for two different dimensionless scan rates and one layer of spheres. (**a**) Schematic depiction of the line at which the concentration profile is calculated. (**b**,**c**) Concentration profiles calculated for scan rates of 5 and 0.1, respectively. Please note that, for the clarity of presentation, voltammograms are modelled within a dimensionless potential range of ± 10 rather than ± 20 as in all other simulations [30]. Reproduced with permission from Appl. Mater. Today.; published by Elsevier, 2019.

They provide simulations of CV features, such as peak heights, peak position, and peak-to-peak separation that are affected by the thickness of the inactive layer, the size of the sphere, and the scan rate. The simulations reveal that, under the limiting conditions (extremely high or low scan rates), one can expect a similar peak-to-peak separation as for the bare electrode. This can be explained by classical voltammetry theory, as at fast scan rates, the diffusion layer does not spread from the electrode surface into the porous layer (due to the time frame of the experiment), which consequently has a similar diffusion profile to the bare electrode. At slow scan rates, there is sufficient time for diffusion from the bulk into the porous layer (and towards the electrode). Accordingly, the diffusion profile becomes similar to that of the bare electrode. Interestingly, substantial variation in the peak-to-peak difference in the intermediate region of the scan rates can be expected. This is a consequence of the interplay between the two aforementioned phenomena. At slightly higher scan rates, a strong decoupling of these two phenomena can be expected, leading to extensive depletion of the electroactive species in the volume between the sphere and the electrode. This is followed by low mass transfer from the bulk toward the electrode surface, resulting in a smaller peak-to-peak separation. At slightly slower scan rates, there is enough time for the diffusion of electroactive species from the bulk toward the electrode surface, which manifests as a higher peak-to-peak separation. Of course, before this diffusion occurs, the depletion of the electroactive species in the small volume of the electrolyte between the sphere and the electrode happens (thin-layer effect). This can be understood as the superposition (weak coupling) of these two processes. These phenomena are strongly dependent on the thickness of the layer. As the layer becomes thicker, less contribution from diffusion in the bulk solution is possible. As a result, at a certain scan rate, the peak-to-peak separation is lesser for a thicker layer. However, diffusion from the bulk affects the signal at slower scan rates (the diffusion path within the pore is longer for a thicker layer). Therefore, the maximum peak-to-peak separation can be expected at slower scan rates as the layer becomes thicker. Additionally, a reduced peak height at intermediate scan rates can be observed for the modified electrode, as a consequence of harnessed mass transfer toward the electrode. This effect becomes more pronounced as the layer thickness increases.

Based on the previously mentioned model, Yang et al. [95] attempted to resolve the intertwined behavior of thin-layer and semi-infinite diffusion by simulating the chronoamperometric response for different thicknesses (number of layers) of the inactive sphere layers. For this purpose, they introduced the previously mentioned diffusion indicator [77,94] to separate the contributions and predict the occurrence of semi-infinite diffusion, the thin-layer effect, as well as the 'bottlenecking' phenomenon. First, the authors made a distinction between this model ('blocked' electrode) and diffusion within a film, as these two cases cannot be considered in the same way due to the fact that diffusion through a film is homogeneous with a reduced (but uniform) diffusion coefficient. In short, the authors divide the chronoamperometric response (e.g., mass transport within a porous structure) into different regimes that occur at different times (Figure 12). The first regime, at short times, is characterized by the chronoamperometric response agreeing with the Cottrellian behavior for the bare electrode. The second regime can be observed after a normalized time of 0.1 and shows identical behavior (deviation from the Cottrellian behavior, as flux drops more rapidly), regardless of the number of sphere layers. Accordingly, the phenomenon occurring in this regime can be a consequence of the perturbation of the concentration of the electroactive species in the first layer. The third regime occurs for electrodes containing two or more layers, where the concentration profile of the electroactive species exhibits an oscillatory character depending on the number of layers. These oscillations indicate the presence of 'bottlenecks' (higher concentration differences in the electroactive species, e.g., highest local flux) in the diffusion process, which occur where the gap between the layers is smallest, at the equator of the sphere. Finally, the fourth regime occurs at longer times and is again characterized by Cottrellian dependence. After this, the author explored the possibility of using the diffusion indicator to infer information observed by chronoamperometric measurements. Analysis by applying diffusion indicator reveals that, in the second regime, 'thin-layer' diffusion occurs to the same extent, regardless of the number of layers. So, this is a phenomenon that occurs within the first layer. At the border of the second and third regimes, the influence of diffusion from the above 'compartment' (the volume between the first and second layer of the sphere, or in the case of one layer, from the bulk) can be noticed. As diffusion toward the electrode is hampered by the funnels created by the spheres, steady-state behavior can be noticed. This can be recognized by a dramatic increase in the diffusion indicator toward positive values. This behavior occurs only twice, regardless of the number of layers, as these funnels, which made serial connection (one above the other), hamper the delivery of the electroactive species into the compartment between the first and second layers, making the maintenance of steady-state diffusion toward the electrode impossible. Finally, as it is an intuitive conclusion, this behavior is more emphasized, as the radius of the sphere is smaller.

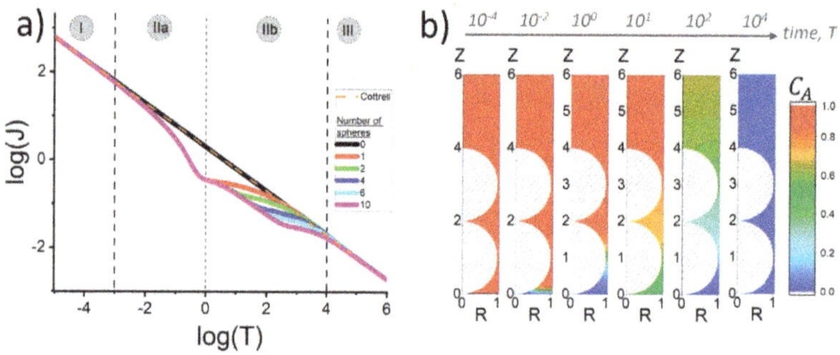

Figure 12. (**a**) Dimensionless flux-time transient to a planar electrode with n layers of surface blocks plotted in log-log form. At $T = 0$, a potential step is applied to the electrode from a potential of zero Faradaic current to a potential of diffusion-limited current to drive the reaction $A + e^- \rightarrow B$. The logarithm is base 10. (**b**) Concentration profiles in the case of two layers of insulating sphere blocks (n = 2) modified on the electrode at different times. Other simulation parameters: $R_{max} = 1.05$, $\Delta X = 1 \times 10^{-12}$, $\Delta T = 1 \times 10^{-12}$, $\omega_{X,T} = 1.02$, and $\varepsilon = 500$ [95]. Reproduced with permission from Appl. Mater. Today.; published by Elsevier, 2021.

All the above-mentioned theoretical and experimental approaches (summarized in Table 1) are based on mass transport phenomena and can be useful in the interpretation of the electrochemical phenomena that take place in various modified plane electrodes, as they consider the various geometry of the porous layers, electrical properties (inactive or active), or even partially conductive layers. It is clear that voltammetric and chronoamperometric responses can be (and usually are) an ensemble of different phenomena, including thin-layer diffusion, semi-infinite diffusion, and even adsorption phenomena, along with the possibility of an electrocatalytic effect, in cases where the porous layer is conductive. Therefore, the main effort should be directed towards separating these phenomena and detecting them.

Table 1. Parameters, electrochemical methods, and analyte used in experimental and simulation studies.

Porous Layer	Conductivity of		Electrochemical Method(s) *		Analyte(s)	Reference
	Substrate	Porous Layer	Simulation	Experimental		
SWCNT	GC (conductive)	N/A	CV, LSV	CV	$[Fe(CN)_6]^{4-}$	[68]
MWCNT	GC (conductive)	N/A	CV	CV, CA	dopamine	[71]
MWCNT	BPPG (conductive)	N/A	CV	CV	nicotine	[2]
Porous	conductive	conductive	DPV, SWV, LSV	-	-	[1]
Cylindrical/micropillar	conductive or non-conductive	conductive, non-conductive, partially conductive	CA	-	-	[73]
	conductive	partially conductive	CV	-	-	[74]
	metallized SiO_2 (conductive)	gold (conductive)	CA, CV	CA, CV	$[Fe(CN)_6]^{4-}$	[75]
	gold sputtered PDMS	drop-casted Pt–Pd/MWCNTs	-	CV, CA	sarccosine	[76]
	non-conductive	partially conductive	CA	-	-	[77]
	conductive	conductive	-	LSV	-	[80]

Table 1. *Cont.*

Porous Layer	Conductivity of		Electrochemical Method(s) *		Analyte(s)	Reference
Cylindrical pores	conductive	non-conductive	-	CV	-	[81]
	conductive	non-conductive	-	CV	-	[83]
	non-conductive	conductive	-	LSV	-	[84]
	no substrate	partially conductive	-	LSV	-	[85]
Spheres	conductive	conductive spheres, reaction inside and on the spheres surface	-	CV	-	[86]
	no substrate	partially conductive spheres conductive, semi-spheres	-	CA	-	[94]
	conductive	non-conductive	-	CV	-	[30]
	conductive	non-conductive	-	CA	-	[95]

* CA—chronoamperometry; CV—cyclic voltammetry; LSV—linear sweep voltammetry; DPV—differential pulse voltammetry, SWV—square wave voltammetry.

6. Single-Entity Measurements

Recently, the Compton group published a series of articles related to the determination of the electrocatalytic behavior of the MWCNTs toward different species, such as bromine/bromide redox pair [29], oxygen reduction [98], the VO^{2+}/VO^{2+} redox pair [99], nitrogen-doped MWCNTs for oxygen reduction [100], as well as studies of the electro-oxidation of the MWCNTs and amino-functionalized MWCNTs at different pH values [101]. The severity of these studies lies in the use of the single-entity approach to resolve the following question: are carbon nanotubes electrocatalytic or not?

A short review and perspective on the mentioned studies are presented in the work of Kumar et al. [26]. The single-entity approach is useful for exploring the electrocatalytic properties of various nanoparticles, including carbon nanotubes. This technique relies on the stochastic behavior (motion and 'adsorption') of the CNTs in their diluted solution toward (for motion) and onto (for 'adsorption') the electrode surface (Figure 13).

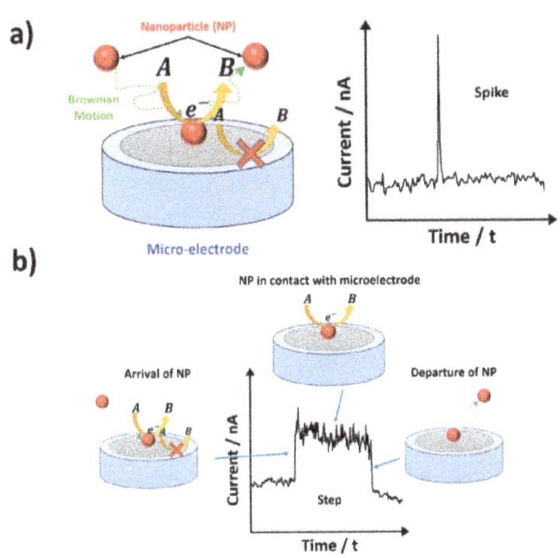

Figure 13. Schematic representation of (**a**) shorter duration impacts, resulting in current spikes and (**b**) a longer residence time at a microelectrode showing current steps. The arrows in yellow denote the mediated one-electron reduction reaction $A(aq) + e^- \rightarrow B(aq)$ [26]. Reproduced from J. Phys. Chem. Lett.; published by American Chemical Society, 2022. This publication is licensed under CC-BY 4.0.

If the CNTs have electrocatalytic properties, then an increase in the current in the presence of the redox couple can be observed in the vicinity of the electrode or adhering to the electrode. By comparing the results obtained with drop-casted MWCNTs and those obtained using the single-entity approach with a pristine glassy carbon electrode, the authors concluded that the electrocatalytic effect of MWCNTs is present to some extent, as the rate constants for the oxidation and reduction in bromine and bromide are approximately one order of magnitude larger than those for the glassy carbon electrode [29]. However, the absolute value of the rate constants was around 10^{-4} and 10^{-3} for reduction and oxidation, respectively, which cannot explain the improved reversibility of the voltammetric responses of the drop-casted MWCNT electrode. Interestingly, for the VO^{2+}/VO^{2+} redox pair [99], improved electron transfer was observed for the oxidation process, but it was retarded for the reduction process. This is not so surprising when considering the different transition states (e.g., changes in geometry, including breaking and forming chemical bonds) of these species. In addition, the redox reactions are complex in the sense of pH and proton transfer. Furthermore, a comparative study that included both the single-entity and drop-casted electrodes of the electrochemical reduction in the oxygen on MWCNTs was reported [98]. As the mechanism of this reduction involves a complex mechanism through quinone surface groups present on the MWCNTs, that issue is beyond the scope of this article. However, it is important to emphasize that, contrary to expectations, oxygen reduction occurs at a more anodic potential for the single-entity technique than for drop-casted MWCNTs. This behavior is explained by the irreversible destruction of the signal caused by the destruction of the quinone group on the MWCNTs when measurements were performed by CV. In contrast, due to the different settings of the single-entity measurements, fresh MWCNTs arrive onto the electrode surface, so the above-mentioned 'poisoning' is not present.

Finally, in a very interesting paper, Kumar et al. [102] demonstrated that using CV and single-entity measurements can be applied to distinguish thin-layer from adsorptive phenomena. They performed electrochemical oxidation of 4-hexylresorcinol on b-MWCNTs (bamboo-like MWCNTs) and revealed that distinguishing between thin-layer and adsorptive phenomena, even for different numbers of b-MWCNT layers, was inconclusive from the data obtained by voltammetry. The authors, based on calculations of the average charge per impact for different 4-hexylresorcinol concentrations, concluded that, if adsorption takes place on MWCNTs, then it must be of the multilayer (4–20 layers) kind of adsorption. When they applied an analysis for the calculation of the diffusion coefficient (D) of 4-hexylresorcinol, with the assumption of the diffusion-controlled process, they obtained a value of D in the order of magnitude of those reported in the literature. Thus, they concluded that voltammetric response is a consequence of thin-layer diffusion rather than an adsorption effect.

Besides the Compton group, other authors have also addressed the electrochemical behavior of the porous electrode. The effect of geometrical structure on the electrochemical behavior of carbon fiber, carbon nanospike, and carbon nanotube yarn electrodes was studied experimentally as well as through simulations by Cao et al. [103]. They demonstrated that an increase in geometry and scan rate results in more pronounced thin-layer behavior. In their simulation models, they successfully separated the CV signals into those belonging to diffusion and thin-layer electrochemistry, thus providing a comprehensive understanding of the influence of electrode geometry on electrochemical behavior.

To assess the role of CNTs in bioanalytical systems, Carrara et al. [4] studied three effects (Nernst, Cottrell, and Randles–Sevcik) using common electrochemical techniques. They demonstrated that, even in situations where CNTs apparently improve the analytical response, some effects, such as the Nernst effect, could be negligible or may even result in opposite trends in similar electrochemical systems. Also, they warned that many authors make the mistake of considering the specific active area as the sole reason for the improved signal. Finally, they concluded that various complex physicochemical processes govern the electrochemical response, and that to understand and enhance the performance of sensors based on CNTs (particularly those not related to H_2O_2 electrochemistry), much work still needs to be completed.

Based on fact that a tremendous number of synthesized compounds are used as electrocatalysists, Garcia et al. [104] contributed to establishing criteria for the characterization of electrochemical reactions occurring under finite diffusion conditions. By theoretical analysis, they presented that, in the case of non-highly efficient electrocatalysis, mass transport plays a crucial role. They also suggested that traditional methods for determining rate constants should be corrected by incorporating other modalities of diffusive fields and spatial domains.

Finally, it should be emphasized that mass transfer problems in various electrochemical systems continue to attract attention, as evidenced by a Special Issue devoted to this subject [105]. All the reviewed articles above demonstrate that alternative approaches can be used to distinguish the phenomena occurring within the porous layer structure, as well as the complexity and time frame of mass transport phenomena, thus posing a challenge to the electrochemical impedance spectroscopy (EIS).

7. Electrochemical Impedance Spectroscopy of the Porous Electrodes

For electrochemical devices, especially those designed for energy storage, porous electrodes dominate. Additionally, many porous materials have been widely used in electrochemical sensing devices. Since the processes occurring in porous electrodes can significantly differ from those at non-porous electrodes, one of the most powerful, non-invasive, and in situ tools for assessing and understanding phenomena in electrochemical systems is electrochemical impedance spectroscopy (EIS). These processes include electrode kinetics, capacitance effects, as well as mass transfer phenomena in solution and other specific processes. Due to the wide application of carbon nanomaterials (CNMs) in electrochemical and electrical systems, either as pristine materials or in composite form, a tremendous number of articles featuring EIS analysis can be found. These studies mostly focus on the development and characterization of microelectronics, electronic components, electroanalysis, energy conversion, energy harvesting, electrical energy storage, fuel cells, hydrogen storage, battery systems, electrochemical sensing, corrosion, composite materials, etc. Due to their porosity, CNTs are indispensable in such investigations. Many general information and data related to the analysis and interpretation of EIS spectra in these systems can be found in dedicated chapters of various scientific books or articles [106–108].

Historically, one of the earliest general works related to the application of EIS for the characterization of porous electrodes was reported by Candy et al. [109]. By comparing the simulated results with those of a Raney gold electrode, they concluded that the impedance technique can be used to evaluate the radius, depth, and number of cylindrical pores when no electrochemical reaction occurs at the electrode surface. Various approaches to the application of impedance methods for the investigation of porous electrodes were further developed by Raistrick et al. [110].

As two examples of the phenomena that take place in the porous structure, the authors related double-layer charging in the porous structure and fuel cell gas diffusion. Also, Lasia [111] simulated impedance spectra by the numerical addition of the impedances of pores and obtained an equation that described the potential and current distribution in cylindrical porous electrodes with semi-infinite and finite lengths. An important contribution to the understanding, interpretation, prediction, and deconvolution of diffusion-related EIS data, considering various pore geometries, was provided by Cooper et al. [112]. They found that impedance spectra can significantly deviate from the commonly used Warburg element in electrical equivalent circuits, potentially leading to the misinterpretation of impedance data. Furthermore, Tröltzsch and Kanoun [113] presented a general perspective on diffusion impedance in porous media using the transmission line model. They demonstrated how various processes—such as diffusion, charge transfer, charge accumulation, conductivity, and porosity—represented by different impedance models reported in scientific literature, are interconnected.

The effect of the pore size distribution on the transmission line model, as a new impedance model of porous materials in order to characterize the electric double-layer capacitor (EDLC), was developed by Song et al. [114]. In fact, they demonstrated the possibility of predicting the impedance of porous materials based on known pore size distributions. Similarly, Trishank Sharma [115] characterized the EIS response for various geometrical parameters of pores, such as radii, depth, density, and electrolyte conductivity. By describing the capacitive behavior of pores using a constant phase element, the authors concluded that these arrays can be represented as the sum of the EIS responses of individual pores. Also, a model for gaining insight into the physical processes, particularly the transient behavior of porous electrodes, was presented by Devan et al. [116]. Using their model, the authors successfully quantified various parameters, including ohmic resistance, charge transfer, and polarization resistance in dependence of various parameters, such as porosity, diffusion coefficient, particle size, etc. An enhanced model, based on the impedance transmission line with two transport channels, was presented by Bisquert [117]. He discussed the blocking of charge carriers at macroscopic boundaries at various frequencies and derived diagnostic criteria to determine whether the dispersion of the constant phase element is caused by the boundary or the inner surface.

As graphite is the most commercially significant anode material for Li-ion batteries, a tremendous number of articles involving EIS analysis of these devices can be found. Among the published articles, some provide a broader perspective on methods and the interpretation of porous electrode impedance spectra, such as those based on the transmission line model [118], or revisiting the physical interpretation of EIS spectra for graphite porous electrodes across different frequencies in the context of ion transport and pseudocapacitive behavior at high and middle frequencies [119]. In view of the extensive research on this topic, it remains a major challenge to compile a comprehensive overview.

Notably, Gruet et al. [120] investigated the EIS response of composite electrodes containing graphite (spherical particles) to assess and understand the degradation processes in Li-ion batteries. The authors presented a model that can predict optimal porosity in order to achieve the best performances upon the composition of the electrode.

One of the excellent reviews that provides an in-depth insight into the challenges, history, and journey of this subject is the review by Huang et al. [121]. One of the early efforts to yield data concerning the structure and composition of porous carbon materials, with an emphasis on the highly accurate determination of the available area through EIS, was carried out by Armstrong et al. [122].

Owing to the fact that carbon materials are widely used in energy storage, mostly in EDLC, Yoo et al. [123] attempted to establish an analytical method to correlate the EIS results with those obtained with the direct current method, such as galvanostatic charge–discharge cycling. They profiled rate capabilities of the porous carbon electrode in light of ionic accessibility and described the distribution of capacitance in respect to the latter. Suss et al. [124] presented both theoretical and experimental characterizations of hierarchical porous carbon aerogels, with varying structures of nanoscale pores integrated into a micron-scale porous network. The focus of the work presented by Abouelamaiem et al. [125] was on the development of a simple EEC model that can be used to describe porous carbon in supercapacitors and on the assessment of the role of pore morphology in improving the electrochemical performance of EDLCs. The authors concluded that, for optimizing porosity, pore size distribution, electrical pore resistance, and capacitance to achieve optimal charging and discharging capabilities, the assessment of relaxation times is crucial. Various inorganic oxide porous structures together with porous carbon were examined as materials for supercapacitors in the study by De et al. [126]. When compared to bulk materials, porous materials showed lower values of equivalent series resistance and charge transfer resistance. Also, the porous structure was found to act as an ion reservoir, stabilizing lattice expansion during cycling. Furthermore, the authors used simple Fick's laws to calculate the diffusion coefficient in the examined porous materials.

Besides the general aspects of EIS analysis on porous structures or porous carbon, many carbon nanomaterials have been characterized by EIS, either purposefully or as part of studies related to specific subjects. Yang et al. [127] studied closed and open CNTs at different potentials to understand phenomena that can occur in lithium batteries, such as intercalation. Differences in morphology (e.g., open vs. closed) can significantly influence double-layer capacitance and diffusion phenomena. Closed nanotubes exhibited spectra that could be explained by adsorption phenomena, while open CNTs showed charge transfer from adsorbed intermediate species. Additionally, diffusion effects in open CNTs occur around both the inner core spaces and the outer surface, leading to faster diffusion processes. In contrast, diffusion in closed CNTs occurs only at the outer surface, resulting in slower diffusion. The authors also calculated the diffusion coefficient of Li ions in CNTs.

Leppänen et al. [128] conducted a comprehensive electrochemical and physicochemical analysis of various carbon nanomaterials, which included cyclic voltammetry and EIS. Their study demonstrated how electrochemical performance can be significantly influenced by small changes in the structural properties of the materials. Several electrochemical parameters were considered, including the point of zero charge, open circuit potential, double-layer capacitance, pseudocapacitance, working potential windows at various pH levels, reaction kinetics for both outer and inner sphere probes, and heterogeneous electron transfer. They found that small capacitance and a wide potential window are closely related to the reactivity of the nanomaterials, particularly for applications in electroanalysis. Interestingly, they applied a simple Randles model for their analysis, probably because they used an equimolar solution of $Rh(NH_3)_3{}^{2+}/{}^{3+}$, and there was no discussion of porosity (at least not for interpreting the EIS data).

The ultramicroelectrodes made of SWCNTs prepared by Dumitrescu et al. [129] showed very reproducible EIS spectra (even for electrodes with different diameters). These spectra were modeled using a simple EEC model and provided valuable data on the double layer capacitance, the diffusion coefficient of $FcTMA^{+/2+}$, and the heterogeneous rate constant. The electrochemical response of twist-spun yarn MWCNTs (with a diameter of around 10 μm) at different potentials in both aqueous and non-aqueous solutions was studied by Mirfakhrai et al. [130]. They aimed to correlate the exponent of the constant phase element (CPE) with the physical structure of the yarn and found that the behavior of the CPE is closely related to the underlying physical and fractal structure of the yarns.

The electrochemical activity of a composite material consisting of SWCNTs and MWCNTs, along with silver, was investigated by Niessen et al. [131] to explain their response. They found that the EIS data were cross-correlated with those obtained from cyclic voltammetry (CV), revealing that more than 90% of the total charge was stored electrostatically (for commercially obtained CNTs). The study also raised doubts about whether CNTs can act as a hydrogen reservoir. Also, various composite materials containing mixtures of 2D and 3D carbon materials [132,133] were characterized using various electrochemical techniques, including EIS, to improve the efficiency of electrodes for high energy density supercapacitors.

Comparative EIS studies on SWCNTs and MWCNTs polymer nanocomposites to optimize the material composition for biosensor applications were presented by Tertis et al. [134].

They proposed a mechanism involving adsorption and relaxation processes. Additionally, Galicia et al. [135] utilized EIS to provide a perspective for characterizing CNT–chitosan scaffolds under flow conditions. They discovered that the transition from single to mixed diffusion and charge transfer mechanisms controls the overall response. As a conclusion, they suggested the potential to use chitosan–CNT composites for applications in flow conditions.

Recently, Han et al. [136] treated SWCNTs with O_2 plasma to introduce oxygen-containing groups for enhanced sensitivity. Using EIS, the authors analyzed spray-coated CNTs on a platinum electrode. As the SWCNT layer thickness increased, they observed an increase in charge transfer resistance due to the reduced effective electrode area, resulting from decreased roughness and the porosity of the layer. In addition, the time constant was lowered due to the improved diffusion, as the charge transfer resistance decreased.

8. Conclusions

This review focuses on addressing the following fundamental question in carbon nanotube (CNT) electrochemistry: What is the role and purpose of incorporating CNTs into electrochemical sensors? Although CNTs are widely used in the development of electrochemical sensors, there is still a lack of clarity regarding the exact purpose of their involvement. To understand this, it is essential to establish the relationship between their physical and electrochemical properties and their influence on electroanalytical signals. Challenges arise from the blurred boundary between physical and electrochemical influences. In fact, the interplay, or even synergy, between physical and electrical properties may occur. This requires a meticulous approach to modeling CNT electrochemical systems, incorporating a variety of physical and electrical parameters (conductivity or non-conductivity of the basic 'particle' and/or substrate). These complex relationships can be studied using diverse electrochemical methods, from amperometry to voltammetry. Consequently, several theoretical models have been proposed to describe the phenomena occurring in CNTs during electrolysis.

Generally, these models, based on chronoamperometric and voltammetric methods, predict that the electrochemical response reflects mass transport processes, which can be described by a combination of semi-infinite and thin-layer diffusion. However, this mixed mass transport regime presents various manifestations, as it depends on the physical properties of the 'basic particles' (such as geometry, shape, orientation, and size, which serve as building blocks of the porous layer), electrical properties (e.g., conductivity), and electrochemical parameters (e.g., the kinetics of heterogeneous electron transfer and reversibility). The latter allows us to distinguish and estimate the electrochemical signal by adjusting the appropriate voltammetric parameters (e.g., scan rates) to obtain differences in the peak-to-peak separation, high peak currents, or even, splitting the voltammetric signals that are related to the different mass transport regime. The reliability of these findings is further improved when chronoamperometry is introduced, providing a diffusion indicator that offers both qualitative and quantitative insights into the nature of diffusion fluxes.

Although all the developed models can be used to elucidate the electrochemical phenomena occurring in CNTs, the simple question of whether carbon nanotubes act as electrocatalysts remains unanswered. Fortunately, single-entity experiments conducted on microelectrodes for selected electroactive species revealed that the reversibility of the electrochemical system cannot be explained by an improved heterogeneous electron transfer due to the presence of CNTs. It is important to note that this statement should be taken 'with a grain of salt', as it as it relevant only to specific cases (see reference [91]). Additionally, single-entity approaches can help differentiate between adsorption and thin-layer diffusion, as the electrochemical results for these two processes are nearly identical.

As presented, it is challenging to draw a definitive conclusion on this subject, as the various theoretical models need to be integrated into a unified theory. This would involve merging and simplifying the mathematical calculations, ultimately streamlining the approach. Additionally, a challenge arises from the fact that many reports on the use of CNTs for electroanalytical purposes do not focus on the role of CNTs in the architecture of the sensors. Moreover, the conclusions drawn from experimental results are often affected by the variability in the properties of CNTs, as they are sourced from different producers, and these properties can vary depending on the specific electroactive species involved. Other complications, such as adsorption within the CNT layer or the 'fouling effect', may result in a 'blocked' electrode. These are issues that future studies will need to address.

The approaches discussed above are designed as 'alternatives' and 'challenges' to electrochemical impedance spectroscopy (EIS), which is extensively used for analyzing and clarifying various electrochemical and physical processes in porous layers.

This review also provides a brief overview of the use of electrochemical impedance spectroscopy (EIS) in understanding mass transport phenomena in porous environments. Although this technique is highly effective for gaining insights into diffusion processes, its interpretation can be challenging. This is due to the necessity of using impedance theory and modeling, which require precise information about the system's geometry, the defined interface between the electrode and electrolyte, and so on. Despite the large number of articles published on EIS and porous structures, limited information is available for clarifying and assessing mass transfer phenomena in carbon nanotubes (CNTs) when they are used in electrochemical sensors. This points to a future perspective on interpreting EIS data, in light of the insights gained through models, simulations, and experiments. Furthermore, the connection between cyclic voltammetry and EIS, as discussed in [137], is crucial for elucidating phenomena in the electrochemical system in real time and simultaneously.

Author Contributions: Conceptualization, J.D. and M.B. (Marijo Buzuk); visualization, J.D. and M.B. (Marijo Buzuk); writing—original draft preparation, J.D. and M.B. (Marijo Buzuk); writing—review and editing, I.Š.R., N.V., M.B. (Maša Buljac) and J.R.; resources, I.Š.R., N.V., M.B. (Maša Buljac) and J.R.; supervision, M.B. (Marijo Buzuk); funding acquisition, I.Š.R., N.V., M.B. (Maša Buljac) and J.R. All authors have read and agreed to the published version of the manuscript.

Funding: This research received no external funding.

Institutional Review Board Statement: Not applicable.

Informed Consent Statement: Not applicable.

Data Availability Statement: The data presented in this study are available within the content of this article.

Conflicts of Interest: The authors declare no conflicts of interest.

References

1. Laborda, E.; Molina, A.; Martínez-Ortiz, F.; Compton, R.G. Electrode modification using porous layers. Maximising the analytical response by choosing the most suitable voltammetry: Differential Pulse vs Square Wave vs Linear sweep voltammetry. *Electrochim. Acta* **2012**, *73*, 3–9. [CrossRef]
2. Sims, M.J.; Rees, N.V.; Dickinson, E.J.F.; Compton, R.G. Effects of thin-layer diffusion in the electrochemical detection of nicotine on basal plane pyrolytic graphite (BPPG) electrodes modified with layers of multi-walled carbon nanotubes (MWCNT-BPPG). *Sens. Actuators B Chem.* **2010**, *144*, 153–158. [CrossRef]
3. Primo, N.E.; Gutiérrez, F.; Rubianes, D.M.; Ferreyra, F.; Rodríguez, L.M.; Pedano, L.M.; Gasnier, A.; Gutierrez, A.; Eguílaz, M.; Dalmasso, P.; et al. Electrochemistry in One Dimension: Applications of Carbon Nanotubes. In *Electrochemistry of Carbon Electrodes*; Alkire, R., Bartlett, P., Lipkowski, J., Eds.; Wiley-VCH Verlag GmbH & Co.: Weinheim, Germany, 2015; Volume 16, pp. 83–120. [CrossRef]

4. Carrara, S.; Baj-Rossi, C.; Boero, C.; De Micheli, G. Do carbon nanotubes contribute to electrochemical biosensing? *Electrochim. Acta.* **2014**, *128*, 102–112. [CrossRef]
5. Banks, C.E.; Crossley, A.; Salter, C.; Wilkins, S.J.; Compton, R.G. Carbon nanotubes contain metal impurities which are responsible for the "electrocatalysis" seen at some nanotube-modified electrodes. *Angew. Chem. Int. Ed.* **2006**, *45*, 2533–2537. [CrossRef]
6. Scott, C.L.; Pumera, M. Carbon nanotubes can exhibit negative effects in electroanalysis due to presence of nanographite impurities. *Electrochem. Commun.* **2011**, *13*, 426–428. [CrossRef]
7. Kosala, W. Conducting Polymer Electrodes for Thermogalvanic Cells. Ph.D. Thesis, Linköping University, Linköping, Sweden, 2018.
8. Hubbard, A.T.; Peters, D.G. Electrochemistry in thin layers of solution. *Crit. Rev. Anal. Chem.* **1973**, *3*, 201–242. [CrossRef]
9. Wang, J. Carbon-nanotube based Electrochemical Biosensors: A Review. *Electroanalysis* **2005**, *17*, 7–14. [CrossRef]
10. Gooding, J.J. Nanostructuring electrodes with carbon nanotubes: A review on electrochemistry and applications for sensing. *Electrochim. Acta* **2005**, *50*, 3049–3060. [CrossRef]
11. Rivas, G.; Rubianes, M.; Rodriguez, M.; Ferreyra, N.; Luque, G.; Pedano, M.; Miscoria, S.; Parrado, C. Carbon nanotubes for electrochemical biosensing. *Talanta* **2007**, *74*, 291–307. [CrossRef]
12. Agüí, L.; Yáñez-Sedeño, P.; Pingarrón, J.M. Role of carbon nanotubes in electroanalytical chemistry. *Anal. Chim. Acta.* **2008**, *622*, 11–47. [CrossRef]
13. Yáñez-Sedeño, P.; Pingarrón, J.M.; Riu, J.; Rius, F.X. Electrochemical sensing based on carbon nanotubes. *TrAC, Trends Anal. Chem.* **2010**, *29*, 939–953. [CrossRef]
14. D'Orazio, P. Biosensors in clinical chemistry–2011 update. *Clin. Chim. Acta* **2011**, *412*, 1749–1761. [CrossRef] [PubMed]
15. Walcarius, A.; Minteer, S.D.; Wang, J.; Lin, Y.; Merkoçi, A. Nanomaterials for bio-functionalized electrodes: Recent trends. *J. Mater. Chem. B* **2013**, *1*, 4878–4908. [CrossRef] [PubMed]
16. Ammam, M.; Fransaer, J. Glucose/O_2 biofuel cell based on enzymes, redox mediators, and Multiple-walled carbon nanotubes deposited by AC-electrophoresis then stabilized by electropolymerized polypyrrole. *Biotechnol. Bioeng.* **2012**, *109*, 1601–1609. [CrossRef]
17. Pang, P.; He, J.; Park, J.H.; Krstić, P.S.; Lindsay, S. Origin of giant ionic currents in carbon nanotube channels. *ACS Nano.* **2011**, *5*, 7277–7283. [CrossRef]
18. Mejri, A.; Arroyo, N.; Herlem, G.; Palmeri, J.; Manghi, M.; Henn, F.; Picaud, F. Impact of single-walled carbon nanotube functionalization on Ion and Water Molecule Transport at the nanoscale. *Nanomaterials* **2024**, *14*, 117. [CrossRef]
19. Secchi, E.; Niguès, A.; Jubin, L.; Siria, A.; Bocquet, L. Scaling behavior for ionic transport and its fluctuations in individual carbon nanotubes. *Phys. Rev. Lett.* **2016**, *116*, 154501. [CrossRef]
20. Yao, Y.-C.; Taqieddin, A.; Alibakhshi, M.A.; Wanunu, M.; Aluru, N.R.; Noy, A. Strong electroosmotic coupling dominates ion conductance of 1.5 nm diameter carbon nanotube porins. *ACS Nano.* **2019**, *13*, 12851–12859. [CrossRef]
21. Daiguji, H. High mobility in tight spaces. *Nat. Nanotechnol.* **2010**, *5*, 831–832. [CrossRef]
22. Samoylova, O.N.; Calixte, E.I.; Shuford, K.L. Selective ion transport in functionalized carbon nanotubes. *Appl. Surf. Sci.* **2017**, *423*, 154–159. [CrossRef]
23. Lee, J.Y.; Kang, J.; Byun, K.; Kang, E.; Hwang, H.; Lee, J.H.; Lee, H.-J.; Kwon, O.; Kim, Y.-M. Field-effect ion-transport devices with carbon nanotube channels: Schematics and simulations. *J. Korean Phys.* **2004**, *45*, 432–437.
24. Joseph, S.; Mashl, R.J.; Jakobsson, E.; Aluru, N.R. Ion channel based biosensors: Ionic transport in carbon nanotubes. In Proceedings of the Nanotech 2003, Nanotechnology Conference and Trade Show, San Francisco, CA, USA, 23–27 February 2003.
25. Cheng, Y.; Zhang, J.; Jiang, S.P. Are metal-free pristine carbon nanotubes electrocatalytically active? *Chem. Commun.* **2015**, *51*, 13764–13767. [CrossRef] [PubMed]
26. Kaliyaraj Selva Kumar, A.; Lu, Y.; Compton, R.G. Voltammetry of carbon nanotubes and the limitations of particle-modified electrodes: Are carbon nanotubes electrocatalytic? *J. Phys. Chem. Lett.* **2022**, *13*, 8699–8710. [CrossRef] [PubMed]
27. Dong, Q.; Nasir, M.Z.; Pumera, M. Semi-conducting single-walled carbon nanotubes are detrimental when compared to metallic single-walled carbon nanotubes for electrochemical applications. *Phys. Chem. Chem. Phys.* **2017**, *19*, 27320–27325. [CrossRef]
28. Scott, C.L.; Pumera, M. Electroanalytical parameters of carbon nanotubes are inferior with respect to well defined surfaces of glassy carbon and EPPG. *Electrochem. Commun.* **2011**, *13*, 213–216. [CrossRef]
29. Kaliyaraj Selva Kumar, A.; Miao, R.; Li, D.; Compton, R.G. Do carbon nanotubes catalyse bromine/bromide redox chemistry? *Chem. Sci.* **2021**, *12*, 10878–10882. [CrossRef]
30. Kätelhön, E.; Chen, L.; Compton, R.G. Nanoparticle electrocatalysis: Unscrambling illusory inhibition and catalysis. *Appl. Mater. Today* **2019**, *15*, 139–144. [CrossRef]
31. Qiu, C.; Jiang, L.; Gao, Y.; Sheng, L. Effects of oxygen-containing functional groups on carbon materials in supercapacitors: A Review. *Mater. Des.* **2023**, *230*, 111952. [CrossRef]
32. Yang, F.; Ma, X.; Cai, W.-B.; Song, P.; Xu, W. Nature of oxygen-containing groups on carbon for high-efficiency electrocatalytic CO_2 reduction reaction. *J. Am. Chem. Soc.* **2019**, *141*, 20451–20459. [CrossRef]
33. Kim, J.-H.; Kim, S.-H.; Kim, B.-J.; Lee, H.-M. Effects of Oxygen-Containing Functional Groups on the Electrochemical Performance of Activated Carbon for EDLCs. *Nanomaterials* **2023**, *13*, 262. [CrossRef]

34. Tojo, T.; Sakurai, K.; Muramatsu, H.; Hayashi, T.; Yang, K.-S.; Jung, Y.C.; Yang, C.-M.; Endo, M.; Kim, Y.A. Electrochemical role of oxygen containing functional groups on activated carbon electrode. *RSC Adv.* **2014**, *4*, 62678–62683. [CrossRef]
35. Matsubara, K.; Waki, K. The effect of the oxygen-containing functional groups on the electrochemical reduction of oxygen of multi-walled carbon nanotubes in acid media. *ECS Trans.* **2010**, *28*, 35–46. [CrossRef]
36. Franklin, G.; Halim, E.-M.; Merlet, C.; Taberna, P.-L.; Simon, P. Insights into the effect of oxygen functional groups on the heterogeneous electron transfer at activated carbons from scanning electrochemical microscopy. *Electrochim. Acta* **2023**, *465*, 142944. [CrossRef]
37. Liu, X.; Wang, Y.; Zhan, L.; Qiao, W.; Liang, X.; Ling, L. Effect of oxygen-containing functional groups on the impedance behavior of activated carbon-based electric double-layer capacitors. *J Solid State Electrochem.* **2011**, *15*, 413–419. [CrossRef]
38. Alemany-Molina, G.; Martínez-Sánchez, B.; Morallón, E.; Cazorla-Amorós, D. The role of oxygen heteroatoms in the surface (electro)chemistry of carbon materials. *Carbon Rep.* **2022**, *1*, 162–174. [CrossRef]
39. Gong, K.; Chakrabarti, S.; Dai, L. Electrochemistry at Carbon Nanotube Electrodes: Is the nanotube tip more active than the sidewall? *Angew. Chem. Int. Ed.* **2008**, *47*, 5446–5450. [CrossRef]
40. Moore, R.R.; Banks, C.E.; Compton, R.G. Basal plane pyrolytic graphite modified electrodes: Comparison of carbon nanotubes and graphite powder as electrocatalysts. *Anal. Chem.* **2004**, *76*, 2677–2682. [CrossRef]
41. Banks, C.E.; Moore, R.R.; Davies, T.J.; Compton, R.G. Investigation of modified basal plane pyrolytic graphite electrodes: Definitive evidence for the electrocatalytic properties of the ends of carbon nanotubes. *Chem. Commun.* **2004**, *16*, 1804–1805. [CrossRef]
42. Banks, C.E.; Compton, R.G. Exploring the electrocatalytic sites of carbon nanotubes for NADH detection: An edge plane pyrolytic graphite electrode study. *Analyst* **2005**, *130*, 1232–1239. [CrossRef]
43. Güell, A.G.; Meadows, K.E.; Dudin, P.V.; Ebejer, N.; Macpherson, J.V.; Unwin, P.R. Mapping nanoscale electrochemistry of individual single-walled carbon nanotubes. *Nano Lett.* **2014**, *14*, 220–224. [CrossRef]
44. Kim, J.; Xiong, H.; Hofmann, M.; Kong, J.; Amemiya, S. Scanning electrochemical microscopy of individual single-walled carbon nanotubes. *Anal. Chem.* **2010**, *82*, 1605–1607. [CrossRef] [PubMed]
45. González-Segura, K.; Cañete-Rosales, P.; del Rio, R.; Yáñez, C.; Ferreyra, N.F.; Rivas, G.A.; Bollo, S. Effect of the dispersing agent on the electrochemical response of glassy carbon electrodes modified with dispersions of carbon nanotubes. *Electroanalysis* **2012**, *24*, 2317–2323. [CrossRef]
46. Kharissova, O.V.; Kharisov, B.I.; de Casas Ortiz, E.G. Dispersion of carbon nanotubes in water and non-aqueous solvents. *RSC Adv.* **2013**, *3*, 24812. [CrossRef]
47. Kharisov, B.I.; Kharissova, O.V.; Dimas, A.V. The dispersion, solubilization and stabilization in "solution" of single-walled carbon nanotubes. *RSC Adv.* **2016**, *6*, 68760–68787. [CrossRef]
48. Sun, G.; Liu, Z.; Chein, G. Dispersion of pristine multi-walled carbon nanotubes in common organic solvents. *Nano* **2010**, *5*, 103–109. [CrossRef]
49. Cheng, Q.; Debnath, S.; Gregan, E.; Byrne, H.J. Ultrasound-Assisted SWNTs Dispersion: Effects of Sonication Parameters and Solvent Properties. *J. Phys. Chem. C* **2010**, *114*, 8821–8827. [CrossRef]
50. Kim, S.W.; Kim, T.; Kim, Y.S.; Choi, H.S.; Lim, H.J.; Yang, S.J.; Park, C.R. Surface modifications for the effective dispersion of carbon nanotubes in solvents and polymers. *Carbon* **2012**, *50*, 3–33. [CrossRef]
51. Ma, P.-C.; Siddiqui, N.A.; Marom, G.; Kim, J.-K. Dispersion and functionalization of carbon nanotubes for polymer-based nanocomposites: A Review. *Compos.-A Appl. Sci. Manuf.* **2010**, *41*, 1345–1367. [CrossRef]
52. Liu, H.; Shen, G. Ordered arrays of carbon nanotubes: From synthesis to applications. *Nano Biomed. Eng.* **2012**, *4*, 107–117. [CrossRef]
53. Zhou, M.; Shang, L.; Li, B.; Huang, L.; Dong, S. The characteristics of highly ordered mesoporous carbons as electrode material for electrochemical sensing as compared with carbon nanotubes. *Electrochem. Commun.* **2008**, *10*, 859–863. [CrossRef]
54. Walcarius, A. Recent trends on electrochemical sensors based on ordered mesoporous carbon. *Sensors* **2017**, *17*, 1863. [CrossRef] [PubMed]
55. Gao, C.; Guo, M.; Liu, Y.; Zhang, D.; Gao, F.; Sun, L.; Li, J.; Chen, X.; Terrones, M.; Wang, Y. Surface modification methods and mechanisms in carbon nanotubes dispersion. *Carbon* **2023**, *212*, 118133. [CrossRef]
56. Alvarez, N.; Nawarathne, C.; Abdul Hoque Pankaj, H.A.; Ruhunage, C.G. Techniques for covalent bonding of carbon nanotubes to substrates. U.S. Patent No. US 2023/0407017 A1, 2023. Available online: https://patents.google.com/patent/US20230407017A1/fr (accessed on 14 December 2024.).
57. Anderson, L.B.; Reilley, C.N. Thin-layer electrochemistry: Use of twin working electrodes for the study of chemical kinetics. *J. Electroanalyt. Chem.* **1965**, *10*, 538–552. [CrossRef]
58. Christensen, C.R.; Anson, F.C. Chronopotentiometry in thin layer of solution. *Anal. Chem.* **1963**, *35*, 205–209. [CrossRef]
59. Hubbard, T.A.; Anson, C.F. Thin-layer chronopotentiometric determination of reactants adsorbed on platinum electrodes. *J. Electroanalyt. Chem.* **1965**, *9*, 163–164. [CrossRef]
60. McClure, J.E.; Maricle, D.L. New dip-type thin-layer electrolysis cell. N-value determinations in nonaqueous systems. *Anal. Chem.* **1967**, *39*, 236–238. [CrossRef]

61. Sheaffer, J.C.; Peters, D.G. Simple electrode for thin-layer electrochemistry. *Anal. Chem.* **1970**, *42*, 430–432. [CrossRef]
62. Hubbard, A.T. Study of the kinetics of electrochemical reactions by thin-layer voltammetry. *J. Electroanal. Chem. Interf. Electrochem.* **1969**, *22*, 165–174. [CrossRef]
63. Cushing, J.P.; Hubbard, A.T. Study of the kinetics of electrochemical reactions by thin-layer voltammetry II. Electro-oxidation of platinum(II) complexes. *J. Electroanal. Chem. Interf. Electrochem.* **1969**, *23*, 183–203. [CrossRef]
64. Lau, A.L.Y.; Hubbard, A.T. Study of the kinetics of electrochemical reactions by thin-layer voltammetry Ill. Electroreduction of the chloride complexes of platinum(II) and (IV). *J. Electroanal. Chem. Interf. Electrochem.* **1970**, *24*, 237–249. [CrossRef]
65. Hubbard, A.T.; Anson, F.C. Study of electrochemistry of chloride and bromide complexes of platinum(II) and (IV) by thin-layer electrochemistry. *Anal. Chem.* **1966**, *38*, 1887–1893. [CrossRef]
66. Lau, A.L.Y.; Hubbard, A.T. Study of the kinetics of electrochemical reactions by thin-layer voltammetry IV. The influence of chemisorbed species on the reactions of platinum complexes at platinum electrodes. *J. Electroanal. Chem. Interf. Electrochem.* **1971**, *33*, 77–93. [CrossRef]
67. Tyurin, R.S.; Lyalikov, Y.S.; Zhdanov, S.I. Thin-layer electrochemistry. *Russ. Chem. Rev.* **1972**, *41*, 1086–1100. [CrossRef]
68. Streeter, I.; Wildgoose, G.G.; Shao, L.; Compton, R.G. Cyclic voltammetry on electrode surfaces covered with porous layers: An analysis of electron transfer kinetics at single-walled carbon nanotube modified electrodes. *Sens. Actuators B Chem.* **2008**, *133*, 462–466. [CrossRef]
69. Šljukić, B.; Banks, C.E.; Compton, R.G. Iron oxide particles are the active sites for hydrogen peroxide sensing at multiwalled carbon nanotube modified electrodes. *Nano Lett.* **2006**, *6*, 1556–1558. [CrossRef]
70. Banks, C.E.; Ji, X.; Crossley, A.; Compton, R.G. Understanding the electrochemical reactivity of bamboo multiwalled carbon nanotubes: The presence of oxygenated species at tube ends may not increase electron transfer kinetics. *Electroanalysis* **2006**, *18*, 2137–2140. [CrossRef]
71. Keeley, G.P.; Lyons, E.G.M. The effects of thin layer diffusion at glassy carbon electrodes modified with porous films of single-walled carbon nanotubes. *Int. J. Electrochem. Sci.* **2009**, *4*, 794–809. [CrossRef]
72. Nicholson, R.S. Theory and application of cyclic voltammetry for measurement of Electrode Reaction Kinetics. *Anal. Chem.* **1965**, *37*, 1351–1355. [CrossRef]
73. Dickinson, E.J.; Streeter, I.; Compton, R.G. Theory of chronoamperometry at cylindrical microelectrodes and their arrays. *J. Phys. Chem. C.* **2008**, *112*, 11637–11644. [CrossRef]
74. Ban, Z.; Kätelhön, E.; Compton, R.G. Voltammetry of Porous Layers: Staircase vs Analog Voltammetry. *J. Electroanal. Chem.* **2016**, *776*, 25–33. [CrossRef]
75. Sánchez-Molas, D.; Esquivel, J.P.; Sabaté, N.; Muñoz, F.X.; del Campo, F.J. High aspect-ratio, fully conducting gold micropillar array electrodes: Silicon micromachining and electrochemical characterization. *J. Phys. Chem. C.* **2012**, *116*, 18831–18846. [CrossRef]
76. Chen, C.; Ran, B.; Wang, Z.; Zhao, H.; Lan, M.; Chen, H.; Zhu, Y. Development of micropillar array electrodes for highly sensitive detection of biomarkers. *RSC Adv.* **2020**, *10*, 41110–41119. [CrossRef] [PubMed]
77. Le, H.; Kätelhön, E.; Compton, R.G. Characterising the nature of diffusion via a new indicator: Microcylinder and microring electrodes. *J. Electroanal. Chem.* **2019**, *855*, 113602. [CrossRef]
78. Aoki, K.; Honda, K.; Tokuda, K.; Matsuda, H. Voltammetry at microcylinder Electrodes. Part II. Chronoamperometry. *J. Electroanal. Chem. Interf. Electrochem.* **1985**, *186*, 79–86. [CrossRef]
79. Szabo, A.; Cope, D.K.; Tallman, D.E.; Kovach, P.M.; Wightman, R.M. Chronoamperometric current at Hemicylinder and band Microelectrodes: Theory and experiment. *J. Electroanal. Chem. Interf. Electrochem.* **1987**, *217*, 417–423. [CrossRef]
80. Henstridge, M.C.; Dickinson, E.J.; Compton, R.G. Mass transport to and within porous electrodes. Linear sweep voltammetry and the effects of pore size: The prediction of double peaks for a single electrode process. *Russ. J. Electrochem.* **2012**, *48*, 629–635. [CrossRef]
81. Menshykau, D.; Compton, R.G. The influence of electrode porosity on diffusional cyclic voltammetry. *Electroanalysis* **2008**, *20*, 2387–2394. [CrossRef]
82. Chevallier, F.G.; Jiang, L.; Jones, T.G.J.; Compton, R.G. Mathematical modelling and numerical simulation of cyclic voltammetry at an electrode covered with an insulating film containing cylindrical micropores. *J. Electroanal. Chem.* **2006**, *587*, 254–262. [CrossRef]
83. Menshykau, D.; Compton, R.G. Electrodes modified with electroinactive layers: Distinguishing through-film transport from Pinhole (Pore) diffusion. *Langmuir* **2009**, *25*, 2519–2529. [CrossRef]
84. Ward, K.R.; Compton, R.G. Quantifying the apparent 'catalytic' effect of porous electrode surfaces. *J. Electroanal. Chem.* **2014**, *724*, 43–47. [CrossRef]
85. Ward, K.R.; Xiong, L.; Lawrence, N.S.; Hartshorne, R.S.; Compton, R.G. Thin-layer vs. semi-infinite diffusion in cylindrical pores: A basis for delineating Fickian transport to identify nano-confinement effects in Voltammetry. *J. Electroanal. Chem.* **2013**, *702*, 15–24. [CrossRef]
86. Barnes, E.O.; Chen, X.; Li, P.; Compton, R.G. Voltammetry at Porous Electrodes: A theoretical study. *J. Electroanal. Chem.* **2014**, *720–721*, 92–100. [CrossRef]

87. Davies, T.J.; Moore, R.R.; Banks, C.E.; Compton, R.G. The cyclic voltammetric response of electrochemically heterogeneous surfaces. *J. Electroanal. Chem.* **2004**, *574*, 123–152. [CrossRef]
88. Davies, T.J.; Banks, C.E.; Compton, R.G. Voltammetry at spatially heterogeneous electrodes. *J. Solid State Electrochem.* **2005**, *9*, 797–808. [CrossRef]
89. Brunauer, S.; Emmett, P.H.; Teller, E. Adsorption of gases in multimolecular layers. *J. Am. Chem. Soc.* **1938**, *60*, 309–319. [CrossRef]
90. McAllister, M.J.; Li, J.-L.; Adamson, D.H.; Schniepp, H.C.; Abdala, A.A.; Liu, J.; Herrera-Alonso, M.; Milius, D.L.; Car, R.; Prud'homme, R.K.; et al. Single sheet functionalized graphene by oxidation and thermal expansion of graphite. *Chem. Mater.* **2007**, *19*, 4396–4404. [CrossRef]
91. Kariim, I.; Abdulkareem, A.S.; Abubakre, O.K. Development and characterization of MWCNTs from activated carbon as adsorbent for metronidazole and levofloxacin sorption from pharmaceutical wastewater: Kinetics, isotherms and thermodynamic studies. *Sci. Afr.* **2020**, *7*, e00242. [CrossRef]
92. Tarazona, P. Free-energy density functional for hard spheres. *Phys. Rev. A* **1985**, *31*, 2672–2679. [CrossRef]
93. Lykiema, J.; Sing, K.S.W.; Haber, J.; Kerker, M.; Wolfram, E.; Block, J.H.; Churaev, N.V.; Everett, D.H.; Hansen, R.S.; Haul, R.A.W.; et al. Reporting Physisorption Data for Gas/Solid Systems with Special Reference to the Determination of Surface Area and Porosity. *Pure Appl. Chem.* **1984**, *57*, 603–619. [CrossRef]
94. Elliott, J.R.; Compton, R.G. Local diffusion indicators: A new tool for analysis of Electrochemical Mass Transport. *J. Electroanal. Chem.* **2022**, *908*, 116114. [CrossRef]
95. Yang, M.; Batchelor-McAuley, C.; Kätelhön, E.; Compton, R.G. A new approach to characterising the porosity of particle modified electrodes: Potential step chronoamperometry and the diffusion indicator. *Appl. Mater. Today* **2021**, *25*, 101249. [CrossRef]
96. Henstridge, M.C.; Dickinson, E.J.F.; Aslanoglu, M.; Batchelor-McAuley, C.; Compton, R.G. Voltammetric selectivity conferred by the modification of electrodes using conductive porous layers or films: The oxidation of dopamine on glassy carbon electrodes modified with multiwalled carbon nanotubes. *Sens. Actuators B Chem.* **2010**, *145*, 417–427. [CrossRef]
97. Chan, H.T.H.; Kätelhön, E.; Compton, R.G. Voltammetry using multiple cycles: Porous electrodes. *J. Electroanal. Chem.* **2017**, *799*, 126–133. [CrossRef]
98. Lu, Y.; Li, X.; Compton, R.G. Oxygen reduction reaction at single entity multiwalled carbon nanotubes. *J. Phys. Chem. Lett.* **2022**, *13*, 3748–3753. [CrossRef]
99. Kaliyaraj Selva Kumar, A.; Compton, R.G. Single-entity "nano-catalysis": Carbon nanotubes and the VO_2^+/VO^{2+} Redox Reaction. *ACS Catal.* **2022**, *12*, 4754–4764. [CrossRef]
100. Lu, Y.; Li, X.; Kaliyaraj Selva Kumar, A.; Compton, R.G. Does nitrogen doping enhance the electrocatalysis of the oxygen reduction reaction by multiwalled carbon nanotubes? *ACS Catal.* **2022**, *12*, 8740–8745. [CrossRef]
101. Lu, Y.; Li, X.; Compton, R.G. Electro-oxidation of amino-functionalized multiwalled carbon nanotubes. *Chem. Sci.* **2022**, *13*, 1355–1366. [CrossRef]
102. Kaliyaraj Selva Kumar, A.; Compton, R.G. Understanding carbon nanotube voltammetry: Distinguishing adsorptive and thin layer effects via "single-entity" electrochemistry. *J. Phys. Chem. Lett.* **2022**, *13*, 5557–5562. [CrossRef]
103. Cao, Q.; Shao, Z.; Hensley, D.K.; Lavrik, N.V.; Venton, B.J. Influence of geometry on thin layer and diffusion processes at carbon electrodes. *Langmuir* **2021**, *37*, 2667–2676. [CrossRef]
104. Martinez-Garcia, A.J.; Hernandez-Tovar, J.V.; Gonzalez, J. Molecular electrocatalysis under finite diffusive mass transport conditions. An analytical approach. *ChemCatChem* **2024**, *16*, e202400267. [CrossRef]
105. Volgin, V.M.; Davydov, A.D. Mass-transfer problems in the electrochemical systems. *Russ. J. Electrochem.* **2021**, *48*, 565–569. [CrossRef]
106. Lasia, A. Impedance of Porous Electrodes. In *Electrochemical Impedance Spectroscopy and its Applications*; Springer: New York, NY, USA, 2014; pp. 203–250. [CrossRef]
107. Lasia, A. Modeling of Impedance of Porous Electrodes. In *Modeling and Numerical Simulations. Modern Aspects of Electrochemistry*; Springer: New York, NY, USA, 2008; Volume 43, pp. 67–137. [CrossRef]
108. Mei, B.-A.; Munteshari, O.; Lau, J.; Dunn, B.; Pilon, L. Physical interpretations of Nyquist plots for EDLC electrodes and devices. *J. Phys. Chem. C* **2018**, *122*, 194–206. [CrossRef]
109. Candy, J.-P.; Fouilloux, P.; Keddam, M.; Takenouti, H. The characterization of porous electrodes by impedance measurements. *Electrochim. Acta* **1981**, *26*, 1029–1034. [CrossRef]
110. Raistrick, I.D. Impedance studies of porous electrodes. *Electrochim. Acta* **1990**, *35*, 1579–1586. [CrossRef]
111. Lasia, A. Impedance of porous electrodes. *J. Electroanal. Chem.* **1995**, *397*, 27–33. [CrossRef]
112. Cooper, S.J.; Bertei, A.; Finegan, D.P.; Brandon, N.P. Simulated impedance of diffusion in porous media. *Electrochim. Acta* **2017**, *251*, 681–689. [CrossRef]
113. Tröltzsch, U.; Kanoun, O. Generalization of transmission line models for deriving the impedance of diffusion and porous media. *Electrochim. Acta* **2012**, *75*, 347–356. [CrossRef]
114. Song, H.-K.; Jung, Y.-H.; Lee, K.-H.; Dao, L.H. Electrochemical impedance spectroscopy of porous electrodes: The effect of pore size distribution. *Electrochim. Acta* **1999**, *44*, 3513–3519. [CrossRef]

115. Sharma, T. Impedance characterization of porous carbon electrodes with controlled pore size distribution: Experimental verification. Master's Thesis, University of British Columbia, Vancouver, CU, Canada, 2019. Available online: https://open.library.ubc.ca/collections/ubctheses/24/items/1.0376556 (accessed on 15 December 2024).
116. Devan, S.; Subramanian, V.R.; White, R.E. Analytical solution for the impedance of a porous electrode. *J. Electrochem. Soc.* **2004**, *151*, A905–A913. [CrossRef]
117. Bisquert, J. Influence of the boundaries in the impedance of porous film electrodes. *Phys. Chem. Chem. Phys.* **2000**, *2*, 4185–4192. [CrossRef]
118. Morasch, R.; Keilhofer, J.; Gasteiger, H.A.; Suthar, B. Methods–understanding porous electrode impedance and the implications for the impedance analysis of Li-ion battery electrodes. *J. Electrochem. Soc.* **2021**, *168*, 080519. [CrossRef]
119. Xu, L.; Xiao, Y.; Yu, Z.; Yang, Y.; Yan, C.; Huang, J. Revisiting the electrochemical impedance spectroscopy of porous electrodes in Li-ion batteries by employing reference electrode. *Angew. Chem. Int. Ed.* **2024**, *63*, e202406054. [CrossRef] [PubMed]
120. Gruet, D.; Delobel, B.; Sicsic, D.; Lucas, I.T.; Vivier, V. On the electrochemical impedance response of composite insertion electrodes–toward a better understanding of porous electrodes. *Electrochim. Acta* **2019**, *295*, 787–800. [CrossRef]
121. Huang, J.; Gao, Y.; Luo, J.; Wang, S.; Li, C.; Chen, S.; Zhang, J. Editors' choice—Review—Impedance response of porous electrodes: Theoretical framework, physical models and applications. *J. Electrochem. Soc.* **2020**, *167*, 166503. [CrossRef]
122. Armstrong, R.D.; Eyre, D.; Race, W.P.; Ince, A. An impedance method for the characterization of porous carbons. *J. Appl. Electrochem.* **1971**, *1*, 179–183. [CrossRef]
123. Yoo, H.D.; Jang, J.H.; Ryu, J.H.; Park, Y.; Oh, S.M. Impedance analysis of porous carbon electrodes to predict rate capability of electric double-layer capacitors. *J. Power Sources* **2014**, *267*, 411–420. [CrossRef]
124. Suss, M.E.; Baumann, T.F.; Worsley, M.A.; Rose, K.A.; Jaramillo, T.F.; Stadermann, M.; Santiago, J.G. Impedance-based study of capacitive porous carbon electrodes with hierarchical and bimodal porosity. *J. Power Sources* **2013**, *241*, 266–273. [CrossRef]
125. Abouelamaiem, D.I.; He, G.; Neville, T.P.; Patel, D.; Ji, S.; Wang, R.; Parkin, I.P.; Jorge, A.B.; Titirici, M.-M.; Shearing, P.R.; et al. Correlating electrochemical impedance with hierarchical structure for porous carbon-based supercapacitors using a truncated transmission line model. *Electrochim. Acta* **2018**, *284*, 597–608. [CrossRef]
126. De, P.; Halder, J.; Gowda, C.C.; Kansal, S.; Priya, S.; Anshu, S.; Chowdhury, A.; Mandal, D.; Biswas, S.; Dubey, B.K.; et al. Role of porosity and diffusion coefficient in porous electrode used in supercapacitors—Correlating theoretical and experimental studies. *Electrochem. Sci. Adv.* **2022**, *3*, e2100159. [CrossRef]
127. Yang, Z.; Wu, H. The electrochemical impedance measurements of carbon nanotubes. *Chem. Phys. Lett.* **2001**, *343*, 235–240. [CrossRef]
128. Leppänen, E.; Akhoundian, M.; Sainio, S.; Etula, J.; Pitkänen, O.; Laurila, T. Structure-property relationships in Carbon Electrochemistry. *Carbon* **2022**, *200*, 375–389. [CrossRef]
129. Dumitrescu, I.; Unwin, P.R.; Macpherson, J.V. Electrochemical impedance spectroscopy at single-walled carbon nanotube network ultramicroelectrodes. *Electrochem. Commun.* **2009**, *11*, 2081–2084. [CrossRef]
130. Mirfakhrai, T.; Oh, J.; Kozlov, M.; Fang, S.; Zhang, M.; Baughman, R.H.; Madden, J.D. Carbon Nanotube Yarn Actuators: An electrochemical impedance model. *J. Electrochem. Soc.* **2009**, *156*, K97–K103. [CrossRef]
131. Niessen, R.A.H.; de Jonge, J.; Notten, P.H.L. The Electrochemistry of Carbon Nanotubes: I. Aqueous Electrolyte. *J. Electrochem. Soc.* **2006**, *153*, A1484. [CrossRef]
132. Cheng, Q.; Tang, J.; Ma, J.; Zhang, H.; Shinya, N.; Qin, L.-C. Graphene and carbon nanotube composite electrodes for supercapacitors with ultra-high energy density. *Phys. Chem. Chem. Phys.* **2011**, *13*, 17615–17624. [CrossRef]
133. Cui, X.; Lv, R.; Sagar, R.U.; Liu, C.; Zhang, Z. Reduced graphene oxide/carbon nanotube hybrid film as high performance negative electrode for supercapacitor. *Electrochim. Acta* **2015**, *169*, 342–350. [CrossRef]
134. Tertiş, M.; Florea, A.; Feier, B.; Marian, I.O.; Silaghi-Dumitrescu, L.; Cristea, A.; Săndulescu, R.; Cristea, C. Electrochemical impedance studies on single and multi-walled carbon nanotubes-polymer nanocomposites for biosensors development. *J. Nanosci. Nanotechnol.* **2015**, *15*, 3385–3393. [CrossRef]
135. Galicia, M.; Li, X.; Castaneda, H. Interfacial characterization of single- and multi-walled CNT-doped chitosan scaffolds under two flow conditions. *J. Electrochem. Soc.* **2014**, *161*, H751–H761. [CrossRef]
136. Han, J.-H.; Hwang, S.; Hyub Kim, J. Electrochemical impedance spectroscopy analysis of plasma-treated, spray-coated single-walled carbon-nanotube film electrodes for chemical and electrochemical devices. *Microchem. J.* **2023**, *188*, 108446. [CrossRef]
137. Pajkossy, T.; Mészáros, G. Connection of CVS and impedance spectra of reversible redox systems, as used for the validation of a dynamic electrochemical impedance spectrum measurement system. *J. Solid State Electrochem.* **2020**, *24*, 2883–2889. [CrossRef]

biosensors

MDPI

Review

Recent Status on Lactate Monitoring in Sweat Using Biosensors: Can This Approach Be an Alternative to Blood Detection?

Leonardo Messina [1,*,†] and Maria Teresa Giardi [1,2]

[1] Microsis srl, Via degli Olmetti, 8a, 00060 Formello, Italy; mt.giardi@biosensor.it
[2] Institute of Cristallography, CNR Area of Research of Rome, 00010 Rome, Italy
* Correspondence: info@microsis.it
† Present address: Biosensor srl; Via degli Olmetti 44, 00060 Formello, Italy.

Abstract: Recent studies have shown that lactate is a molecule that plays an indispensable role in various physiological cellular processes, such as energy metabolism and signal transductions related to immune and inflammatory processes. For these reasons, interest in its detection using biosensors for non-invasive analyses of sweat during sports activity and in clinical reasons assessments has increased. In this minireview, an in-depth study was carried out on biosensors that exploited using electrochemical methods and innovative nanomaterials for lactate detection in sweat. This detection of lactate by biosensors in the sweat method seems to be feasible and highly desirable. From this commentary analysis, we can conclude that the correlation between lactate concentrations in sweat and blood is not yet clear, and studies are needed to clarify some key issues essential for the future application of this technology.

Keywords: lactate detection; wearable biosensors; blood/sweat correlation; electrochemical biosensors

1. Introduction and Literature Search Methods

Recently, the possibility of non-invasive and effective methodologies for biological studies has markedly increased; these methods play a prominent role in scientific research, especially in biosensor applications. Measuring sweat components is beneficial for disease diagnosis, health care, and exercise management, and it is expected to play a crucial role in various fields, including point-of-care testing [1,2]. Lactate is an important molecule present in the biological fluids of our bodies as the core of metabolic regulation in vivo. Our bodies produce it through glycolysis in anaerobic conditions. Its presence in abnormal concentrations can be a symptom of a pathology or a specific psychophysical state in an individual. The accumulation of lactate in the body is more dangerous than other molecular fuels, and an increase in serum lactate can lead to lactic acidosis [3]. Its determination is important to be able to evaluate the presence of disorders, such as shock or respiratory and heart attacks [4]. Lactate regulates the breakdown of fat and sugar for energy in athletes. However, it is also associated with muscular fatigue, and it is a major limitation in athletic performance. Lactate concentrations vary depending on the biological fluid; in blood, under normal conditions, its concentration ranges from 0.6 to 2.0 mM, while in sweat, it ranges from 4 to 25 mM [5]. Under pathological conditions, such as pressure-induced ischemia, or during intense physical activity, it can increase to 70 mM [6]. Lactate detection using sweat necessitates the development of a non-invasive and rapid methodology that differs from those already present in the market. In fact, this metabolite is usually studied via blood analysis using invasive methods. Scientific research in this field has focused on obtaining a biosensor capable of evaluating lactate concentrations in sweat in real time. Recently, due to the possibility of detecting lactate in sweat, many researchers have sought to analyze the main characteristics and features of this methodology through reviews. For example, Yang et al. [7] investigated the impact of lactate in sports, focusing on the description of electrochemical biosensors, all based exclusively on the use of enzymes. The remarkable study by Pérez and Orozco [8], on the other hand, focused on the anatomical description of the sweat glands, which are involved in lactate metabolism, and on the biochemical processes that occur during sweating. Shen et al. [9] described, in detail, the entire range of biosensors that can be applied for the detection of sweat lactate, discussing self-powered semiconductors, as well as optical and electrochemical biosensors. Compared to the above studies, this minireview exclusively focuses on a detailed analysis of recent research on electrochemical biosensors, based on both enzymatic and non-enzymatic processes, ending with the application of MIPs with an emphasis on new nanomaterials. For a complete analysis of biosensors used for lactate detection, a targeted bibliographic search was carried out, and the findings of this in-depth study reveal the extent to which the challenges in lactate detection are not addressed, as most studies are limited to the technical results obtained without discussing the possibility of applying electrochemical biosensors for the detection of this analyte in sweat. For this reason, after a careful analysis of articles, we performed a targeted screening of only the most recent and innovative articles on the use of nanomaterials for electrochemical biosensors. A flowchart of the literature search and identification of relevant articles for this minireview is depicted in Figure 1. Of the 21 studies included, 15 articles were categorized and reviewed (Table 1). For the analysis of the correlation between lactate concentration in sweat and blood in Section 3, the remaining six articles are considered. Furthermore, a market analysis was carried out to gain insight into the state of the art of these biosensor technologies (Table 2).

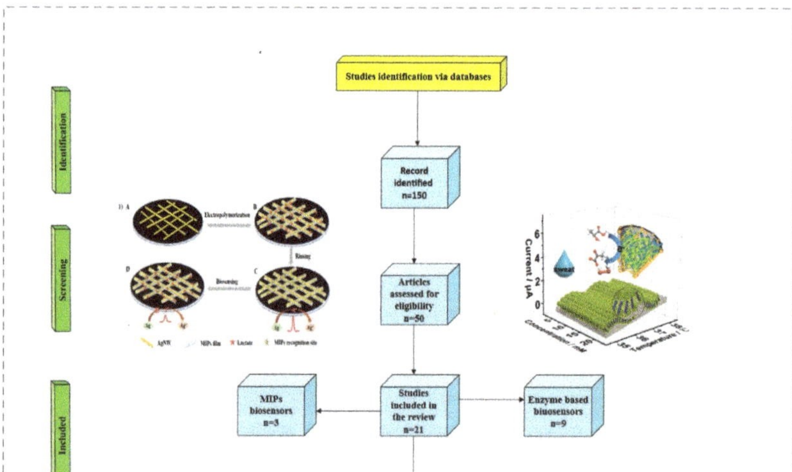

Figure 1. Flowchart of study selection for this minireview. Images reprinted with permission from [10–13]. Copyright © 2024 American Chemical Society; copyright 2024 Elsevier; copyright 2020 Elsevier; copyright © 1999–2024 John Wiley & Sons, Inc., or related companies.

2. Electrochemical Biosensors and Nanomaterials for Lactate Analysis in Sweat

Many types of biosensors can be used for lactate detection, including optical, spectroscopic, and electrochemical biosensors. Electrochemical devices may involve different techniques, such as the use of enzymes, which are capable of providing high selectivity and high sensitivity under external environmental conditions. Inspired by the metal architecture in nanomaterials, scientists have developed artificial biocatalysts or functionalized nanozymes that are more stable and specific and ensure effective electrochemical response with little dependence on detection conditions, such as pH and temperature. Nanozymes involve the formation of a non-enzymatic substrate that undergoes a redox reaction with the analyte of interest. For example, using nickel-based nanozymes, it is possible to develop NiOOH, a well-known and efficient electrocatalyst for lactate oxidation [14]. Another type of widely used electrochemical biosensor involves MIPs, which are polymers capable of selectively recognizing the analyte of interest and a superior alternative to the enzymatic method.

2.1. Enzymatic Electrochemical Biosensors

The electrochemical detection method using lactate concentration in sweat is perhaps the most widely used approach in the existing literature. In this method, enzymes such as lactate oxidase (LOx) and lactate dehydrogenase (LDH) are commonly employed, and electronic mediators, such as Prussian blue [15], are used for effective electrochemical response. Lactate dehydrogenase is a very stable enzyme and can efficiently convert lactate into pyruvate with nicotinamide adenine dinucleotide (NAD) as a coenzyme. Kumar et al. [16] demonstrated that it is possible to detect lactate in sweat even in the absence of a redox probe, with a wide concentration range (0.1–100 mM). In this method, lactate dehydrogenase is immobilized on the surface of a working electrode using an EDC–NHS crosslinker (1-ethyl-3-(3-dimethylaminopropyl carbodiimide and *N*-hydroxysuccinimide). Prior to this, the SPCE is pretreated with oxygen plasma, followed by ECX electrochemical oxidation using 10% acrylic acid with a DC voltage of 1.8 V. In the case of LOx, lactate is recognized and treated enzymatically through an oxidation reaction in the presence of oxygen, generating pyruvate and hydrogen peroxide as products. The peroxide then undergoes a redox reaction induced by the electronic mediator, which leads to the passage of electrons inside the electrochemical cell, thus generating a detectable response. It should be noted that in these enzymatic methods, difficulties arise in obtaining linearized data. This is due to variations in external conditions such as pH and temperature, parameters that modify enzymatic activity and compromise its integrity. Enzyme-based sensors are more susceptible to pH changes, and it is possible to identify the underlying mechanisms by studying lactate oxidase activity under various reaction conditions. Importantly, the study by Ping Xu et al. [17] underlines that lactate oxidase achieves its maximum enzymatic activity in a relatively narrow pH range, with its peak activity at pH = 7.7. When moving towards acidic or basic pH, a rapid reduction in enzymatic activity occurs, thus compromising lactate detection in biological matrices. pH variations during lactate detection are associated with the redox reaction between the substrate and the electronic mediator, which generates either protons, thus acidifying the reaction environment, or hydroxyl ions, which have the opposite effect. For lactate detection, electrochemical biosensors can be combined with other types of sensors that provide important information on physical conditions such as temperature and pH. This also allows one to better control the measurement conditions of the metabolite, thus achieving its precise detection in sweat. In the study by Wu et al. [10], an electrochemical sensor was used to simultaneously measure the lactate concentration and the temperature by analyzing the transmembrane current and the current directly generated via the redox reaction. The sensor was composed of gold nanoelectrode arrays fabricated on the nanoporous polycarbonate (PC) membrane by encapsulating lactate oxidase (LOx) in chitosan (CS) hydrogel (Figure 2). Flexible gold nanoporous electrodes not only increased the electrode area but also provided a nanoconfined space to accelerate the catalytic reaction of LOx and control the substrate's concentration on the surface of LOx to reduce substrate inhibition.

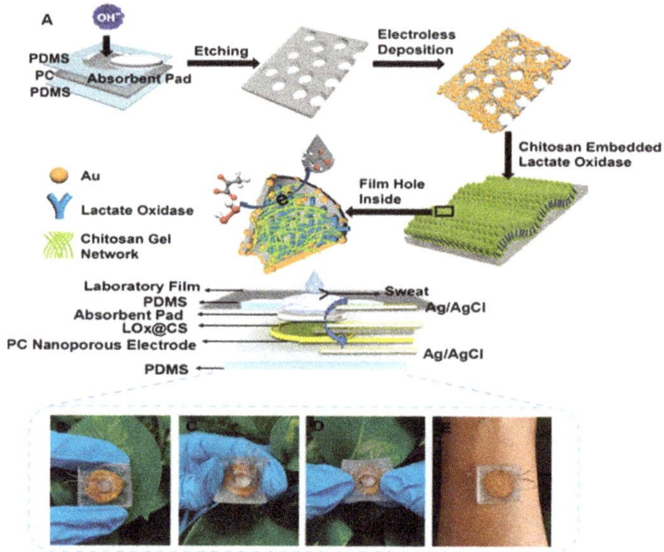

Figure 2. Schematic illustration of the preparation process of the stretchable LO_x@CS PC sensor: (**A**) preparation of the LO_x@CS PC sensor; (**B–E**) photograph of the prepared sensors for lactate detection in sweat. Reprinted with permission from [10]. Copyright © 2024 American Chemical Society.

Prussian blue is not the only electronic mediator used in this field. Khan and Andreescu investigated a new material, called MXCeO2 (M represents an "early" transition metal and X represents carbon or/and nitrogen), which is obtained from the combination of cerium oxide with Mxenes and serves as a catalytic amplifier for electrochemical biosensors. They demonstrated that this substance undergoes an efficient catalytic action towards hydrogen peroxide reduction and is selective for oxidases (e.g., LOx) [18]. In another study, Xuan Weng [19] used a combination of nanozymes with peroxidase-like properties. The combination of cerium oxide with molybdenum sulfide led to an increase in active catalytic sites, resulting in enhanced catalytic efficiency. Following the redox reaction with hydrogen peroxide, a significantly intense electrochemical signal was obtained, which allowed for the accurate identification of lactate concentrations in sweat. In some methods, an electronic mediator may not be used. For instance, for the continuous monitoring of athletic performance, Shuling Deng used a working electrode made of oxidized graphene nanosheets and obtained a response with high sensitivity and a wide detection range, between 0.1 and 80 mM [20]. Enzyme immobilization using a working electrode is also very important since poor immobilization can lead to the inaccurate detection of the analyte. Thus, chemical compounds that enable optimal enzyme immobilization are often used. In Liang Tian's study, lactate oxidase and peroxidase were anchored to the electrode via $NiCo_2O_4$ microspheres, coated with carbon nanotubes [21]. The microspheres were placed on the working electrode, thus improving its conductivity. Ferrocene–methanol has also been used as a redox mediator and is capable of lowering the reaction potential due to the effect of HRP. In this approach, the reaction byproducts are minimized due to the high redox potential of the process. A very recent study by Qingrong He et al. [22] also highlighted the possibility of achieving enzyme self-immobilization on the surface of the working electrode without the use of external crosslinkers. In this interesting study, the structure of lactate oxidase was modified through the modification of phenazine. Through this modification, they obtained accurate results for lactate detection in sweat, as phenazine has excellent electron transport properties, which reduce the detection potential for lactate. Enzyme immobilization is key in the successful development of enzyme-based biosensors. It must be highlighted that LOX is an unstable enzyme, and thus an appropriate immobilization technique is crucial for the fabrication of a robust biosensor. Several immobilization techniques have been investigated, among which a new technique using electrospray deposition at room temperature (ESD) (Figure 3) resulted in biosensors with superior performance. These devices have a high storage capability of up to 90 days and are not affected by low temperatures; moreover, they can be reused for up to 24 measurements on a new as well as a three-month-old electrode [23].

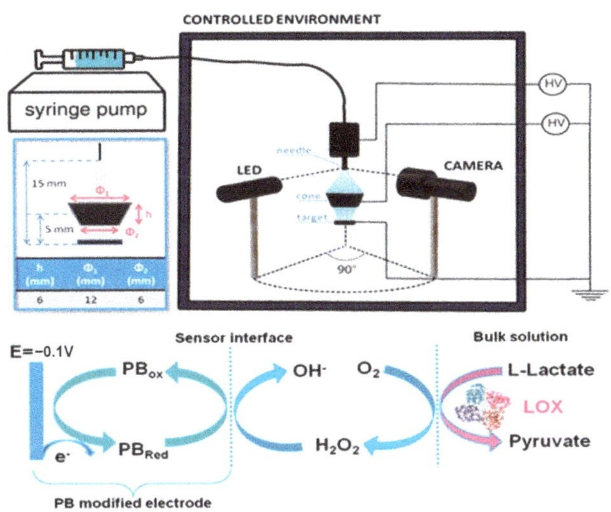

Figure 3. ESD setup (top right panel) and enlarged scheme of the deposition region (top left panel) where the dimensions and the relative distance between the needle of immobilization, the cone target, and the SPE are depicted. Schematic of the redox reactions that occur on the electrode surface during amperometric measurements (bottom panel). Reprinted with permission from [23]. Copyright 2024 Royal Society of Chemistry.

2.2. *Enzyme-Free Electrochemical Biosensors—Semiconductors*

In these biosensors, enzymes are not used to detect lactate. Instead, this method involves the use of certain chemical–physical characteristics of inorganic compounds to "capture" the analyte through specific interactions. These biosensors have superior sensitivity for the detection of electrochemical signals; however, their poor selectivity remains the main unresolved challenge. The multiple analytes present in sweat can interfere with measurement; therefore, the detection methods using these biosensors are not as precise as those based on enzymes. Ling-Yu Chang [24] studied a typical non-enzymatic biosensor using a nickel–cobalt layered double hydroxide (LDH) capable of interacting with the lactate at its numerous electrochemical sites, thus inducing its oxidation. Furthermore, the biosensor involves the use of N-doped graphene quantum dots, which lower the resistive phenomena on the working electrode. The authors also demonstrated that the selective detection of lactate in a matrix such as sweat was still possible despite numerous interferences. To evaluate the selectivity of NiCo-based layered double hydroxides with N-doped graphene quantum dots (NGQD/m-NiCo LDH-based SPCE), an interference test was performed with a series of chemical compounds present in human sweat, such as glucose, ascorbic acid, and uric acid. The interference test results reveal a distinct response with lactate, whereas only minor fluctuations were observed after the addition of uric acid, ascorbic acid, and glucose. These results confirmed the superior selectivity of NGQD/m-NiCo LDH-based SPCE for lactate detection. Given its excellent electrochemical properties, nickel is widely used in non-enzymatic biosensors. Gao et al. employed nickel oxide nanoparticles placed on a graphene support. Nickel is present in its two oxidation states +2 and +3, and in $NiOOH/Ni(OH)_2$, there is a crystalline core of metallic nickel inside a shell of NiO, with a crystalline/amorphous surface. Graphene was used to enhance the electrochemical sensitivity of the generated biosensor, which also yielded good results regarding linearity in lactate detection [14]. This device facilitates the accurate identification of electrochemical responses, with a wide detection range. Pei Li [12] developed a working electrode via the glancing angle deposition (GLAD) of nickel oxide on a gold surface. GLAD is a physical vapor deposition process that involves depositing a material onto a substrate oriented at oblique angles relative to a vapor source, typically accompanied by substrate rotation. This results in the formation of unique nanostructured thin films with columnar architectures, which increase the number of active sites available for catalysis, leading to a high surface-to-volume ratio. The increased surface area allows for a greater interaction between the analyte and catalyst, in which Ni in the oxidated state undergoes a redox reaction with lactate, resulting in the oxidation of lactate to pyruvate, while Ni is reduced to its +2 form (Figure 4), thus enhancing the sensitivity and lowering the detection limit of the sensor. Oxidized nickel holds promise as an alternative to lactate oxidase, yielding excellent results. It stands out among the different nanozymes (Co, Zn, and Cu) due its low toxicity, high catalytic activity, and economic feasibility.

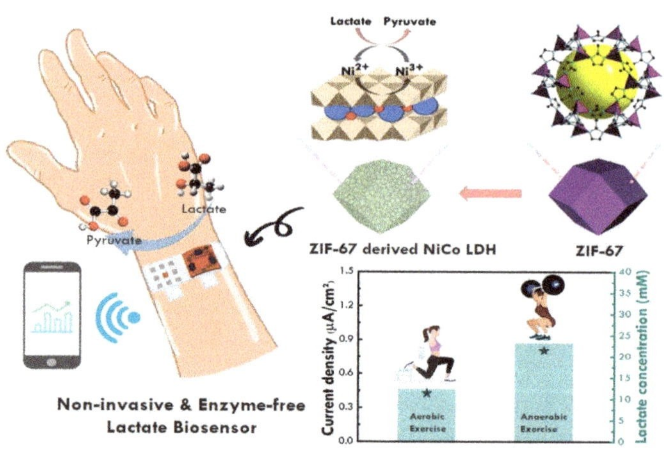

Figure 4. Schematic illustration of enzyme-free electrochemical biosensors, based on Ni derivatives. Ni in the oxidated state undergoes a redox reaction with lactate, resulting in its oxidation into pyruvate and the reduction of Ni to its +2 form. Reprinted with permission from [25]. Copyright 2024 Elsevier.

2.3. *MIPs Electrochemical Biosensors*

To date, the use of molecularly imprinted polymers seems to be a valid alternative to the use of the enzymatic approach. It is characterized by the direct electropolymerization of specific compounds onto the working electrode, together with a template molecule that must then be detected. Following the removal of the template (for example, through an overoxidation reaction), the polymer layer comprises highly specific areas for the target molecule, which are captured and subsequently detected through electrochemical reactions (Figure 5). This type of biosensor has very high selectivity and detection sensitivity; moreover, due to the absence of enzymes, the system is not affected by changes in pH and temperature. Additionally, electronic mediators can be used to increase their sensitivity in detecting electrochemical signals. In the study by Grace Dykstra [26], a working electrode was prepared using a silk-screened carbon electrode, with a deposited layer of Prussian blue which acted as the electronic transfer mediator. Subsequently, an MIP composed of aminophenylboronic acid and pyrrole molecules was developed via electropolymerization. In another recent study using MIPs with Prussian blue (PB) as an electronic mediator, lactate was used as the template molecule, and pyrrole (Py) served as the functional monomer. PB was embedded into the MIP as a built-in redox probe, thus eliminating the need for an additional probe and facilitating the simultaneous quantification of lactate concentration. Moreover, the MIP-doped platinum nanoparticles (PtNPs) enhanced the electron transfer ability, further improving the sensitivity of the sensors [27]. As demonstrated by Yangyang Chen [28], these biosensors can be easily integrated with the textile fabric, making the device suitable for wearing. They built a device with excellent electrochemical properties, which was implanted on a textile surface and could undergo many detection cycles.

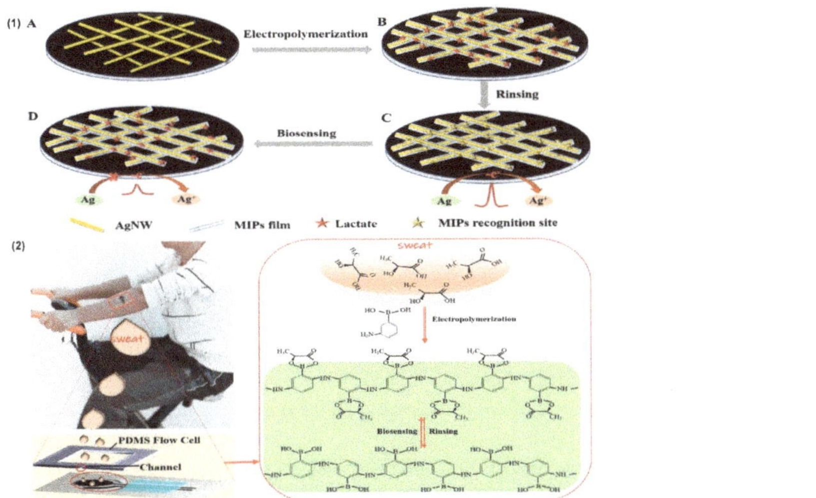

Figure 5. (**1**) The fabrication process of the MIP–AgNWs electrochemical biosensors for the epidermal monitoring of lactate; (**2**) the application of a screen-printed three-electrode biosensor chip on a male volunteer's arm and the principles of MIP formation and biosensing. Reprinted with permission from [11]. Copyright 2020 Elsevier.

3. Comparative Analysis of Non-Invasive Electrochemical Biosensors for the Detection of Lactate in Sweat

From a careful analysis of the biosensors included in this review, it can be concluded that enzyme-free devices have a very wide range of linear detection, due to the lack of any direct dependence on variations in pH and temperature during the analysis of lactate concentrations in sweat.

Table 1. Comparative analysis of non-invasive electrochemical biosensors for lactate detection in sweat.

Working Electrodes	Measurement Techniques	Typology	Sensitivity	Linearity	Detection Limits	Ref.
Nano–Au/CS PC	Amperometry	LOx enzyme	0.8 μA/mMcm^2	0.01–35 mM	0.144 μM	[10]
Au/GLAD NiO	Amperometry	Enzyme-free	80 μA/mMcm^2	1–65 mM	16 μM	[12]
NiOOH/Ni(OH)$_2$	Amperometry	Enzyme-free	80 μA/mMcm^2	0.02–53 mM	0.13 μM	[14]
SPCE/PB	Amperometry	LOx enzyme	−0.01 μA/mMcm^2	1–20 mM	0.2 mM	[15]
SPCE	EIS	LDH enzyme	N/A	0.100 mM	0.1 mM	[16]
SPCE/MXCeO$_2$	Amperometry	LOx enzyme	N/A	0.01–12 mM	0.4 mM	[18]
CeO$_2$/MoS$_2$AuNPs	Amperometry	LOx enzyme	0.027 μA/mMcm^2	0–100 mM	N/A	[19]
GO	EIS	LOx enzyme	N/A	0.1–80 mM	N/A	[20]
NiCo$_2$O$_4$/SWCNTs	Amperometry	LOx enzyme	N/A	0–30 mM	39.9 μM	[21]
SPGE	Amperometry	LOx enzyme	25.58 μA/mMcm^2	0–5 mM	0.135 mM	[22]
NGQD/NiCo LDH	Amperometry	Enzyme-free	62.63 μA/mMcm^2	0–15 mM	0.3 mM	[24]
SPCE/PB/MIPs	DPV/EIS	MIPs	N/A	1–35 mM	0.62 mM	[26]
SPCE/PtNPs/Pt@MI	Amperometry	MIPs	N/A	0–1.5 mM	1.1 mM	[27]
MCF/PBPPY	Amperometry	MIPs	0.11 μA/mMcm^2	0.01–25 mM	N/A	[28]
SPCE	Amperometry	LOx enzyme	0.9 μA/mMcm^2	0.03–0.5 mM	N/A	[29]

The only enzymatic biosensor that is not affected by these variations is the device proposed by Xing Xuan et al. [15], which operates via a lipophilic membrane placed externally on the working electrode. This not only reduces the flow of sweat reaching the biosensor but also counteracts diffusion phenomena within the fluid, thus reducing the loss of linearity caused by pH changes. S. Deng obtained the best results in terms of detection range, but their study was carried out on simulated sweat and therefore does not fully consider real-world conditions, such as, for instance, the presence of interference [20]. This analysis also reveals that it is possible to replace enzymatic biosensors with MIPs, which exhibit superior selectivity and good linearity in the obtained data [26].

4. Blood and Sweat Lactate Analysis Using Electrochemical Biosensors

Sweating involves a very complex mechanism [13] (Figure 6), a phenomenon that can be identified through large lactate oscillations produced by the sweat glands. In fact, blood filtering is not involved in the passage of lactate into sweat, and its production is not linked to biochemical signals; rather, it is more related to the psychological, nervous, mechanical, and hormonal signals sent directly to the glands themselves [30].

Firstly, the correlation between the concentrations of lactate in blood and sweat is not yet clear. This is a major challenge since the data obtained from the detection of lactate in sweat could be significant for the analysis of the physiological state of an individual. A more in-depth review of the literature revealed that lactate detection in sweat is complex. Many factors should be considered in the development of a reliable biosensor. The direct generation of the metabolite by the sweat glands leads to large variations in the lactate concentration in sweat, which does not seem to be directly related to that of the blood [30]. The extent to which this phenomenon is dependent on the metabolism of the glands and how it can be predicted using mathematical algorithms remain unexplored. In a previous study, Garcia et al. developed a methodology for the accurate prediction of blood lactate values using a biosensor, which was able to reliably and accurately predict absolute blood lactate values, with only 0.3 mM of accumulated error compared to portable blood lactate meters, the current gold standard for sports clinicians [31].

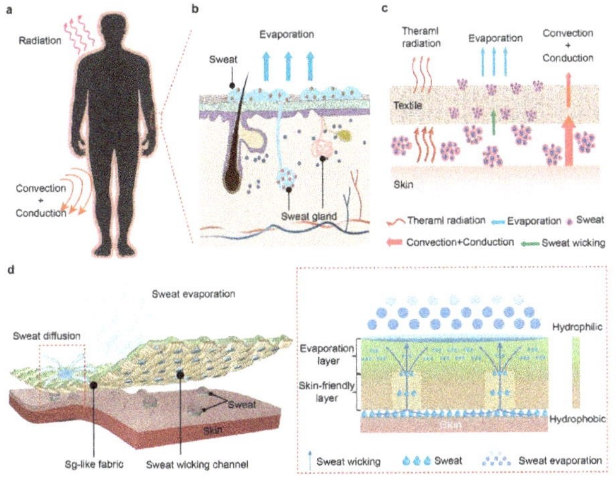

Figure 6. Schematic illustration of the complex mechanism of sweating from human skin with sudoriferous gland participation. (**a**) Illustration of heat transfer between the human body and the surrounding environment. (**b**) Sweat evaporation via sweat glands. (**c**) Schematic illustrating the heat transfer process from the human body covered with a textile to the surrounding environment. (**d**) Schematic illustration of the Sg-like fabric. Reprinted with permission from [13]. Copyright © 1999–2024 John Wiley & Sons, Inc., or related companies.

Through targeted bibliographic studies, we explored this research gap in depth. Of the six articles analyzed, only two highlighted a correlation between lactate levels in blood and sweat. However, in analyzing the values obtained in these studies, we found significant differences. Sakharov [32] highlighted that lactate concentration is divided in capillary blood, venous blood, and sweat. The final correlation between lactate levels in venous blood and sweat is determined via equation C (lactate in venous blood) as 0.73 × C (lactate in sweat), which is quite different from the relation experimentally determined by Karpova [33] for the working muscle area and latent muscle area (angular coefficient = 5.7 and 2.6, respectively). As such, even if the correlation is mild or good ($R^2 > 0.95$), there is no agreement between the different findings in the literature. This, therefore, suggests a degree of randomness in the values found when analyzing the two biological fluids, a hypothesis supported by the remaining four articles, in which the correlation is quite low, $R^2 < 0.9$. This highlights that the correlation between the different concentrations of lactate in sweat is undetermined. In particular, Klous [30] and Maeda [34] revealed that lactate concentration in sweat does not change with variations in its concentration in the blood and vice versa (Figure 7).

Furthermore, studies should also focus on the dilution of lactate in sweat due to an increase in sweating, and the effect of variations in pH and temperature on this phenomenon, as well as understanding how it influences the detection of the analyte. In fact, it has been noted that all of these processes lead to a loss of linearity in biosensor responses used for lactate detection. Kudo sought to solve this problem by designing an "atypical" biosensor. To prevent measurement interference due to the reabsorption or evaporation of sweat, bacterial contamination, and pH variation, a wristwatch biosensor using PBS as a continuous carrier for sweat collection was designed [29]. Additionally, appropriate mathematical adjustments were made to correct measurements in real time, such that the integrity and reliability of the data could be preserved.

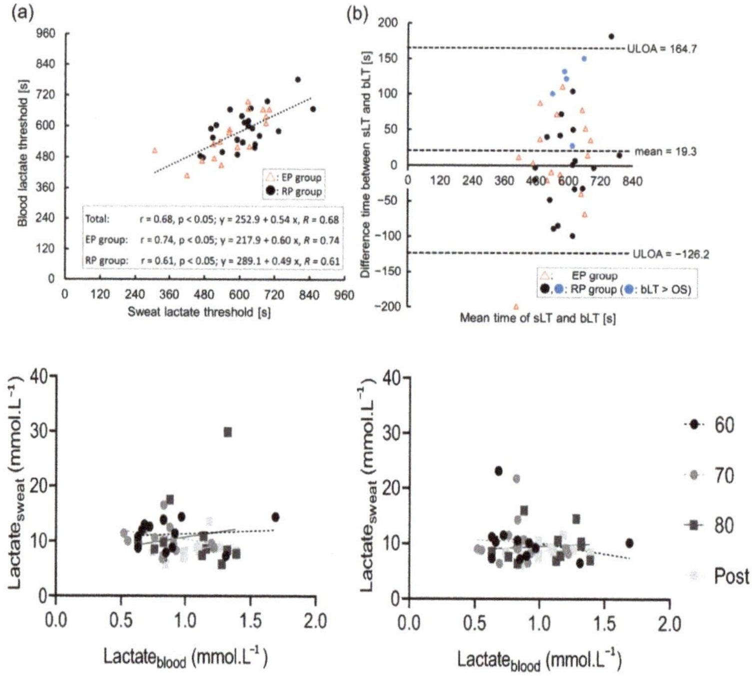

Figure 7. The relationship between the sweat and blood lactate thresholds (sLT and bLT): (**a**) scatter plot and approximation line between the sLT and bLT; (**b**) validity testing using Bland–Altman plots. The figure describes the correlations between lactate concentrations in blood plasma and upper arm sweat (left panels) and upper back sweat (right panels) during incremental cycling in the heat (33 °C, 65% relative humidity RH; n = 12). Black circles with a dashed regression line represent 60% maximum heart rate (HRmax (60)); grey circles with a dashed line represent 70% maximum heart rate (HRmax (70)); dark grey squares with a solid line represent 80% maximum heart rate (HRmax (80)); and light grey squares with a solid line represent post-exercise (post). Reprinted with permission from [30,34]. Springer Nature 2020 copyright. MDPI 2023 copyright.

Moreover, to achieve improvements in biosensor responses, it is necessary to focus directly on the device parameters, among which the loss of linearity in biosensor responses considering the influences of pH, temperature, and the flow of sweat reaching the working electrode are the primary factors. Using a membrane that limits diffusion, placed directly above the working electrode, results in excellent biosensor responses at the expense of a reduction in sensitivity [15]. Nevertheless, this paves the path towards the application of enzymatic electrochemical biosensors. Furthermore, it is possible to integrate lactate oxidase with an inorganic compound, thus improving the immobilization of the enzyme on the electrode and enhancing the electrochemical response [21]. Another solution to reducing or eliminating the dependence of detection results on the pH and temperature conditions of the system could be to use different biosensors. The use of MIPs may avoid this problem, as they simply involve interactions between the polymer layer and the lactate itself. However, even in MIP applications, whether the generated interactions are affected by these variations remains unknown.

5. Conclusions

In future research, a more in-depth study of the physiological mechanism determining the presence of lactate in sweat is needed.

As can be seen from Table 2, there are still no biosensors in the market that can be used for lactate detection in sweat. The only exception is the IST device, but this is currently only capable of detecting levels in simulated sweat and is still under development. It uses a buffer to guarantee pH neutrality during the measurement; this achieves strong linearity in the obtained data and therefore success in lactate analysis.

Table 2. Current progress in the development of non-invasive devices for lactate sensing.

Company	Sample Type	Stage of Development	Measuring Range	Sample Volume	Testing Time	Operating Temperature
Zimmer & Peacock	Sweat	Research	N/A	N/A	N/A	N/A
Nova Niomedical	Blood	Market	0.5–25 mM	0.5 μL	10 s	10–45 °C
PKvitality	Sweat	Research	N/A	N/A	N/A	N/A
IST	Sweat	Market	0.05–25 mM	N/A	<90 s	15–42 °C
Cosmed	Blood	Market	0.5–25 mM	0.3 μL	15 s	N/A
APExBIO	Blood	Market	0.7–22.2 mM	3 μL	45 s	10–40 °C

Currently, lactate detection in sweat may not yield the same results as those obtained with simulated sweat. The rest of the devices available on the market are non-wearable, invasive, and involve detection through blood analysis (Figure 8). The detection parameters of these devices are all very similar, and the lactate detection range reaches a maximum of 25 mM.

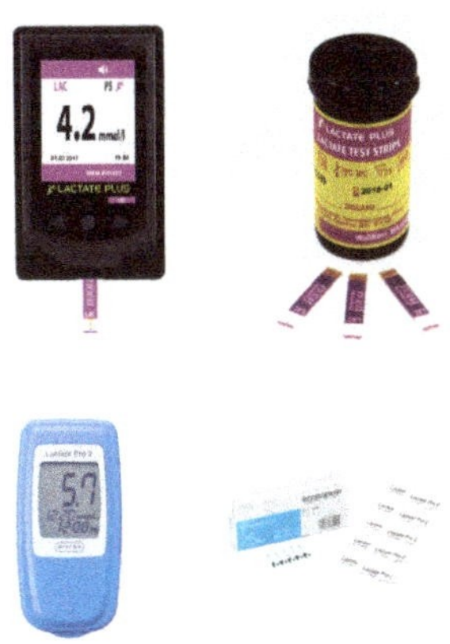

Figure 8. Examples of devices present on the market for lactate sensing. On the top, Lactate Plus is presented with lactate test strips, made by Nova Biomedical; below is Lactate Pro 2 with its strips made by Arkray. Reprinted with permission from Nova Biomedical Italy srl and from Arkray, Inc. © 2023 Nova Biomedical V1.128. Copyright© 2024 Arkray, Inc.

In conclusion, the development of biosensors with sensitivity and linear detection range compatible with sweat lactate concentrations seems more feasible than ever. The main problem remains the necessity of finding a biomedical correlation between lactate levels present in the blood with those present in sweat. For this purpose, a new European-supported project, called H2TRAIN, has been established to seek a definitive solution to the problems listed in this minireview. H2TRAIN employs a new cutting-edge biosensor for the detection of various parameters in sweat, including lactate, cortisol, and C-reactive protein. These detectors are based on graphene oxide, a 2D material that is functionalized as a detection biomarker, complemented by edge-cloud AI. The edge-cloud AI continuum is a novel concept, and this project will endow biosensors with smart properties for monitoring the levels of different compounds in sweat [35]. The project encompasses the entire supply chain, from R&D and the manufacturing of microsystems to electronics assembly, as well as physical integration in wearables and textiles and the integration of smart systems (implementation of software and AI algorithms). The key issues raised by this technology will be addressed. The design, testing, and validation of the proposed solutions will be carried out in medical applications, as well as from the perspective of sports sciences, with a substantial contribution from the social sciences and humanities. Guidance offered by end-user cohorts and external organizations related to different stages of the development cycle will help in advancing future applications of the technology [36]. The problem of finding a correlation between lactate levels present in the blood with those present in sweat could be addressed, first and foremost, through artificial intelligence algorithms [personal communication by Prof. Juan A. Montiel-Nelson, University of Las Palmas, Spain]. In fact, data on lactate concentrations, collected through measurements using blood as a matrix, could be analyzed and compared with data related to the oscillations in lactate concentrations in sweat. Through in-depth analysis and post-measurement correction, it would perhaps be possible to fill the research gap that continues to exist regarding the correlation between lactate concentrations in the two biological matrices. Finally, lactate concentrations in sweat could be used to depict patients' pathological conditions.

Author Contributions: Conceptualization, L.M. and M.T.G.; writing—original draft preparation, L.M. and M.T.G.; writing—review and editing, L.M. and M.T.G.; supervision, M.T.G. All authors have read and agreed to the published version of the manuscript.

Funding: This project has been co-funded by the Italian Ministry MIMIT and the European Union's Horizon Europe Research and Innovation Programme [HORIZON-KDT-JU-2023-1-IA-Topic-1]—[GA 101140052]. Site Web: https://www.h2train-project.eu/ (accessed on 23 December 2024).

Institutional Review Board Statement: Not applicable.

Informed Consent Statement: Not applicable.

Data Availability Statement: As this is a review article, no data were generated.

Conflicts of Interest: Microsis srl is a company that may have a commercial interest in the results of this research and technology. The potential individual conflict of interest played no role in the study design; in the collection, analysis and interpretation of data; in the writing of the manuscript, and in the decision to publish the results. The authors have no other relevant affiliations or financial involvement with any organization or entity with a financial interest in or financial conflict with the subject matter or materials discussed in the manuscript apart from those disclosed.

References

1. Sonner, Z.; Wilder, E.; Heikenfeld, J.; Kasting, G.; Beyette, F.; Swaile, D.; Sherman, F.; Joyce, J.; Hagen, J.; Kelley-Loughnane, N.; et al. The microfluidics of the eccrine sweat gland, including biomarker partitioning, transport, and biosensing implications. *Biomicrofluidics* **2015**, *9*, 031301. [CrossRef] [PubMed]
2. Gubala, V.; Harris, L.F.; Ricco, A.J.; Tan, M.X.; Williams, D.E. Point of Care Diagnostics: Status and Future. *Anal. Chem.* **2011**, *84*, 487–515. [CrossRef] [PubMed]
3. Robergs, R.A.; Ghiasvand, F.; Parker, D. Biochemistry of exercise-induced metabolic acidosis. *Am. J. Physiol. Integr. Comp. Physiol.* **2004**, *287*, R502–R516. [CrossRef]
4. Zymliński, R.; Biegus, J.; Sokolski, M.; Siwołowski, P.; Nawrocka-Millward, S.; Todd, J.; Jankowska, E.A.; Banasiak, W.; Cotter, G.; Cleland, J.G.; et al. Increased blood lactate is prevalent and identifies poor prognosis in patients with acute heart failure without overt peripheral hypoperfusion. *Eur. J. Heart Fail.* **2018**, *20*, 1011–1018. [CrossRef]
5. Green, J.M.; Pritchett, R.C.; Crews, T.R.; McLester, J.R.; Tucker, D.C. Sweat lactate response between males with high and low aerobic fitness. *Eur. J. Appl. Physiol.* **2003**, *91*, 1–6. [CrossRef]

6. García, E.T. Development of a Flexible Biosensor for the Monitoring of Lactate in Human Sweat for Its Medical Use in Pressure Ischemia. 2014. Available online: http://dspace.lib.cranfield.ac.uk/handle/1826/9254 (accessed on 26 July 2024).
7. Yang, G.; Hong, J.; Park, S.-B. Wearable device for continuous sweat lactate monitoring in sports: A narrative review. *Front. Physiol.* **2024**, *15*, 1376801. [CrossRef]
8. Pérez, D.; Orozco, J. Wearable electrochemical biosensors to measure biomarkers with complex blood-to-sweat partition such as proteins and hormones. *Microchim. Acta* **2022**, *189*, 127. [CrossRef]
9. Shen, Y.; Liu, C.; He, H.; Zhang, M.; Wang, H.; Ji, K.; Wei, L.; Mao, X.; Sun, R.; Zhou, F. Recent Advances in Wearable Biosensors for Non-Invasive Detection of Human Lactate. *Biosensors* **2022**, *12*, 1164. [CrossRef]
10. Wu, Z.-Q.; Cao, X.-Q.; Hua, Y.; Yu, C.-M. A Bifunctional Wearable Sensor Based on a Nanoporous Membrane for Simultaneous Detection of Sweat Lactate and Temperature. *Anal. Chem.* **2024**, *96*, 3087–3095. [CrossRef]
11. Zhang, Q.; Jiang, D.; Xu, C.; Ge, Y.; Liu, X.; Wei, Q.; Huang, L.; Ren, X.; Wang, C.; Wang, Y. Wearable electrochemical biosensor based on molecularly imprinted Ag nanowires for noninvasive monitoring lactate in human sweat. *Sens. Actuators B Chem.* **2020**, *320*, 128325. [CrossRef]
12. Li, P.; Kalambate, P.K.; Harris, K.D.; Jemere, A.B.; Tang, X. Robust and flexible electrochemical lactate sensors for sweat analysis based on nanozyme-enhanced electrode. *Biosens. Bioelectron. X* **2024**, *17*, 100455. [CrossRef]
13. Zhang, Q.; Wang, M.; Chen, T.; Chen, Z.; Liu, D.; Zhang, Z.; Zhuo, L.; Wang, Y.; Xiao, X.; Zhu, B.; et al. Sweat Gland-Like Fabric for Personal Thermal-Wet Comfort Management. *Adv. Funct. Mater.* **2024**, 2409807. [CrossRef]
14. Gao, W.; Guan, X.; Wang, J.; Heinig, N.F.; Thomas, J.P.; Zhang, L.; Ding, K.; Leung, K.T. Supported NiO_x Nanocatalysts on Graphene for Nonenzymatic Lactate Sensing. *ACS Appl. Nano Mater.* **2024**, *7*, 7162–7171. [CrossRef]
15. Xuan, X.; Chen, C.; Molinero-Fernandez, A.; Ekelund, E.; Cardinale, D.; Swarén, M.; Wedholm, L.; Cuartero, M.; Crespo, G.A. Fully Integrated Wearable Device for Continuous Sweat Lactate Monitoring in Sports. *ACS Sens.* **2023**, *8*, 2401–2409. [CrossRef]
16. Kumar, N.; Lin, Y.-J.; Huang, Y.-C.; Liao, Y.-T.; Lin, S.-P. Detection of lactate in human sweat via surface-modified, screen-printed carbon electrodes. *Talanta* **2023**, *265*, 124888. [CrossRef]
17. Xu, P.; Yano, T.; Yamamoto, K.; Suzuki, H.; Kumagai, H. Characterization of a lactate oxidase from a strain of gram negative bacterium from soil. *Appl. Biochem. Biotechnol.* **1996**, *56*, 277–288. [CrossRef]
18. Khan, R.; Andreescu, S. Catalytic $MXCeO_2$ for enzyme based electrochemical biosensors: Fabrication, characterization and application towards a wearable sweat biosensor. *Biosens. Bioelectron.* **2023**, *248*, 115975. [CrossRef]
19. Weng, X.; Li, M.; Chen, L.; Peng, B.; Jiang, H. A wearable nanozyme–enzyme electrochemical biosensor for sweat lactate monitoring. *Talanta* **2024**, *279*, 126675. [CrossRef]
20. Deng, S. Application of graphene oxide nanosheet lactate biosensors in continuous assessment of athlete fitness. *Alex. Eng. J.* **2024**, *88*, 31–35. [CrossRef]
21. Tian, L.; Cai, L.; Ding, Z.; Zhou, Y.; Zhang, Y.; Liu, Q.; Ge, X.; Yu, C. Sweat lactate biosensor based on lactate oxidase immobilized with flower-like $NiCo_2O_4$ and carbon nanotubes. *Microchem. J.* **2024**, *200*, 110417. [CrossRef]
22. He, Q.; Wang, C.; Jain, R.; Byrnes, J.; Farquhar, E.R.; Reed, E.; Berezovsky, E.; Chance, M.R.; Lodowski, D.; An, R. An engineered lactate oxidase based electrochemical sensor for continuous detection of biomarker lactic acid in human sweat and serum. *Heliyon* **2024**, *10*, e34301. [CrossRef] [PubMed]
23. Castrovilli, M.C.; Scognamiglio, V.; Tempesta, E.; Chiarinelli, J.; Parracino, M.; Frisulli, V.; Giardi, M.T.; Avaldi, L.; Rossi, D.; Cartoni, A. Improved reuse and storage performances at room temperature of a new environmentally friendly lactate oxidase biosensor prepared by ambient electrospray immobilization. *Green Chem.* **2023**, *25*, 5257–5266. [CrossRef]
24. Chang, L.-Y.; Rinawati, M.; Guo, Y.-T.; Lin, Y.-C.; Chang, C.-Y.; Su, W.-N.; Mizuguchi, H.; Huang, W.-H.; Chen, J.-L.; Yeh, M.-H. Nitrogen-Doped Graphene Quantum Dots Incorporated into MOF-Derived NiCo Layered Double Hydroxides for Nonenzymatic Lactate Detection in Noninvasive Biosensors. *ACS Appl. Nano Mater.* **2024**, *7*, 14431–14442. [CrossRef]
25. Wang, Y.-X.; Tsao, P.-K.; Rinawati, M.; Chen, K.-J.; Chen, K.-Y.; Chang, C.; Yeh, M.-H. Designing ZIF-67 derived NiCo layered double hydroxides with 3D hierarchical structure for Enzyme-free electrochemical lactate monitoring in human sweat. *Chem. Eng. J.* **2021**, *427*, 131687. [CrossRef]
26. Dykstra, G.; Chapa, I.; Liu, Y. Reagent-Free Lactate Detection Using Prussian Blue and Electropolymerized-Molecularly Imprinted Polymers-Based Electrochemical Biosensors. *ACS Appl. Mater. Interfaces* **2024**, *16*, 66921–66931. [CrossRef]
27. Pei, S.; Ji, W.; Yang, Y.; Liu, T.; Yang, S.; Wu, J.; Dai, J.; Hou, X.; Wu, Q.; Li, L. Flexible Dual-Channel Molecularly Imprinted Electrochemical Sensor for Simultaneously Monitoring Sweat Cortisol and Lactate Levels. *Anal. Sens.* **2024**, *4*, e202400003. [CrossRef]
28. Chen, Y.; Hu, X.; Liang, Q.; Wang, X.; Zhang, H.; Jia, K.; Li, Y.; Zhang, A.; Chen, P.; Lin, M.; et al. Large-Scale Flexible Fabric Biosensor for Long-Term Monitoring of Sweat Lactate. *Adv. Funct. Mater.* **2024**, *34*, 2401270. [CrossRef]
29. Konno, S.; Kudo, H. Fundamental Study of a Wristwatch Sweat Lactic Acid Monitor. *Biosensors* **2024**, *14*, 187. [CrossRef]

30. Klous, L.; de Ruiter, C.J.; Scherrer, S.; Gerrett, N.; Daanen, H.A.M. The (in)dependency of blood and sweat sodium, chloride, potassium, ammonia, lactate and glucose concentrations during submaximal exercise. *Eur. J. Appl. Physiol.* **2020**, *121*, 803–816. [CrossRef]
31. Rabost-Garcia, G.; Colmena, V.; Aguilar-Torán, J.; Galí, J.V.; Punter-Villagrasa, J.; Casals-Terré, J.; Miribel-Catala, P.; Muñoz, X.; Cadefau, J.; Padullés, J.; et al. Non-Invasive Multiparametric Approach To Determine Sweat–Blood Lactate Bioequivalence. *ACS Sens.* **2023**, *8*, 1536–1541. [CrossRef]
32. Sakharov, D.A.; Shkurnikov, M.U.; Vagin, M.Y.; Yashina, E.I.; Karyakin, A.A.; Tonevitsky, A.G. Relationship between Lactate Concentrations in Active Muscle Sweat and Whole Blood. *Bull. Exp. Biol. Med.* **2010**, *150*, 83–85. [CrossRef] [PubMed]
33. Karpova, E.V.; Laptev, A.I.; Andreev, E.A.; Karyakina, E.E.; Karyakin, A.A. Relationship Between Sweat and Blood Lactate Levels During Exhaustive Physical Exercise. *ChemElectroChem* **2020**, *7*, 191–194. [CrossRef]
34. Maeda, Y.; Okawara, H.; Sawada, T.; Nakashima, D.; Nagahara, J.; Fujitsuka, H.; Ikeda, K.; Hoshino, S.; Kobari, Y.; Katsumata, Y.; et al. Implications of the Onset of Sweating on the Sweat Lactate Threshold. *Sensors* **2023**, *23*, 3378. [CrossRef] [PubMed]
35. Teunissen, H. H2TRAIN. Inside. Available online: https://www.inside-association.eu/post/h2train (accessed on 8 October 2024).
36. PONS IP. H2TRAIN, the Largest Project Funded Under the European Chips Act for the Application of Digital Technologies in Smart Sports and Health Care, Has Been Launched with the Participation of PONS IP. Available online: https://ponsip.com/en/ip-news/news/h2train-the-largest-project-funded-under-the-european-chips-act-for-the-application-of-digital-technologies-in-smart-sports-and-health-care-has-been-launched-with-the-participation-of-pons-ip/ (accessed on 8 October 2024).

MDPI AG
Grosspeteranlage 5
4052 Basel
Switzerland
Tel.: +41 61 683 77 34

Biosensors Editorial Office
E-mail: biosensors@mdpi.com
www.mdpi.com/journal/biosensors

www.ingramcontent.com/pod-product-compliance
Lightning Source LLC
Chambersburg PA
CBHW040745250926
51850CB00007BB/1415
* 9 7 8 3 7 2 5 8 7 9 6 1 8 *